Soldiers Steadfast and Loyal

Soldiers Steadfast and Loyal

Medal of Honor and Distinguished Service Cross
Recipients of the 4th Division in World War One

Michael D. Belis

Deeds Publishing| Athens

Copyright © 2022 — Michael D. Belis

ALL RIGHTS RESERVED—No part of this book may be reproduced in any form or by any electronic or mechanical means, including information storage and retrieval systems, without permission in writing from the authors, except by a reviewer who may quote brief passages in a review.

Published by Deeds Publishing in Athens, GA
www.deedspublishing.com

Printed in The United States of America

Cover design by Michael D. Belis.

ISBN 978-1-950794-78-2

Books are available in quantity for promotional or premium use. For information, email info@deedspublishing.com.

First Edition, 2022

10 9 8 7 6 5 4 3 2 1

Dedicated to the Soldiers of the 4th Infantry Division, past, present, and future.

CONTENTS

Preface	xi
Acknowledgements	xix
4th Division Order of Battle World War One	1
Campaigns of the 4th Division in World War One	3

Medal of Honor	7	George N. Brigham	71
William Shemin	9	George Brown	74
		Raymond Buma	76
Distinguished Service Cross	15	William J. Cahill	79
Samuel P. Adkisson	18	Alberis Callewaert	82
August Aibner	22	Willis M. Campbell	85
Tom F. Barto	24	Charles F. Carbaugh	88
Joseph Bassi	27	Marion H. Cardwell	91
William B. Beach	31	James B. Carpenter	94
Harold V. Beal	36	Joseph H. Carvo	96
Harold Bedolfe	38	Claude E. Cherry	99
James P. Behan	40	Arthur I. Clark	102
Elwyn L. Berwick	43	Joseph T. Clement	105
Fred E. Billman	46	Herbert Arnold Cohn	109
Edward S. Blackman	48	James Conway	112
Homer J. Bleau	51	Frank B. Cook Jr.	114
Guy W. Boardman	53	Fred E. Cullen	117
Frank C. Bolles	57	Walter Currie	120
Lawrence Boop	62	Earl W. Curtis	123
Gustav J. Braun	64	Clinton Day	125
Olaf Brekke	68	Charles E. Deleuw	128

Walter H. Detrow	131	James P. Growdon	223
Albert Dietz	134	Leonard E. Guy	226
Lester C. Dill	137	Harold Lamonte Hall	229
Joseph Dilworth	140	Arthur M. Hamilton	232
Frank J. Downs	143	William H. Hammond	235
Thomas D. Drake	146	James W. Hanbery	239
Charles T. Dunbar	150	Mathias Willoughby Haney	244
Charles B. Duncan	153	Samuel H. Hanna	248
Peter W. Ebbert	157	Roy Harris	251
Andrew Edmiston	161	Edward D. Haskew	254
Emil J. Eklund	165	Morrison Hayes	257
Edwin A. Elliott	169	William Herren	260
Harold W. Enright	173	William B. Hook	265
Charles H. Epler	175	Samuel H. Houston	267
Etienne Escudier	178	Henry Howard	270
Charles H. Evans	180	William L. Hunter	273
Daris V. Ford	183	Albert L. J. Ihrke	276
Pietro Formica	186	Homer S. Jarvis	279
Ernest Fosnes	189	Frank Jaworski	282
Earl R. Fretz	193	Reuben L. Johnson	285
Henry J. Garst	196	Thomas W. Kearns	287
Isaac Gataino	199	Charles Kelly	290
Reuben L. George	201	Orval Kline	292
Charles Glenn	204	Max S. Kos	296
Cornelius T. Glynn	207	Jacob Kreis	299
Arthur J. Goetsch	210	Ralph E. Ladue	303
David S. Grant	212	Edward R. Lawless	308
Frank B. Gresham	216	John Legnosky	311
Glenn M. Grove	219	Fred Adcook Lieuallen	314

Name	Page	Name	Page
Joe Limon	317	Morton Osborn	429
Luther E. Lindahl	320	Paul J. Pappas	432
Clyde H. Lindsey	323	James K. Parsons	436
Joseph Longowski	326	Arthur Paulson	440
Charles J. Love	328	Donald A. Pegg	443
Murray R. MacKall	331	John C. Persons	446
Robert A. Madden	334	Oscar W. Peterson	450
John J. Madore	337	James J. Pirtle	453
Arthur H.G. Mallet	340	Richard G. Plumley	457
James Manning	343	Benjamin A. Poore	461
Richard Marcella	347	Ernest R. Potter	466
Robert G. Marshall	350	John H. Pratt Jr.	470
Cecil N. Martin	353	Fred N. Rapp	473
Henry F. Martin	356	Carl Rasmussen	476
Forrest L. Martz	360	Lee M. Ray	479
Roy E. Mathews	364	James T. Rice	482
Edward McAndrew	367	Stefano Riggio	485
Arnot L. McArthy	370	Lowell H. Riley	488
Howard C. McCall	373	Charles C. Rismiller	491
George C. McCelvey	378	Edward D. Ritchie	493
Arno S. McClellan	381	James H. Roberts	497
Joseph McCollum	385	Leo D. Roberts	501
William H. McGinnis	388	Raymond D. Robertson	504
Earl M. McKinley	392	Alvan C. Sandeford	507
Jean L. Meurisse	395	Otto A. Schwanke	511
Arthur M. Miller	397	Louis Scionti	514
Manton C. Mitchell	399	William A. Shea	517
Hans E. Morgan	403	Anthony F. Shedlock	520
Robert H. Murdoch	406	Albert B. Simpson	523
Martin Nelson	409	Ralph Slate	527
Francis K. Newcomer	412	Ford D. Smith	530
John H. Norton	416	Joe Smith	532
John W. Norton	420	Raymond R. Smith	535
Robert William Norton	423	Rutherford H. Spessard	538
Cornelius J. O'Brien	426	Dave W. Stearns	542

Wallis H. Sturtevant	545	Harrison B. Webster	576
Clark O. Tayntor	549	John Samuel Weimer	580
Henry Tudury	552	CARROLL B. WEST	583
Henry D. Turner	556	Stephen J. Weston	586
Thomas Vander Veen	560	Columbus Whipple	590
John C. Vann	563	Gilbert W. Wilcox	593
Paul Von Krebs	566	Merle R. Winsor	596
Emmett W. Waltman	569	William J. Wood	599
Arthur H. Warfield	572	Robert L. Worden	601
Joseph Waskiewic	574		

Appendix	605
Endnotes	613
Sources	617
About the Author	653

PREFACE

World War One began in Europe in August 1914. In April 1917, the United States entered the war. The first elements of the American Expeditionary Forces (A.E.F.) sailed to France in June 1917 and by October 1917 small numbers of American soldiers were in the front-line trenches supporting British and French units. The first offensive attack made by units of an American Division occurred in May 1918 and was fought by the 1st Division at the Battle of Cantigny.

The 4th Division of the United States Army was formed in December 1917, and during the month of May 1918 most of its units arrived in France, with the last few units arriving in early June. The 4th Division fought in its first offensive battle on July 18, 1918 and participated in combat in three major Offensives and two Defensive Sectors. On October 19, 1918, all of its units were pulled out of the front lines, except for the 4th Artillery Brigade and the 4th Ammunition Train, which remained in action until the Armistice of November 11, 1918.

From December 1918 until July 1919 the Division served on occupation duty in Germany. In July 1919 the Division returned to the United States. It was at Camp Dodge, Iowa (with a Provisional Regiment at a camp at Gary, Indiana) as it began the demobilization process after the war. The Division was officially inactivated on September 21, 1921 at Camp Lewis, Washington, leaving one Brigade on active duty to maintain a Divisional presence in the Army.

The Division was reactivated as a full-strength organization on June 1, 1940 at Fort Benning, Georgia. During World War Two it was the first American Division to land on the beaches in Normandy on D-Day, fought in five campaigns in Europe, and served on occupation duty in Germany. In July 1945, the Division returned to the United States at Camp Butner, North Carolina, where it was inactivated on March 12, 1946.

The Division was reactivated at Fort Ord, California on July 15, 1947 and has remained on active duty ever since. It served in Germany in the 1950's as part of the NATO deterrent to possible Soviet aggression. The Division served in eleven campaigns in the Vietnam War. In the Global War On Terror, it has served in Iraq and Afghanistan. Today it continues to deploy units overseas as deemed necessary by the Department of Defense.

The 4th Division was nicknamed the Ivy Division, the name taken from the pronunciation of the Roman numeral for four (IV). The Army approved shoulder sleeve insignia for the 4th Division is traced back to a design attributed to the first Commanding General of the Division, Major General George H. Cameron. In contemplating the designation "IV" and translating that to read "Ivy" Cameron came up with a design of four Ivy leaves joined together at a central disc. The design was officially approved for wear as a shoulder sleeve insignia by the American Expeditionary Forces on October 30, 1918. (The shape and size of the Ivy leaves in the design were standardized by the Adjutant General on June 20, 1922 and differ slightly from the insignia worn at the end of World War One.) Thus, the Ivy vine and its leaves came to be associated with the Division. In the language of flowers Ivy signifies fidelity. That concept was used to create the official motto of the Division which is "Steadfast and Loyal".

During World War One, the official designation of the Division was the 4th Division. In the 1920's, the Division was officially renamed the 4th Infantry Division. From 1940 to 1943, it was officially known as the 4th Motorized Division. From 1943 to 1970, it was officially known as the 4th Infantry Division. From 1970 to 2018, it was officially known as the 4th Infantry Division (Mechanized). In 2018, it once again officially became the 4th Infantry Division.

Through its service, it has been known by the nicknames of the Ivy Division, The Rolling Fourth, The Famous Fourth, The Famous Fighting Fourth, and the Ironhorse Division. In 2018, it once again reclaimed its original nickname of the Ivy Division. It served in combat in World War One, World War Two, in Vietnam, in Iraq, and Afghanistan. It also served in peacetime between the World Wars and during the Cold War.

During World War One there were only two medals authorized for the United States Army to award its soldiers as recognition for acts of valor, the Medal of Honor and the Distinguished Service Cross. The Army Medal of Honor was established by an Act of Congress on July 12, 1862 for enlisted men, amended on March 3, 1863 to include officers, and is the highest decoration for bravery on the battlefield which the Army can bestow. The Distinguished Service Cross was established by an Act of Congress on July 9, 1918 for acts of valor of a lesser degree than that required for the Medal of Honor, which are performed by any soldier. The Medal of Honor is the highest decoration, and the Distinguished Service Cross is the second highest decoration of any kind in the Army.

The same Act which established the Distinguished Service Cross also established the Citation Star for gallantry, but that award was not worn on the uniform as a full-size medal. It was displayed as a 3/16-inch-tall silver star device worn on the ribbon of the Campaign Medal which for World War One was the Victory Medal. When the Silver Star Medal was created in 1932, living recipients of the Citation Star were allowed to apply to have their award converted to the Silver Star Medal. At the time, no such provision was made for posthumous recipients of the Citation Star.

During the first World War however, and for fourteen years after it, the only full-size medals with which the Army could reward its soldiers for valorous acts that went above and beyond the normal duties expected of them were the Medal of Honor and the Distinguished Service Cross.

One hundred eighty-six Distinguished Service Crosses were awarded to soldiers for deeds they performed while serving in the 4th Division during World War One. One of those awards was later upgraded to the Medal of Honor. Forty-one of those Distinguished Service Crosses were awarded posthumously.

The Department of Defense is in the process of creating lists of all recipients of the highest awards for valor and making those lists available to the public. However, those lists only give name, rank, and conflict, and do not give the individual Citations. The Citations for recipients of the Distinguished Service Cross in World War One are spread out among a number of publications and documents spanning the years 1918 to 1941. In those publications and documents the format and language used in presenting the Citations differs.

Most of the Citations are contained in AMERICAN DECORATIONS A List of Awards of the Congressional Medal of Honor the Distinguished-Service Cross and the Distinguished-Service Medal AWARDED UNDER THE AUTHORITY OF THE CONGRESS OF THE UNITED STATES 1862-1926 which was compiled by the Office of the Adjutant General in 1926 and published in 1927. Five supplements were published for that volume, one in 1937, one in 1939, two in 1940, and one in 1941, each of which contain more Citations.

Most Citations in the records of the Adjutant General and other government documents for the Distinguished Service Cross do not begin with the more formal introductory phrase "The President of the United States of America, authorized by Act of Congress, July 9, 1918, takes pleasure in presenting the Distinguished Service Cross to ...". Actually, most Citations in those records and documents have no introductory phrase at all and simply give a description of the action performed. For the sake of continuity, that is how the Citations will be presented in this volume. The one exception will be the Citation for the Medal of Honor to William Shemin which is presented here using the formal language and phrasing of how it was issued in 2015.

Army Citations are famous for their misspellings, bad grammar, and incorrect use of punctuation marks, and while most of the Citations have been "cleaned up" in that respect for presentation here, the basic language of the original Citation describing the act for which the medal was awarded remains unchanged. Soldier's names are presented in the headings of the Citations here in the manner they were presented in the original Citation, with the exception that names which were misspelled in the original Citations have been corrected for presentation on these pages. Middle initials were not always indicated, though in some cases the full middle name was spelled out. In the text of most of the original Citations, pronouns were used, although many Citations did specify proper names in the text. In the text of the Citations presented here, proper names have been substituted in many instances where pronouns were used in the original.

In a great many of the Citations recorded by the Adjutant General, serial numbers (Army Service Numbers) for enlisted men were not included. They have been researched for this volume and all which could be found are presented here. Serial numbers for officers were almost never given, since officers were not issued numbers until starting in 1920 and are not presented here. Date and place of birth, home of record, and even complete unit were not always given in the original Citations, though most of those details were listed in the AMERICAN DECORATIONS volumes mentioned above. In many Citations the home of record given was that of the soldier's emergency contact, which may or may not have been the soldier's actual residence. Therefore, home of record is only mentioned here in cases where it could be verified as actual residence.

Complete units were not always given in the original Citations but have been added here where they could be determined. Rank is spelled out in the Citation headings but may be abbreviated in the profile text since it was abbreviated in most other documents. General Orders numbers for the awards have been researched and all which could be found are presented here. Medal numbers were not given in most of the government publications. Those which could be found are presented here.

It is hoped that the profiles presented here will give the

reader a better understanding of the men who received these awards and of the incredible courage and sense of duty they possessed on that European battlefield a century ago. They and all other men and women of the 4th (Infantry) Division for the past one hundred years have been and today still are Soldiers Steadfast and Loyal.

—**Michael D. Belis**
Carencro, Louisiana March 31, 2022

ACKNOWLEDGEMENTS

Thanks to my wife Margaret who corrected major grammatical errors in the early draft of the text. She keeps me on the "straight and narrow" and puts up with me spending much of my waking hours researching and writing military history. During the times when I'm buried for hours in the computer room, she'll stick her head in the doorway and ask, "How are the dead army people doing?" I'd be lost without her.

Thanks to Robert O. Babcock, CEO of Deeds Publishing, 4th Infantry Division veteran, 4th Infantry Division President and Historian, and past long time President of both the 4th Infantry Division Association and the 22nd Infantry Regiment Society. Bob has been a close friend for over twenty years. His devotion to preserving the history of the 4th Infantry Division is unequalled. It was at Bob's suggestion that this work on the Medal of Honor and Distinguished Service Cross recipients of the 4th Division in World War One was begun. Thanks to Bob, Jan, Mark, Matt, and everyone from Deeds Publishing for their hard work and dedication.

Thanks to Eric Lowman for his help in assembling research on 4th Division soldiers from Indiana and whose committment

Michael D. Belis

to honoring the military history of the people of the State of Indiana is quite impressive.

A special debt of gratitude is owed to Gary Mitchell, whose diligence in researching and compiling the assigned numbers of the Distinguished Service Cross of the United States during the World War One era will never be surpassed. Gary took raw data gathered by Charles P. Hey and Colonel Albert F. Gleim and refined it into a presentation of several volumes which are of great value to scholars, historians, and collectors. Because of Gary's work, 164 assigned medal numbers of the Distinguished Service Cross were identified for soldiers of the 4th Division, out of the 186 such medals awarded to members of the Division. Only two additional numbers were found from other sources, so it is Gary's efforts that resulted in such a high percentage of the Distinguished Service Crosses of the 4th Division being identified as to their assigned medal numbers. This may be the highest percentage of discovered medal numbers for any Division in the American Expeditionary Forces. Conversations with Gary were especially enlightening and informative. He was patient with my many questions and imparted a great deal of knowledge on the subject of awards of all kinds to American soldiers in World War One. His generosity in sharing his research and his expertise is greatly appreciated.

4TH DIVISION ORDER OF BATTLE WORLD WAR ONE

7th Infantry Brigade
39th Infantry Regiment
47th Infantry Regiment
11th Machine Gun Battalion

8th Infantry Brigade
58th Infantry Regiment
59th Infantry Regiment
12th Machine Gun Battalion

4th Field Artillery Brigade
13th Field Artillery Regiment (155mm)
16th Field Artillery Regiment (75mm)
77th Field Artillery Regiment (75mm)
4th Trench Mortar Battery

Divisional
Headquarters Company
10th Machine Gun Battalion
4th Engineer Regiment
8th Field Signal Battalion

Michael D. Belis

Trains
4th Train Headquarters and Military Police
4th Ammunition Train
4th Supply Train
4th Engineer Train
4th Sanitary Train (Ambulance Companies and Field Hospitals 19, 21, 28, 33)

CAMPAIGNS OF THE 4TH DIVISION IN WORLD WAR ONE

The Army Center of Military History in its Lineage and Honors section for the 4th Infantry Division lists the following five campaigns for the Division's participation in World War One in this order:

Aisne-Marne
St. Mihiel
Meuse-Argonne
Champagne 1918
Lorraine 1918

However, the Army Center of Military History in its separate official list of World War One Campaigns of the U.S. Army does not include either Champagne 1918 or Lorraine 1918 as campaigns.

Army Regulation 600-8-22 denotes the clasps authorized for the World War I Victory Medal and in that Regulation there are no clasps authorized for Champagne 1918 or for Lorraine 1918.

The terms of Champagne 1918 and Lorraine 1918 used by

the Center of Military History apparently were originally intended to refer to the two defensive sectors in which the 4th Division served, with Champagne indicating the Vesle Sector and Lorraine indicating the Toulon Sector. Those two Sectors should not however be seen as campaigns, because they are not. The Center of Military History still inaccurately indicates that the 4th Division served in five campaigns and this mistake is repeated in other accounts and descriptions of the Division in World War One.

Brief Histories of Divisions, U.S. Army 1917-1918 Prepared in the Historical Branch, War Plans Division, General Staff June 1921 which is on file at the Army Command & General Staff College Combined Arms Research Library at Fort Leavenworth, Kansas gives descriptions of the U.S. Army Divisions that participated in World War One, including a chronology of combat service for most of the units in each Division. Since this study was completed by the Historical Branch of the General Staff immediately after the war in order to create an official synopsis of the U.S. Army Divisions in the war, it can be considered the authority for such.

In that publication under the heading of "COMBAT SERVICE, 4TH DIVISION" the following campaigns and defensive sectors in which the 4th Division participated in the order they occurred are listed as:

Aisne-Marne Offensive
Vesle Sector
Toulon Sector
St. Mihiel Offensive
Meuse-Argonne Offensive

BATTLE PARTICIPATION OF ORGANIZATIONS OF THE AMERICAN EXPEDITIONARY FORCES IN FRANCE, BELGIUM AND ITALY 1917-1918 published by the War Office (War Department) in 1920 lists the campaign participation credit for the various units of the Army in World War One. It also uses the names of Aisne-Marne Offensive, Vesle Sector, Toulon Sector, St. Mihiel Offensive and Meuse-Argonne Offensive in reference to the operations of the 4th Division.

The *4th Division Summary of Operations in the World War prepared by the American Battle Monuments Commission* published by the United States Government Printing Office in 1944 is the official history of the front-line action of the Infantry and Machine Gun units of the 4th Division in World War One. That study also uses the names of Aisne-Marne Offensive, Vesle Sector, Toulon Sector, St. Mihiel Offensive and Meuse-Argonne Offensive in reference to the operations of the 4th Division.

Thus it can be seen that the 4th Division did not serve in five campaigns but instead served in three campaigns and two defensive sectors. The above campaigns and sectors from the General Staff publication of 1921, the War Office publication of 1920, and the American Battle Monuments Commission publication of 1944 are therefore used in the following pages to denote the service of the soldiers presented here.

The Division Colors today carry five streamers for service in World War One, including streamers for Champagne and Lorraine. During a reorganization of campaign streamers by the War Department's Heraldic Program Office in 1945, those two streamers were discontinued as official issue. Use of such outdated and discontinued streamers is allowed by The Institute of Heraldry, to units formed previous to the 1945 streamer reorganization, however, as a matter of carrying on tradition. Thus, the two streamers of Champagne and Lorraine are still flown from the 4th Infantry Divison Colors to represent the Division's service in the Vesle Sector and the Toulon Sector.

The Victory Medal clasps authorized for soldiers of the 4th Division in World War One are listed immediately below in the order worn. Regardless of how many different defensive sectors in which a soldier participated, only one clasp for service in a defensive sector is authorized and is designated by the term "Defensive Sector".

The authorized clasps for the 4th Division in World War One are:

Aisne-Marne
St. Mihiel
Meuse-Argonne
Defensive Sector

MEDAL OF HONOR

Michael D. Belis

William Shemin

The President of the United States of America, authorized by Act of Congress, March 3, 1863, has awarded in the name of Congress the Medal of Honor to
Sergeant William Shemin, United States Army
Army Service Number 558172
Company G 47th Infantry Regiment 4th Division American Expeditionary Forces
War Department General Orders 5 (November 15, 1918) (Amended June 2, 2015)

CITATION

For conspicuous gallantry and intrepidity at the risk of his life above and beyond the call of duty:

Sergeant Shemin distinguished himself by acts of gallantry and intrepidity above and beyond the call of duty while serving as a Rifleman with G Company 2d Battalion 47th Infantry Regiment 4th Division American Expeditionary Forces in connection with combat operations against an armed enemy on the Vesle River near Bazoches, France from August 7 to August 9, 1918. Sergeant Shemin left cover and crossed open space, repeatedly exposing himself to heavy machine gun and rifle fire to rescue wounded. After Officers and Senior Noncommissioned Officers had become casualties, Sergeant Shemin took command of the platoon and displayed great initiative under fire until wounded on August 9. Sergeant Shemin's extraordinary heroism and selflessness, above and beyond the call of duty are in keeping with the highest traditions of the military service and reflect great credit upon himself, his unit and the United States Army.

Michael D. Belis

DETAILS

William Shemin was born in Bayonne, Hudson County, New Jersey on October 14, 1896, the son of Harris and Bessie Shemin. He studied for a year at Syracuse University New York State College of Forestry, worked as Assistant City Forester of the City of Bayonne, New Jersey and was a forester on the New York State Reservation at Niagara Falls, New York.

Shemin enlisted in the Regular Army as a Private on October 2, 1917 and was assigned to Company G 47th Infantry at Camp Greene, North Carolina. He was promoted to Corporal on April 20, 1918. He sailed to France as a Corporal in Company G 47th Infantry 4th Division aboard the troopship *U.S.S. Princess Matoika* (seized German liner *Prinzess Alice*) on May 10, 1918. As his emergency contact, he listed his father Harry in Bayonne, New Jersey.

Shemin was promoted to Sergeant on June 20, 1918 and served in the Aisne-Marne Offensive. He was awarded the Distinguished Service Cross (Medal # 604) in War Department General Orders 5 (1920) for his actions in the Vesle Sector. On August 7, 1918 Shemin's 2nd Battalion 47th Infantry was the attacking element in the advance across the Vesle River at the village of St. Thibaut, with the objective being the village of Bazoches which was situated across the river. The 4th Engineers constructed and emplaced a footbridge about twenty feet wide for the Infantry to dash across and repaired it as German artillery damaged it throughout the day. Some soldiers of 2nd Battalion got across the river on that footbridge, some got across using trees felled at the river and some got across by swimming and wading. By five o'clock in the afternoon nearly all of Companies G and H and one platoon of Company F, a total of about 350 men, got across the river and established defensive positions.

Shemin was one who made it across. He and his fellow soldiers were under consistent artillery fire from the enemy north of the river. They also came under heavy machine gun fire from German positions before the village of Bazoches, the village itself, and the high ground behind the village. An enemy counterattack failed to dislodge the Americans from their hastily dug positions in front of Bazoches. Shemin ventured out onto the battleground under fire three times to rescue wounded soldiers. When all officers and Sergeants senior to him became casualties he took command of his platoon and directed the platoon in the defense of its position until he was wounded on August 9, when a bullet struck his helmet and entered his head behind his left ear. He also was wounded by shrapnel in his back and spent the next three months in hospital.

Shemin also received a Citation Star. He served on occupation duty in Germany with the 4th Division where he was prominent on the sports teams of the 47th Infantry. He returned to the United States with his Regiment aboard the troopship *U.S.S. Mobile* (seized German liner *Cleveland*) leaving Brest, France July 16, 1919 and arriving at Hoboken, New Jersey July 27, 1919. He was discharged on August 2, 1919.

The 1920 Census showed Shemin living with his parents in Bayonne, New Jersey where he was employed as a gardener for the government. On April 22, 1921 he was awarded the Conspicuous Service Cross (Medal #467) by the State of New York for his actions in World War One. Shemin finished his studies at the Syracuse University New York State College of Forestry and played on the 1922 Syracuse University football team. He married Bertha Schiffer on June 29, 1928 and started his own greenhouse business. The 1930 Census showed Shemin living with his wife, his mother and sister in the Bronx, New York City where he was employed as a florist and indicated he was a veteran of the World War. After 1932 he was awarded the Purple Heart for the wounds he received in World War One. The 1940 Census showed him living with his wife, two daughters and son in the Bronx where he was still employed as a florist. He was a member of the Legion of Valor and the Jewish War Veterans of the United States. His wife Bertha died in 1971. William Shemin died at the age of 76 on August 15, 1973 and is buried in Baron Hirsch Cemetery, Staten Island, Richmond County (Staten Island), New York.

In the early 2000's, Shemin's daughter Elsie Shemin-Roth began what would turn out to be a thirteen-year endeavor to have her father's Distinguished Service Cross upgraded to the Medal of Honor. Thanks in great part to the efforts of U.S. Representative Blaine Luetkemeyer and Retired Army Col. Erwin Burtnick, Shemin's case was reviewed. It was believed that discrimination may have played a part in Shemin not receiving the Medal of Honor. His award of the Distinguished Service Cross was therefore upgraded to the Medal of Honor. On June 2, 2015, in a ceremony at the White House in Washington, D.C., the daughters of William Shemin were presented with the Medal of Honor by President Barrack Obama. William Shemin is the only soldier of the 4th Division awarded the Medal of Honor for actions in World War One.

Representing the 4th Infantry Division during the Medal of Honor ceremony were Sergeant Major of the Army Dan Dailey who had fought with the 4th Infantry Division on four tours in Iraq and Robert Babcock, 4th Infantry Division Association president and historian, a Vietnam veteran of 4th Infantry Division, and Michael May, the Staff Judge Advocate of the 4th Infantry Division Association, also a Vietnam veteran of 4th Infantry Division.

DISTINGUISHED SERVICE CROSS

Publisher Note: Pictures were only found for the one DSC recipient whose award was upgraded to the MOH, and for 62 of the remaining 185 DSC recipients from WWI. Quality of the pictures is understandably not equal to today's photographic standards since they were taken in the early years of the 1900s. We have made them as high quality as we can and decided to include them even though some do not meet the standards of today's photography.

Samuel P. Adkisson

Second Lieutenant, United States Army
Company D 39th Infantry Regiment 4th Division
Date of Action: October 10, 1918
Location: Near Septsarges, France
War Department General Orders 44 (1919)
Medal # 3176

CITATION

Leading his platoon through an unusually heavy barrage Lieutenant Adkisson filled a gap on the right flank which was until then exposed. From this point he attacked and captured several machine guns and 20 prisoners. During an attack he was badly gassed and his platoon reduced in strength to six men but he held his position under a murderous crossfire of artillery and machine guns until relieved three days later.

DETAILS

Samuel Parker Adkisson was born in Louisville, Kentucky on October 25, 1895, the son of Samuel A. and Louise Adkisson. He attended Officers Training Camp at the Presidio in San Francisco, California and was commissioned a 2nd Lieutenant of Infantry (National Guard) in 1917. At the time of his service his home of record was listed as Los Angeles, California. He sailed to England as a 2nd Lieutenant Infantry officer replacement, coming from the California National Guard Camp Kearny June Automatic Replacement Draft (Infantry) Company #12, aboard the British troopship *Lapland* on June 28, 1918. As his emergency contact, he listed his mother Louise in Los Angeles, California. From England he was sent to France and was eventually assigned to the 39th Infantry 4th Division. He carried out the duties of Regimental Assistant Adjutant until being attached in August to Company D 1st Battalion 39th Infantry as a platoon leader. He served in the Toulon Sector and the St. Mihiel Offensive.

Adkisson was awarded the Distinguished Service Cross for his actions in the Meuse-Argonne Offensive on October 10-13, 1918. His Citation gives the location as near the village of Septsarges. The actual location was about four miles north of Septsarges in the wooded area known as the Bois de Forêt. By October 9, 1918, 3rd Battalion 39th Infantry had led the advance toward the Meuse River as far as the wooded area known as the Bois de Fays. On October 10, 2nd Battalion 39th Infantry led the attack, moving through 3rd Battalion's positions, and occupied the wooded area known as the Bois de Peut de Faux, with 1st Battalion 39th Infantry including Adkisson's Company in support. Following close behind 2nd Battalion Adkisson and his platoon moved through heavy enemy fire into a position from which they could protect the exposed right flank of the Regimental advance. Adkisson and his men then assaulted and eliminated several enemy machine guns that had been bypassed by the attacking elements, and captured twenty prisoners. On October 11 the three Battalions of the 39th Infantry including Adkisson and his platoon took the German positions in the next wooded area known as the Bois de Forêt in a bayonet attack. An enemy counterattack was repulsed that afternoon.

On October 12, the day was spent consolidating the gains made and another counterattack was defeated. Early in the morning of October 13, the enemy counterattacked again but the 39th Infantry in the Bois de Forêt held the line they had established. As an indication of how intense the battle was in the area in which Adkisson was commanding his platoon, during the four days mentioned in his Citation of October 10th, 11th, 12th, and 13th his Regiment suffered the loss of more than one hundred killed and five hundred wounded and missing. Adkisson was one of the wounded, suffering from the effects of being gassed at some point in the attack, but remained in command of his platoon. By October 13, his platoon was down from a normal strength of about fifty men to exactly six effective soldiers yet under his leadership they still held their part of the line until relieved in late morning of that day by elements of the 4th Infantry 3rd Division.

Adkisson returned to the United States ahead of his Regiment as a 1st Lieutenant of Infantry in a list of casual sick and wounded officers in Casual Detachment #4 of the 91st Division destined for Camp Upton, New York and immediate discharge aboard the transport *U.S.S. Virginian* leaving St. Nazaire, France April 7, 1919 and arriving at Hoboken, New Jersey April 20, 1919. The date of his discharge could not be found.

By 1921 Adkisson had settled down into being an office clerk in Los Angeles, California and then became a cotton broker. On October 5, 1923 he married Frances Thompson and in 1925 they had a daughter. The 1930 Census showed him living with his wife and daughter and his wife's parents in Los Angeles where he was employed as a Stockbroker, and indicated he was a veteran of the World War. The 1940 Census showed him living with his wife, daughter, and mother-in-law in Los Angeles where he was employed as a secretary for a land company.

When he registered for the draft in 1942, he indicated his home of record as Red Bluff, California and that he was employed by the El Camino Land Corporation in Gerber, California. Adkisson was back in uniform during World War Two and the Korean War and served as an officer in the Army Transportation Corps, most likely in the continental United States eventually reaching the rank of Lieutenant Colonel and being separated from this service in 1955. In 1956 he and his wife Frances moved to Reno, Nevada where he worked in the real estate business. Samuel P. Adkisson died at the age of 72 on May 6, 1968 and is buried in Golden Gate National Cemetery in San Bruno, San Mateo County, California.

August Aibner

Sergeant, United States Army
Army Service Number: 560341
Company M 58th Infantry Regiment 4th Division
Date of Action: October 4, 1918
Location: Near Bois de Fays, France
War Department General Orders 2 (1920)
Medal # 6153

CITATION

Sergeant Aibner advanced in the midst of an enemy barrage and rescued two of his comrades who were lying wounded in advance of our front line, and under heavy enemy fire.

DETAILS

August Aibner was born in Coal Creek, Colorado on February 6, 1893. At the time he entered the service he was employed as a butcher in Thermopolis, Wyoming. He entered the Army on February 19, 1918. Aibner sailed to England as a Corporal with Company M 58th Infantry 4th Division aboard the British troopship *Themistocles* on May 11, 1918. As his emergency contact, he listed his mother Josephine in Walsenburg, Colorado. From England he and his Regiment were sent over to France. At some time in France before October 4, 1918, he was promoted to Sergeant. During the Aisne-Marne Offensive Aibner was wounded on August 5, 1918 and later returned to duty.

Aibner was awarded the Distinguished Service Cross for his actions in the Meuse-Argonne Offensive. On October 4, 1918 the 8th Brigade was the leading element of the 4th Division in the attack toward the Meuse River with the 58th Infantry spearheading the assault. Aibner and 3rd Battalion 58th Infantry left their foxholes in the northwestern edge of the wooded area known as the Bois de Brieulles and headed across open ground toward the wooded area known as the Bois de Fays. A heavy fog concealed much of the attacking force but when the fog cleared, the Germans opened up with artillery and machine gun fire, creating heavy casualties among the American ranks. During this phase of the battle, Aibner ventured out into the artillery barrage and brought two of his wounded fellow soldiers back to safety and medical treatment.

Aibner served on occupation duty in Germany with the 4th Division and returned to the United States as Mess Sergeant of Company M 58th Infantry aboard the troopship *U.S.S. Mount Vernon* (captured German liner *Kronprinzessin Cecilie*) leaving Brest, France July 24, 1919 and arriving at Hoboken, New Jersey August 1, 1919. He was discharged on August 11, 1919.

On November 18, 1924 he married Blanche Shifflette in Sulphur Springs, Colorado. Their daughter was born two years later. The 1930 Census showed Aibner living with his wife and daughter in Denver, Colorado where he was employed as a butcher and indicated he was a veteran of the World War. August Aibner died in Colorado at the age of 70 on November 2, 1963.

Tom F. Barto

Corporal, United States Army
Army Service Number: 568752
Company D 4th Engineer Regiment 4th Division
Date of Action: August 11, 1918
Location: On the Vesle River near Ville Savoye, France
War Department General Orders 129 (1918)
Medal # No assigned number discovered

CITATION

Corporal Barto volunteered to go into Ville Savoye at a time when it was under a heavy bombardment to rescue a wounded officer.

DETAILS

Thomas Franklin "Tom" Barto was born in Neosho, Newton County, Missouri on September 16, 1891, the son of Alexander and Mary Barto. By 1910, he and his family were living in Bellingham, Washington. He entered the Army at Butte, Montana on April 26, 1917. Barto sailed to France as a Corporal in Company D 4th Engineers aboard the troopship *U.S.S. Martha Washington* (seized Austrian liner) on April 30, 1918. As his emergency contact, he listed his mother Mary in Bellingham, Washington. Barto served in the Aisne-Marne Offensive.

He was awarded the Distinguished Service Cross for his actions in the Vesle Sector. He was killed attempting to rescue 2nd Lieutenant Frank B. Cook of Company D 4th Engineers from the village of Ville Savoye which at the time was under a heavy artillery barrage. Barto was accompanied by Private 1st Class Gilbert W. Wilcox, also of Company D 4th Engineers. The two soldiers charged into the deadly artillery fire to get their Lieutenant who lay seriously wounded in the shell torn village.

Though Barto was killed during the action, Wilcox managed to bring out Lieutenant Cook, who recovered from his wounds and returned to the United States with his Regiment when it sailed from France in the summer of 1919. Barto was awarded the Distinguished Service Cross for his bravery and sacrificing his life in the rescue. Wilcox was awarded the Distinguished Service Cross for his bravery in rescuing Lieutenant Cook under heavy fire. Cook was awarded the Distinguished Service Cross for his own actions at the Vesle River that day before going into the village and getting wounded in the horrific bombardment.

Barto was buried in Grave # 11 in the battlefield cemetery # C-64 at the village of Ville Savoye.

His Distinguished Service Cross was presented to his father Alexander H. Barto.

Michael D. Belis

On April 4, 1919 Barto was reinterred in Plot 1 Section G Grave # 53 in the American Cemetery # 617 at the village of Fismes. On March 15, 1921, his body was disinterred for preparation and shipment. His remains were returned to the United States as one of 2,794 deceased American soldiers who came home aboard the United States Army Transport *U.S.A.T. Wheaton* leaving Antwerp, Belgium April 26, 1921 and arriving at Hoboken, New Jersey on May 18, 1921. Tom F. Barto was laid to his final rest in Bayview Cemetery, Bellingham, Whatcom County, Washington on June 19, 1921.

Joseph Bassi

Private, United States Army
Army Service Number: 562059
Company I 59th Infantry Regiment 4th Division
Date of Action: October 4 - 7, 1918
Location: Near Bois de Fays, France
War Department General Orders 71 (1919)
Medal # 3435

CITATION

Showing marked personal courage, Private Bassi repeatedly crossed ground swept by heavy artillery and machine gun fire to deliver important messages. He volunteered for dangerous missions, his example being an inspiration to the other runners.

DETAILS

Joseph Bassi was born in Bassignana, Italy on August 31, 1895. He immigrated to the United States aboard the German liner *S.S. Berlin* leaving Genoa, Italy on February 8, 1912 and arriving at New York City February 22, 1912. Before entering the Army he was a salesman living at 415 Beale Avenue, Memphis, Tennessee. Bassi entered the Army on March 8, 1918.

He sailed to England as a Private 1st Class in Company I 59th Infantry 4th Division aboard the armed British troopship *RMS Olympic* (sister ship of the *Titanic*) on May 5, 1918. As his emergency contact, he listed his cousin Joseph in Memphis, Tennessee. From England he and his Company were sent over to France. Bassi served in the Aisne-Marne Offensive, the Vesle Sector, the Toulon Sector and the St. Mihiel Offensive.

Michael D. Belis

He was awarded the Distinguished Service Cross for his actions in the Meuse-Argonne Offensive. The period of October 4-7, 1918 in and around the wooded area known as the Bois de Fays was one of the most intense with regard to enemy fire of all types and constituted some of the most horrendous of conditions faced by the 4th Division in the war.

The conditions in the Bois de Fays are best described in the following passage from the official history of the 4th Division in World War One:

> "The infantry and machine gun units in the Bois de Fays were under a terrific strain. A scattered line of precarious holes, dug amid the seared and shattered trees of the wood, afforded the only possible shelter. It rained nearly every day and the dampness and cold, the mud and the darkness, made life almost unbearable. But they had to protect a salient open to attack from three sides and to suffer a continuous fire from every form of weapon that could throw a projectile. No fires could be built. Hot food could only be brought up from the Bois de Brieulles at night. Gas reeked through the woods. Aeroplanes bombed them at frequent intervals. The stench of dead bodies became unendurable, and as many as could be found in the thick underbrush were carried to the southern edge of the woods and later buried. Seldom has there been such a combination of horrors and seldom have the supreme qualities of endurance of mind and body and quiet heroism been in greater demand."[1]

Bassi voluntarily performed the job of runner carrying vital communications between units on the battlefield in the days before field radios, repeatedly leaving shelter and moving through the entanglements of the wooded areas and across open areas consistently subjected to enemy machine gun and artillery fire.

He volunteered for the most dangerous of missions, setting an example of courage and devotion to duty that inspired others.

Bassi served on occupation duty in Germany with the 4th Division and was promoted to the rank/position of Mechanic. He returned to the United States as a Mechanic in Company I 59th Infantry and part of Company "K" Third Army Composite Regiment aboard the troopship *U.S.S. Leviathan* (seized German liner *Vaterland*) leaving Brest, France September 1, 1919 and arriving at Hoboken, New Jersey September 8, 1919.

The Third Army Composite Regiment was formed to represent the American Expeditionary Forces in a few parades and celebrations held in Europe at the end of World War I. Five hundred men from each of the 1st, 2nd, 3rd, 4th, 5th and 6th Divisions were organized to make a "Regiment" of 3,000 American soldiers to take part in the ceremonies. Each Division provided their 500 men split into two companies. The soldiers from the 4th Division made up Companies I and K of the Composite Regiment. In the 4th Division the men were selected from the four Infantry Regiments and the three Machine Gun Battalions. Officers and enlisted men were picked based on their appearance, conduct, and record. They had to have served in at least one major engagement of the Division and most of them wore at least one wound stripe indicating they had been wounded at least once.

The Third Army Composite Regiment led the formation of Allied Troops in the parade through Paris witnessed by over two million people on la Fête nationale (The National Celebration) the major national holiday of France on July 14, 1919.

Bassi was discharged on September 25, 1919.

On November 7, 1919, Joseph Bassi filed a petition for naturalization to become an American citizen based on the Act of Congress of May 9, 1918. This act provided that any alien serving in the Armed Forces of the United States during the war could file a petition for naturalization and have certain requirements waived because of his service. An endorsement in his petition folder dated September 1921 indicated his petition had been denied because he failed to prosecute his petition to a final hearing within a reasonable length of time. Further naturalization records for him could not be discovered. When he registered for the draft in 1942, Bassi indicated he lived in Los Angeles, California and was unemployed. Joseph Bassi died at the age of 71 on May 22, 1967.

William B. Beach

Sergeant First Class, United States Army
Army Service Number: 568500
Company C 4th Engineer Regiment 4th Division
Date of Action: August 6 and 8, 1918
Location: Near St. Thibaut, France
War Department General Orders 46 (1919)
Medal # 1303

CITATION

Being a member of a covering detachment sent out to protect a detail which was constructing a bridge over the Vesle River, Sergeant Beach voluntarily left his squad and fought his way alone down the river in order to locate an enemy machine gun nest. The flashes from his automatic rifle drew fire from the enemy and he was forced to jump into the river for protection. Swimming back to his squad, he organized a detail and led it in a successful attack on the hostile position. Two nights later, after this bridge had been destroyed, this soldier with three others volunteered to rebuild the bridge. Under continuous fire from the enemy, he swam the river several times and set the posts for the bridge thereby making possible the infantry attack on the following morning.

DETAILS

William Burdett Beach was born in Cooksville, McLean County, Illinois on July 7, 1893, the son of Frank and Laura Beach. He enlisted in the Regular Army on January 1, 1916, was on Mexican Border Service at Camp Stephen D. Little, Arizona and took part in General John J. Pershing's Punitive Expedition into Mexico. He sailed to France as a Sergeant with Company C 4th Engineers 4th Division aboard the troopship *U.S.S. Martha Washington* (seized Austrian liner) on April 30, 1918. As his emergency contact, he listed his sister Hazel (Mrs. Orville Sims) in Colorado Springs, Colorado. At some time after reaching France he was promoted to Sergeant 1st Class. Beach served in the Aisne-Marne Offensive.

Of the 21 Distinguished Service Crosses awarded to the 4th Engineers during World War One, 15 of them were awarded for actions either at or near the Vesle River. The river was an obstacle which had to be crossed during the Aisne-Marne Offensive. The official history of the 4th Engineers in World War One describes conditions at the Vesle River:

> "We soon learned that we were on the heights above the Vesle River, a little, slow-moving stream flowing between wooded hills, the hills on the German or east bank being steeper and higher than that on our side. At the side we occupied was an open space some half mile wide between the hills and the river, this space was quite swampy, and had no trees or cover of any kind, while on the eastern bank the wooded slopes extended almost to the water. Fringing the river itself was a row of tall poplar trees... The river itself was only four or five feet deep, and between thirty-five and forty-five feet wide. The Germans had stretched a good deal of barbed wire in the bed of the river, as well as entanglements on both sides, and had placed many machine guns under cover of the trees." [2]

For most of its length, the river's current was not overly strong, and the water was not over a man's head but in numerous places the river narrowed and deepened which made the current flow faster and the threat of drowning was very real for a soldier burdened with pack, rifle, and ammunition. The Engineers felled trees to make at least eight-foot bridges and a few bridges large enough to pull the American operated 75mm artillery guns across. The artillery bridges were emplaced near the village of Ville Savoye. Most of the bridges were destroyed or heavily damaged by German artillery and the Engineers rebuilt them again and again, working under machine gun fire, artillery, and gas attacks. Much of the work was carried out at night. Though the night work was slow and it was difficult to see in the darkness, working by night meant German artillery spotters could not effectively call in their artillery fire so a good deal of the work was done at that time. The Engineers suffered their heaviest casualties of the war working at the Vesle River. The 4th Engineers history continued the narrative:

> "The work at the river was continued night after night. Some of the footbridges were wrecked by direct hits of shells, and the big artillery bridge near Viller Savey [sic] was struck again and again. At times the work was utterly impossible owing to the furious shelling, but each time after the firing had died down, they went back to their jobs. The list of dead and wounded grew longer and longer, and every day there were fresh casualties. The constant presence of poisonous gas in the air affected the survivors so that many were hardly able to walk." [3]

> "Thru it all, the Engineers worked, standing at times shoulder deep in water. Trees were dropped into the current and dragged by man strength to the other bank, as two were placed side by side, cross pieces were lashed to them to make a footing, man after man was hit, but others took their places, and the work went on. Gas shells

rained in, and the air was rank with the deadly phosgene and chlorine vapors, but the work in gas masks was impossible and the men risked the danger to finish the job."[4]

Beach was awarded the Distinguished Service Cross for his actions in the Vesle Sector on two separate dates. The first date was August 6, 1918. On that date Beach was part of a detail which had forded the Vesle River and were on the German side, covering the work of his Company which was building and emplacing a footbridge at night across the river. The bridge was to be used by soldiers of the 7th Brigade 4th Division to get across the river near the village of St. Thibaut. An enemy machine gun was hindering the work. Beach took his French Chauchat automatic rifle and moved along the river searching for the German gun. As he fired at the enemy, the muzzle flashes of his automatic rifle drew fire and he was forced to move back into the river for concealment. He swam back to his squad, gathered a few men, returned to the enemy position, led his men in an attack against the gun and silenced it.

The second date in his Citation occurred a couple of nights later on August 8 after the footbridge had been destroyed by enemy artillery. Beach and a few others volunteered to rebuild the bridge. He swam back and forth across the river under fire numerous times, resetting the posts of the bridge so it could be used by the Infantry the next day.

Beach was one of three soldiers of the 4th Division who received the French Military Medal (Médaille Militaire) during World War One and one of only 304 soldiers of the American Expeditionary Forces to be given that award. In addition, Beach was twice awarded the French Croix De Guerre.

He served in the Toulon Sector, the St. Mihiel Offensive

and the Meuse-Argonne Offensive. Beach served in Germany on occupation duty with the 4th Division and returned home as 1st Sergeant of Company C 4th Engineers aboard the troopship *U.S.S. Von Steuben* (seized German liner *Kronprinz Wilhelm*) leaving Brest, France July 21, 1919 and arriving in the United States on July 29, 1919. The January 1920 Census showed him as a 1st Sergeant with the skeleton organization of the 4th Division at the camp at Gary, Indiana. The date of his discharge could not be found.

On May 10, 1921 Beach enlisted as a Private in the United States Marine Corps. He served on Guam for several years, eventually being promoted to Corporal and apparently ended his enlistment in the Marines in 1926 or 1927. At some time between 1920 and 1925 he married Adalien Pitts. They had a daughter in 1925. The 1930 Census showed Beach living with his wife and daughter in Corte Madera, Marin County, California where he was employed as a Civil Engineer for a Railroad Company and indicated he was a veteran of the World War. The 1940 Census showed him living with his wife, daughter, and two sons in Union, Humboldt County, Nevada where he was employed as a Civil Engineer in Government Service.

Beach went back into military service during WWII being commissioned an officer in the Army Corps of Engineers on May 26, 1942 and served on active duty in the United States. He was discharged as a Major on August 18, 1953. William B. Beach died of coronary problems at the age of 61 on August 30, 1954 and is buried in Arlington National Cemetery.

Harold V. Beal

Corporal, United States Army
Army Service Number: 562801
Battery A 13th Field Artillery Regiment 4th Division
Date of Action: August 13, 1918
Location: Near Chery-Chartreuve, France
War Department General Orders 89 (1919)
Medal # 2154

CITATION

Corporal Beal displayed unusual courage in repairing shattered telephone lines during a heavy barrage under direct observation by the enemy. He was repeatedly knocked down by concussion of shells and he was painfully wounded in the shoulder by a bursting shell but he continued at his work until it was completed without seeking medical aid.

DETAILS

Harold Vest "H.V." Beal was born in Oak Ridge, Cape Girardeau County, Missouri on January 8, 1899, the son of Dock and Emma Beal. He was drafted into the Army on May 9, 1917 and was assigned to Headquarters Company 13th Field Artillery. On September 10, 1917 he was transferred to Battery A 13th Field Artillery. Beal sailed to France as a Private 1st Class in Battery A 13th Field Artillery 4th Division aboard the troopship *U.S.S. Great Northern* on May 22, 1918. As his emergency contact, he listed his father Dock in Oak Ridge, Missouri. In France in June 1918 he was promoted to Corporal. He served in the Ainse-Marne Offensive.

Beal was awarded the Distinguished Service Cross along with Private James Behan, his fellow soldier from Battery A for their actions on August 13, 1918 in the Vesle Sector north of the village of Chery-Chartreuve. On that date their Battery and the Artillery Brigade of the 4th Division were temporarily attached to the 77th Division. The Infantry units of the 4th Division were relieved in place along the Vesle River by elements of the 77th Division on the night of August 11 but the artillery units of the 4th Division were held in the front lines until the artillery of the 77th Division could be brought up, which did not fully take place until August 17. On August 13, Beal and Behan worked to repair field telephone lines that were cut by enemy artillery even though they were wounded by shrapnel and knocked down repeatedly by concussion from German shell fire. They continued their work until they got the lines repaired and only then did they head to the aid station to see about their wounds.

Beal served in the Toulon Sector, the St. Mihiel and Meuse-Argonne Offensives and then on occupation duty in Germany with the 4th Division. He returned to the United States with Battery A 13th Field Artillery aboard the transport *U.S.S. Zeelandia* leaving Brest, France July 18, 1919 and arriving at Brooklyn, New York July 31, 1919. He was discharged on August 6, 1919.

Beal went to live with his father and mother in Apple Creek, Missouri. In June 1922 he married Viola McLane. The 1940 Census showed Beal and his wife living in Cape Girardeau, Missouri where he was a salesman. Harold V. Beal died at the age of 64 on January 23, 1963 and is buried in Cape County Memorial Park Cemetery, Cape Girardeau, Cape Girardeau County, Missouri.

Harold Bedolfe

Sergeant First Class, United States Army
Army Service Number: 568256
Company B 4th Engineer Regiment 4th Division
Date of Action: August 5, 1918
Location: Near St. Thibaut, France
War Department General Orders 5 (1920)
Medal # 5684

CITATION

Sergeant First Class Bedolfe went forward, exposed to intense rifle, machine gun, and artillery fire, and carried on his back, with the assistance of another soldier, a badly wounded comrade to a shell hole, thus saving the life of the wounded soldier. In the performance of this act Sergeant Bedolfe was severely wounded.

DETAILS

Harold Fauntleroy Bedolfe was born in San Francisco, California on September 17, 1886, the son of Maitland and Sarah Bedolfe. When he registered for the draft in June 1917 he was living in Billings, Montana and working as an Insurance Lister with American Casualty Company. He entered the Army on August 11, 1917. Bedolfe sailed to France as a Private in Company B 4th Engineers 4th Division aboard the troopship *U.S.S. Martha Washington* (seized Austrian liner) on April 30, 1918. As his emergency contact, he listed his brother Robert in Tacoma, Washington. At some time after reaching France he was promoted to Sergeant 1st Class though it is not clear if he was promoted to that rank before or after his award actions.

Bedolfe was awarded the Distinguished Service Cross for his actions in the Aisne-Marne Offensive during the heavy fighting at the Vesle River near the village of St. Thibaut. For several days the Infantry had been trying to cross the river, with the Engineers giving as much help as they could by building and emplacing footbridges upon which the Infantry could cross in different locations along the banks of the river. As the Germans destroyed the bridges the Engineers rebuilt them again and again. The Engineers were under constant fire from the enemy the entire time and took heavy casualties. On August 5, 1918 Bedolfe went out forward of his position and with another soldier succeeded in rescuing a wounded comrade under heavy fire and was himself wounded in the act.

After recovering from his wounds he served with the 4th Division on occupation duty in Germany. Bedolfe returned to the United States with a detachment from the 4th Engineers aboard the troopship *U.S.S. Agamemnon* (seized German liner *Kaiser Wilhelm II*) leaving Brest, France on September 17, 1919 and arriving at Hoboken, New Jersey September 26, 1919. He was discharged on October 7, 1919.

The 1920 Census showed Bedolfe living in Buxton, Washington County, Oregon where he was employed as a railroad construction foreman. The 1930 Census showed him married to Ethel Threlkel and living with her family at Placer, California where he was the manager of a farm and indicated he was a veteran of the World War. The 1940 Census showed him and his wife living in Placer, California and working on his own farm. Harold F. Bedolfe died at the age of 65 at San Francisco, California on June 1, 1951 and is buried in Newcastle Cemetery, Newcastle, Placer County, California.

James P. Behan

Private, United States Army
Army Service Number: 562804
Battery A 13th Field Artillery Regiment 4th Division
Date of Action: August 13, 1918
Location: Near Chery-Chartreuve, France
War Department General Orders 89 (1919)
Medal # 2156

CITATION

Private Behan displayed unusual courage in repairing shattered telephone lines during a heavy barrage under direct observation by the enemy. He was repeatedly knocked down by concussion of shells and his helmet was smashed by a bursting shell but he continued at his work until it was completed, without seeking medical aid.

DETAILS

(His middle initial is given as P in all Citation lists of the Distinguished Service Cross and in both passenger lists of the ships on which he sailed to and from France. His registration for Selective Service during World War Two gives his middle name as Patrick. His grave marker and obituary give his middle initial as F and middle name as Franklin.)

James Behan was born in New Orleans, Louisiana on November 2, 1899, the son of William and Mattie Behan. He enlisted in the Regular Army on December 6, 1917 and was assigned to Headquarters Company 13th Field Artillery. On January 28, 1918 he was transferred to Battery A. He sailed to

France as a Private in Battery A 13th Field Artillery 4th Division aboard the troopship *U.S.S. Great Northern* on May 22, 1918. As his emergency contact, he listed his mother in New Orleans, Louisiana. He served in the Ainse-Marne Offensive.

Behan was awarded the Distinguished Service Cross along with Private 1st Class Harold Beal, his fellow soldier from Battery A for their actions on August 13, 1918 in the Vesle Sector north of the village of Chery-Chartreuve. On that date their Battery and the Artillery Brigade of the 4th Division were temporarily attached to the 77th Division. The Infantry units of the 4th Division were relieved in place along the Vesle River by elements of the 77th Division on the night of August 11 but the artillery units of the 4th Division were held in the front lines until the artillery of the 77th Division could be brought up, which did not fully take place until August 17. On August 13 Behan and Beal worked to repair field telephone lines that were cut by enemy artillery even though they were wounded by shrapnel and knocked down repeatedly by concussion from German shell fire. They continued their work until they got the lines repaired and only then did they head to the aid station to see about their wounds.

Behan served in the Toulon Sector, the St. Mihiel and Meuse-Argonne Offensives and then on occupation duty with the 4th Division in Germany. On March 1, 1919 he was promoted to the rank/position of Wagoner. He returned to the United States as a Wagoner with Battery A 13th Field Artillery aboard the transport *U.S.S. Zeelandia* leaving Brest, France July 18, 1919 and arriving at Brooklyn, New York July 31, 1919. He was discharged on August 14, 1919.

Behan returned to New Orleans, Louisiana where he married Jennie Gregory in June 1926. They had a daughter. He worked for Waterman's Steamship Company and then with the New Orleans Fire Department, eventually reaching the rank of Lieutenant with the Fire Department. The 1940 Census showed him living with his wife Jennie and daughter in New Orleans where he was employed as a Lieutenant with the Fire Department. James P (F). Behan died at the age of 65 on July 2, 1965 and is buried in St. Patrick Cemetery #1, New Orleans, Louisiana. His grave marker indicates additional service during World War Two. A 1945 directory for the city of New Orleans indicated he was employed by the U.S. Navy in New Orleans but no details of that service could be found.

Elwyn L. Berwick

Corporal, United States Army
Army Service Number: 563217
Battery C 13th Field Artillery Regiment 4th Division
Date of Action: August 13, 1918
Location: Near Chery-Chartreuve, France
War Department General Orders 27 (1920)
Medal # 2287

CITATION

When an enemy shell struck his battery position, setting fire to a powder drum and killing or wounding 30 men, Corporal Berwick, though himself wounded went into the burning dump at imminent risk to his life and assisted in extinguishing the flames. He then assisted in removing the other men before securing aid for himself. Refusing to be evacuated, he reported back to his battery with one arm in a sling and resumed his place as gunner.

DETAILS

Elwyn Lee Berwick was born in Oakland, Alameda County, California on July 3, 1895, the son of James and Kate Berwick. He entered the Army on May 31, 1917. Berwick sailed to France as a Corporal with Battery C 13th Field Artillery on the troopship *U.S.S. Great Northern* on May 22, 1918. As his emergency contact, he listed his sister Mrs. James Gunter in Oakland, California. He served in the Ainse-Marne Offensive.

Berwick was awarded the Distinguished Service Cross for his actions in the Vesle Sector. The 4th Division was pulled out of the front lines for a rest on August 11, 1918 but the Artillery Brigade of the Division remained at the front and was attached to the 77th Division. On August 13, German heavy artillery fire struck Berwick's Battery, killing five men, wounding twenty-seven men including Berwick and damaging two of the 155mm guns of the Battery. Ignoring his own wounds, Berwick helped to put out the fire the shelling had caused and assisted in the carrying of wounded men to the forward aid station. After getting his wounds bandaged, he reported back for duty with his Battery which had continued to fire at the enemy with the remaining guns that had not been damaged. He continued his duties as a gunner even though he had one arm in a sling.

Berwick was also awarded the French Croix De Guerre. He served on occupation duty in Germany with the 4th Division and returned to the United States with Battery C 13th Field Artillery aboard the transport *U.S.S. Zeelandia* leaving Brest, France July 18, 1919 and arriving at Brooklyn, New York on July 31, 1919. He was discharged on August 11, 1919.

Berwick returned to California and married Mable Robinson on January 1, 1923. The 1930 Census showed him living in Oakland, Alameda County, California where he was employed as a salesman for a creamery and indicated he was a veteran of the World War. After 1932 he was awarded the Purple Heart for the wounds he received in World War One. When he registered for the draft in 1942, he indicated that he was unemployed. Elwyn L. Berwick died at the age of 60 on April 15, 1956 and is buried in Golden Gate National Cemetery, San Bruno, San Mateo County, California. His report of interment indicated that at the time of his death he had a wife named Dorothy.

Fred E. Billman

Private, United States Army
Army Service Number: 570919
Medical Detachment 47th Infantry Regiment 4th Division
Date of Action: July 29 - 30, 1918
Location: At Sergy, France
War Department General Orders 35 (1919)
Medal # 1847

CITATION

Private Billman displayed conspicuous bravery by administering first aid to wounded soldiers in areas swept by shell and machine gun fire.

DETAILS

Fred Elias Billman was born in New Tripoli, Lynn Township, Lehigh County, Pennsylvania on January 11, 1899, the son of Harry and Rosalie Billman. He enlisted in the Regular Army on July 10, 1917 at Fort Slocum, New York and was assigned to the Medical Detachment of the 47th Infantry at the Camp at Syracuse, New York on August 2, 1917. Billman sailed as a Private with the Medical Detachment of the 47th Infantry to France aboard the Italian troopship *Caserta* on May 10, 1918. As his emergency contact, he listed his mother Rose in Wind Gap, Pennsylvania.

Billman was awarded the Distinguished Service Cross for his actions in the Ainse-Marne Offensive. First and Third Battalions of the 47th Infantry were engaged in battle in and around the town of Sergy from July 29 - August 1, 1918. The

Americans were under intense fire by German artillery and aggressive attacks by enemy aircraft. During four days at Sergy the two Battalions of the 47th Infantry lost 27 officers and 462 enlisted men killed, wounded or gassed and 6 enlisted men missing in action. As a medic, Billman constantly moved across the battlefield at Sergy on July 29 and 30, attending to wounded soldiers while under heavy fire, saving lives and serving as an inspiration of courage to his fellow countrymen.

Billman was severely wounded by a gas attack in the Vesle Sector on August 7, 1918, spent time in hospital, returned to duty and served in the Meuse-Argonne Offensive.

Billman served with the 4th Division on occupation duty in Germany and returned to the United States with the Sanitary Detachment (Medical Detachment) of the 47th Infantry aboard the troopship *U.S.S. Mobile* (seized German liner *Cleveland*) leaving Brest, France July 16, 1919 and arriving at Hoboken, New Jersey July 27, 1919. He was discharged at Camp Dix, New Jersey on August 2, 1919.

Billman married Rosalie Mae Blue on June 12, 1925 and resided with her in Indiana. The 1930 Census showed Fred and Rosalie living in Indianapolis, Indiana where he was employed as a salesman for a phone company and indicated he was a veteran of the World War. By the 1970's they were living in Sarasota, Florida where Rosalie died in 1979. Fred E. Billman died at the age of 87 at Sarasota, Florida on October 27, 1986.

Michael D. Belis

Edward S. Blackman

Corporal, United States Army
Army Service Number: 572390
Company G 59th Infantry Regiment 4th Division
Date of Action: on or about October 5, 1918
Location: In the Bois de Fays, France
War Department General Orders: None discovered

Note: No War Department General Order numbers could be found for this award. The Citation is listed in DECORATIONS UNITED STATES ARMY 1862-1926 SUPPLEMENT II published by the Adjutant General's Office in 1939 with dates of awards presented as being from the period of July 1, 1937 - June 30, 1938.
Medal # 4158

Note: No issue records exist for the medal number range of 3800-5500. Assignment of the number 4158 to Edward S. Blackman is based on a medal found with that number officially stamped on it and his name engraved on the back.

CITATION

Corporal Blackman repeatedly volunteered to carry messages through intense barrages of artillery fire in order to obtain essential information and to keep his battalion commander informed as to the progress of the battle. While performing this mission he was severely wounded and died later as a result of wounds received during this action.

DETAILS

Edward S. Blackman was born in Harvey, Cook County, Illinois on March 6, 1893, the son of Samuel and Ida Blackman. He and his family were living in Lowell City, Massachusetts in 1900 and by the time Blackman entered the service he and his family had moved to Providence, Rhode Island. When he registered for the draft in June 1917, he indicated he was employed as a Tool-maker-machinist for the E. Jenckes Knitting Machine Company in Pawtucket, Rhode Island. He sailed to England as a Private in Company G 59th Infantry 4th Division aboard the armed British troopship *RMS Olympic* on May 5, 1918. As his emergency contact, he listed his father Samuel in Providence, Rhode Island. From England he and his Regiment were sent over to France. Blackman served in the Aisne-Marne Offensive, the Vesle Sector, the Toulon Sector and the St. Mihiel Offensive. At some time in France, between May and October 1918, he was promoted to Corporal.

Blackman was awarded the Distinguished Service Cross for his actions in the Meuse-Argonne Offensive. On October 5, 1918 while the rest of his Regiment was held in reserve, 2nd Battalion 59th Infantry including Blackman and his Company were sent into the wooded area known as the Bois de Fays as reinforcements for the 58th Infantry. As the furthermost penetration by the 4th Division into the Bois de Fays as of October 5 his location was subjected to intense German machine gun and artillery fire and considerable air attacks. Though he could have delegated the task to a junior rank, Blackman took upon himself the dangerous job of runner in order to maintain communications in the Bois de Fays. He was seriously wounded while carrying messages across this fire swept battleground on or about October 5 and died of his wounds October 15, 1918.

(See the entry for Joseph Bassi for a description of the battlefield conditions in the Bois de Fays upon which Blackman ventured to carry his messages.)

Edward S. Blackman was buried in cemetery # C-230 at the village of Brieulles-sur-Meuse. On June 10, 1919 he was reinterred in Plot 3 Section 109 Grave # 126 at the Argonne American Cemetery # 1232. On September 19, 1921 he was reinterred and laid to his final rest in Plot H Row 39 Grave 4 in the Meuse-Argonne American Cemetery and Memorial, Romagne-sous-Montfaucon, Departement de la Meuse, Lorraine, France.

From 1930-1933 the United States Government operated a program under which mothers or widows of American soldiers killed in World War One and buried in cemeteries in England, Belgium, or France could visit their loved one's grave at the expense of the Federal Government. The program was known as the Pilgrimages of Gold Star Mothers and Widows, or more commonly the Gold Star Pilgrimages. During the life of the program over 6,000 "Gold Star Pilgrims" visited the graves of their sons or husbands. The government paid for everything including transportation, lodging, a bouquet to lay on the grave, and an official photographer to take a photo of each woman standing at their loved one's grave.

In 1930 Mrs. Ida M. Blackman the mother of Edward S. Blackman declined the offer to visit her son's grave in France under the U.S. Government program of Gold Star Pilgrimages.

Blackman's father died in 1932. His mother died in 1937.

His Distinguished Service Cross was presented to his sister Mrs. Harriet McMillan.

Homer J. Bleau

Sergeant, United States Army
Army Service Number: 560708
Company A 59th Infantry Regiment 4th Division
Date of Action: September 29, 1918
Location: Near Brieulles, France
War Department General Orders 89 (1919)
Medal # 7617

CITATION

When his company was held up by heavy artillery and machine gun fire, Sergeant Bleau displayed exceptional bravery and devotion to duty in leading his platoon across an open field in an attack upon an enemy machine gun nest. Even after receiving a wound, from the effects of which he died next morning, he remained with his men, encouraging them on and inspiring them by his fortitude.

DETAILS

Homer Joseph Bleau was born in Canada on September 22, 1892, the son of Auguste and Amanda Bleau. The 1910 Census showed him living at the Industrial School for Boys at Lansing, Michigan. Though the actual date he entered the Army could not be found, when he entered the Army he listed his home of record as Munising, Michigan. He sailed to England as a Corporal in Company A 59th Infantry 4th Division aboard the British troopship *Megantic* on May 3, 1918. As his emergency contact, he listed his uncle Adolph Monnette in Munising, Michigan. From England he and his Regiment were sent over

Michael D. Belis

to France. At some time in France between May and September 1918 he was promoted to Sergeant. Bleau served in the Aisne-Marne Offensive, the Vesle Sector, the Toulon Sector and the St. Mihiel Offensive.

He was awarded the Distinguished Service Cross for his actions in the Meuse-Argonne Offensive. On September 29, 1918 the 59th Infantry as part of the 4th Division's 8th Brigade attacked and together with the 58th Infantry cleared the wooded area known as the Bois de Brieulles. Heavy enemy fire prevented the Brigade from securing the next objective, which was the next wooded area known as the Bois de Fays. With the officers of his platoon all wounded, Bleau, as a Sergeant, led the platoon across the open ground between the two wooded areas in an attack against a German machine gun nest. During the attack Bleau was severely wounded but remained in command. He died of his wounds the next day September 30, 1918.

Bleau was buried in Plot # 3 Grave # 99 in the American battlefield cemetery # 702 at the village of Septsarges.

His Distinguished Service Cross was presented to his brother Charles Bleau.

On March 20, 1919 Homer J. Bleau was reinterred in Plot # 3 Section 65 Grave # 111 in the Argonne American Cemetery # 1232. He was reinterred and laid to his final rest on October 22, 1921 in Plot F Row 18 Grave 26 in the Meuse-Argonne American Cemetery and Memorial, Romagne-sous-Montfaucon, Departement de la Meuse, Lorraine, France.

Guy W. Boardman

Private, United States Army
Army Service Number: 2280987
Company A 59th Infantry Regiment 4th Division
Date of Action: July 19, 1918
Location: Near Courchamps, France
War Department General Orders 108 (1919)
Medal # 7564

CITATION

Though he had been wounded in the ankle, Private Boardman crawled out from a shell hole under heavy machine gun fire and made several trips to a small stream 100 yards away for the purpose of filling the canteens of his wounded comrades, until he was ordered to the rear for medical aid. He was later killed in action while charging an enemy machine gun nest.

DETAILS

Guy Wallace Boardman was born in Kenton, Hardin County, Ohio on November 13, 1894, the son of William and Sadie Boardman. In his registration for the draft in 1917 he listed his home of record as Hughson, California and his occupation as farmer. He entered the army on September 21, 1917. Boardman sailed to England as a Private in Company A 59th Infantry 4th Division aboard the British troopship *Megantic* on May 3, 1918. As his emergency contact, he listed his father William in Hughson, California. From England he and his Regiment were sent over to France.

Boardman was awarded the Distinguished Service Cross for his actions in the Aisne-Marne Offensive. Attached to and under direct command of the French 164th Division the 59th Infantry led the advance on July 19, 1918 in the darkness of early morning in a line east of the village of Courchamps. German artillery, machine gun, and small arms fire took a heavy toll on Boardman's 1st Battalion. The Battalion Commander was wounded, and seven other officers of the Battalion were either killed or wounded. The French were amazed at how the American soldiers of the 59th Infantry in their first engagement of the war continued the attack despite their ranks being depleted by men falling everywhere as bullets and shrapnel cut them down. Their ammunition dwindling, many of the doughboys advanced with fixed bayonets on their empty rifles. The attack was stopped by the intensity of the enemy resistance and the Battalion pulled back to the Courchamps-Priez road where they dug in and held their position.

His Citation states he was wounded in the ankle. Another report indicated that during the advance Boardman had been shot through both legs, yet still remained on the battlefield and performed the actions which earned him the Distinguished Service Cross. Ignoring his wounds, he crawled through enemy machine gun fire several times to a stream one hundred yards away to fill canteens with water for his fellow wounded comrades. After being ordered to the rear for medical aid he was admitted to a hospital where he remained until rejoining his Regiment on September 20, 1918.

At some time after his award action, he was promoted to Private 1st Class. Eleven days after returning to the front lines from the hospital Boardman was killed in action on October 1, 1918 in the Meuse-Argonne Offensive near the wooded area known as the Bois de Brieulles. He and his Company had reached the German trenches on Hill 281, north of the woods where they were stopped by heavy fire from across the Meuse River. Boardman's Company organized a defensive position and carried out patrols. In this environment, Boardman was killed by enemy fire in an attack against a machine gun nest.

Guy W. Boardman was buried in Plot # 1 Grave # 17 in the American battlefield cemetery # 989 at the village of Brieulles-sur-Meuse.

His Distinguished Service Cross was presented to his father William Boardman.

On June 6, 1919 Boardman was reinterred in Plot # 1 Section 50 Grave # 25 in the Argonne American Cemetery # 1232. Guy W. Boardman was reinterred and laid to his final rest on November 19, 1921 in Plot C Row 41 Grave 35 in the Meuse-Argonne American Cemetery and Memorial, Romagne-sous Montfaucon, Departement de la Meuse, Lorraine, France.

On June 3, 1931 Mrs. Sadie G. Boardman, the mother of Guy W. Boardman sailed aboard the *U.S.A.T. Republic* (seized

German liner *S.S. President Grant*) and visited her son's grave in France under the Gold Star Pilgrimage program. See the entry for Edward S. Blackman for an explanation of the Gold Star Pilgrimages.

Frank C. Bolles

Colonel, United States Army
Commanding Officer 39th Infantry Regiment 4th Division
Date of Action: September 26 and September 28, 1918
Location: Near Septsarges, France and near Bois de Fays, France
War Department General Orders 14 (1923)
Medal # 3177

CITATION

On September 26 Colonel Bolles personally directed the assaulting battalion of his regiment when the line was temporarily held up by hostile fire, leading the attacking troops forward to their objective. After reaching the objective, terrific hostile fire caused many casualties, and the line was beginning to waver when Colonel Bolles assisted in the reorganization of the line and by his personal example of courage and fearlessness encouraged his men to hold in the face of the withering machine gun and artillery fire until the flank division had advanced abreast. On September 28, he rallied his men under the sweeping fire of machine guns, minenwerfer and artillery and although painfully wounded, personally assisted in the reorganization of the positions.

DETAILS

Frank Crandall Bolles was born in Elgin, Illinois on September 25, 1872, the son of Elisha and Harriet Bolles. He entered the U.S. Military Academy in June 1891. He was at or near the bottom of his class for three years and in 1894 he was discharged from the Academy, reappointed, and repeated his second-class year (3rd year.) He graduated from the Academy June 12, 1896 number 69 out of 73 graduates and was commissioned a 2nd Lieutenant of Infantry. Between 1898 and 1906 he served three deployments to the Philippines during the War with Spain, the Philippine Insurrection, and the Moro Wars. During those deployments he was wounded twice, awarded a Citation for gallantry, and recommended for the Medal of Honor.

By 1910 he was married to Irene Pettit and had a son. By 1916 he had risen through the ranks to Major. In 1917 he was appointed to the wartime temporary rank of Colonel. His and his wife had a daughter born in February 1918. He was assigned command of the 39th Infantry 4th Division on March 8, 1918 at Camp Greene, North Carolina. Bolles sailed to France as commander of the 39th Infantry aboard the Italian troopship *Duca D'Aosta* on May 10, 1918. As his emergency contact, he listed his wife in Hempstead, Long Island, New

York. Bolles commanded his Regiment in the Aisne-Marne Offensive, the Vesle Sector, the Toulon Sector and the St. Mihiel Offensive.

He was awarded the Distinguished Service Cross for his actions on two separate dates in the Meuse-Argonne Offensive. The first date was September 26, 1918, the first day of the Offensive. On that date, rather than remain at the command post, he personally led 1st Battalion 39th Infantry in the attack against Septsarges. Throughout September 26 and September 27, he was side by side with his men in the front line, infecting them with his courage. When German artillery shattered the 1st and 3rd Battalions and they began to fall back, Bolles assisted in stopping the retreat and helped to organize the disheartened men into defensive positions. His Executive Officer Lieutenant Colonel William E. Holliday was killed in the battle on September 27.

The second date in his Citation was September 28, 1918 when Bolles led the 39th Infantry in yet another advance and personally kept a part of the line from falling back under heavy fire. He was wounded that day by a machine gun bullet to his right leg and forced to retire from the battlefield. The official history of the 39th Infantry in World War One described the effect he had upon the soldiers he commanded:

> "His presence on the front line, however, was much missed by all who had become accustomed to seeing him where the fighting was the fiercest, and whereby his indifference to personal safety he inspired his subordinates with a fearlessness and a determination to go forward."[5]

After being wounded, Bolles was out of the action and in the hospital from September 28 until November 12, 1918 when he resumed command of the Regiment. He served on occupation duty in Germany with the 4th Division. On March 18, 1919, Bolles was decorated with the Distinguished Service Cross by the Commanding General of the American Expeditionary Forces John J. Pershing in a 4th Division awards ceremony at Büchel, Germany.

Bolles brought the 39th Infantry home aboard the troopship *U.S.S. Leviathan* (seized German liner *Vaterland*) leaving Brest, France July 30, 1919 and arriving at Hoboken, New Jersey August 6, 1919. For his actions in France he was also awarded the Distinguished Service Medal, the French Officer of the Legion of Honor (Officier de la Légion d'Honneur) and the French Croix De Guerre with gilt star. He was one of 20 soldiers of the 4th Division and one of only 399 soldiers of the American Expeditionary Forces to be awarded the Italian War Merit Cross (Croce al Merito di Guerra).

Bolles had been recommended for the Medal of Honor and issued a Citation for gallantry for his bravery and command as a 2nd Lieutenant of Company F 18th Infantry in charge of a Gatling gun detachment in battle on the Island of Panay in 1899. The recommendation had never been acted upon and in 1923 was fulfilled by the award of a Distinguished Service Cross. The Citation for that award read:

> "To Second Lieutenant 18th Infantry Frank C. Bolles, United States Army during the attack on Jaro, Panay, Philippine Islands, February 12, 1899. Second Lieutenant Bolles exhibited conspicuous bravery and skill in handling his detachment and directing the fire of his piece. Even after he was seriously wounded in the leg he continued to encourage his men and could scarcely be prevailed upon to

desist from attempting mounting his horse when so crippled as to be unable to do so."

War Department General Orders Number 14 dated 1923 rescinded a previous Citation for the Distinguished Service Cross to Frank C. Bolles and awarded him a Distinguished Service Cross and an oak leaf cluster. In General Orders Number 14 the first award was for his actions in the Philippine Insurrection in February 1899 and the second or oak leaf cluster was for his actions in France of September 26 and 28, 1918. Though Bolles had already been presented with the Medal in a ceremony in Germany for his actions during the Meuse-Argonne Offensive General Orders Number 14 officially made that award the Oak leaf cluster to the medal.

Bolles continued his Army career, being promoted to Brigadier General in 1928. In 1932 he was awarded the Purple Heart with two oak leaf clusters for being wounded twice in the Philippines and once in World War One. Also in 1932, his Citation from the Philippines was converted to the Silver Star Medal. He was promoted to Major General in 1934. In 1935 he commanded the 2nd Infantry Division at Fort Sam Houston, Texas. He retired from the Army on September 30, 1936. In retirement he was President of the Union State Bank in San Antonio, Texas. Frank C. Bolles died at the age of 82 on July 14, 1955 and is buried in Arlington National Cemetery.

Michael D. Belis

Lawrence Boop

Private, United States Army
Army Service Number: 2659181
Company A 59th Infantry Regiment 4th Division
Date of Action: September 29, 1918
Location: Near the Bois de Brieulles, France
War Department General Orders 39 (1920)
Note: General Orders 39 (1920) rescinded a previous Citation for the Distinguished Service Cross issued to Lawrence Boup [sic] in General Orders 46 (1919)
Medal # 6111

CITATION

After all communication with the company on the left had been broken by intense machine gun and artillery fire, Private Boup volunteered and re-established liaison with the flank company, successfully performing this mission by going a distance of over 300 yards through terrific artillery and machine gun fire.

DETAILS

Lawrence Sylvester Boop was born in Glen Iron, Pennsylvania on November 7, 1895, the son of Alvah and Sadie Boop. Before entering the Army, he was married to Edna Walters and worked as a pipe fitter in Girard, Ohio. He was drafted into the Army as a Private on May 26, 1918 and inducted at Niles, Ohio. He took his training at Camp Gordon, Georgia. He sailed to England as a Private and replacement in Company #9 of Infantry of the Camp Gordon July Automatic Replacement Draft aboard the British troopship *RMS Carmania* on July 22, 1918.

As his emergency contact, he listed his wife Edna in Girard, Ohio. From England he was sent over to France where he was assigned to Company A 59th Infantry 4th Division on August 19, 1918. He served in the Toulon Sector and the St. Mihiel Offensive.

Boop was awarded the Distinguished Service Cross for his actions in the Meuse-Argonne Offensive. On the morning of September 29, 1918 Boop's Regiment was part of the attack of the 8th Brigade of the 4th Division out of the wooded area known as the Bois de Brieulles across open ground in an attempt to occupy the next wooded area known as the Bois de Fay. The open ground was swept by German machine gun, trench mortar and artillery fire and Boop made his way across that maelstrom of death to successfully re-establish communications with the unit on his Company's flank. For his bravery he was also awarded the French Croix De Guerre.

Boop returned to the United States several months ahead of his Regiment aboard the battleship *U.S.S. South Carolina* as part of a group of nearly 1000 soldiers from different units sent home under the heading of Casual Companies leaving Brest, France March 5, 1919 and arriving at Newport News, Virginia March 18, 1919. He was discharged on March 29, 1919.

The 1920 Census showed him living with his wife and son in Youngstown, Mahoning County, Pennsylvania where he was employed as a line driver at a steel mill. The 1930 Census showed him living with his wife, son and four daughters in Trumbull, Liberty County, Ohio where he was employed as a foreman of a steel mill and indicated he was a veteran of the World War. Lawrence Boop lived the rest of his life in Trumbull, Ohio and died at the age of 90 at St. Elizabeth Hospital in Youngstown, Mahoning County, Ohio on September 9, 1986.

Michael D. Belis

Gustav J. Braun

Captain, United States Army
47th Infantry Regiment 4th Division
Date of Action: July 29 - 30, 1918
Location: Near Sergy, France
War Department General Orders 46 (1919)
Medal # 1013

CITATION

No medical officer or first-aid man being present, Captain Braun, then First Lieutenant and Battalion liaison officer, established a first-aid station and worked throughout the day and night dressing the wounded. On both days he repeatedly went out himself in the most intense shell fire and carried wounded men to shelter. When the water supply was exhausted, he made several trips through heavy machine gun fire and filled canteens at water holes and a creek in front of the line.

Note: Braun had been promoted to the wartime temporary rank of Captain over a month before his award actions. The statement in his Citation of him being a First Lieutenant at the time refers to his Regular Army permanent rank.

DETAILS

Gustav Joseph Braun was born in Buffalo, Erie County, New York on January 15, 1895, the son of Gustav and Anna Braun. He joined the New York National Guard in 1910 and served on Mexican Border Service in 1916. He was commissioned a 2nd Lieutenant in the Regular Army on August 9, 1917, was immediately promoted to 1st Lieutenant, assigned to the 47th Infantry at the camp at Syracuse, New York and went to Camp Greene, North Carolina with his Regiment as the 4th Division was created. He sailed to France as a 1st Lieutenant in Company I 47th Infantry 4th Division aboard the Italian troopship *Caserta* on May 10, 1918. As his emergency contact, he listed his father Gustav in Buffalo, New York. He was promoted to Captain (temporary) in France on June 17, 1918.

Braun was awarded the Distinguished Service Cross for his actions in the Aisne-Marne Offensive. Two Battalions of the 47th Infantry fought for four days to capture and hold the village of Sergy and for the attack were attached to units of the 42nd Division. At the time Braun was assigned as Battalion liaison officer for 3rd Battalion 47th Infantry and for the fight at Sergy his Battalion was attached to the 168th Infantry Regiment. Braun's Battalion along with 1st Battalion 47th Infantry suffered heavy casualties during the battle. By the time they were relieved, the two Battalions had lost 27 officers and 462 enlisted men killed, wounded, or gassed and 6 enlisted men missing in action. Braun was slightly wounded on July 28, 1918 but remained at the front.

On July 29-30, 1918 with no medical personnel available at his location, Braun took it upon himself to set up an aid station and tend to the wounded. He ventured out on the battlefield to bring in wounded soldiers, saving several lives. He made repeated trips under heavy fire to obtain water for the wounded. On July 30, 1918 Braun became the fifth commander (temporary) of 3rd Battalion in two days after three previous commanders had been wounded and a fourth killed.

On July 31, 1918, Braun was gassed and pulled out of the front lines. The date of his return to the Regiment could not be found. He served on occupation duty in Germany with the 4th Division and returned to the United States in command of Company I 47th Infantry aboard the troopship *U.S.S. Mobile* (seized German liner *Cleveland*) leaving Brest, France July 16, 1919 and arriving at Hoboken, New Jersey July 27, 1919. He reverted back to his permanent rank of 1st Lieutenant on June 30, 1920.

Braun was promoted to the permanent rank of Captain on July 1, 1920. He married Anna Krieger on July 5, 1922. He continued to serve on active duty in the Army including an eight-year assignment as an instructor at the Infantry School at Fort Benning and several years of service in China. He was promoted to Major in 1935, Lieutenant Colonel in 1940 and Colonel (temporary) in 1942.

During WW2 he served in various positions in the 34th Infantry Division beginning in February 1943. In Headquarters Fifth U.S. Army General Orders 175 (1944) he was awarded an oak leaf cluster to the Distinguished Service Cross (indicating a second award of the medal) for his command of the 133rd Infantry Regiment of the 34th Division in combat in Italy in 1944. In February 1945 he was promoted to Brigadier General.

On March 17, 1945 as Assistant Division Commander of the 34th Infantry Division Braun was being flown in a light observation aircraft on a reconnaissance mission over the battlefield in Italy when his plane was shot down by enemy gunfire and he was killed. He was awarded the Silver Star Medal.

Braun was buried in the U.S. Military Cemetery # 5267, Pietramala Cemetery, Mount Beni, Italy. After the war his remains were returned to the United States and on December 20, 1948 Gustav J. Braun was laid to his final rest in Arlington National Cemetery.

Olaf Brekke

Private, United States Army
Army Service Number: 1424692
Company C 58th Infantry Regiment 4th Division
Date of Action: October 2 - 5, 1918
Location: Near Nantillois, France
War Department General Orders 81 (1919)
Medal # 7467

CITATION

Though wounded in the chest by shrapnel he refused to be evacuated, continuing his duties as runner for three days till his organization was relieved. He showed marked personal heroism in performing dangerous missions, exposing himself to heavy artillery and machine gun fire.

DETAILS

Olaf Brekke was born in Grafton, North Dakota on June 18, 1899, the son of Johannes and Lena Brekke. He enlisted in the Minnesota National Guard on March 1, 1917 and was later transferred to the National Army. He sailed overseas as a Private 1st Class of Infantry and replacement in Company #12 of the Camp Cody June Automatic Replacement Draft of Infantry aboard the transport *U.S.S. Titan* on June 28, 1918. As his emergency contact, he listed his father in Fergus Falls, Minnesota. At some time after reaching France, he was assigned to Company C 58th Infantry 4th Division. He was wounded in the Vesle Sector on August 8, 1918 and on an unknown date returned to duty.

Brekke was awarded the Distinguished Service Cross for his actions spanning several days in the Meuse-Argonne Offensive when he was wounded again yet continued his duties as a runner for three days in spite of his wounds. On October 2-3, 1918 Brekke's Company was consolidating its position on the reverse slope of Hill 295 north of the village of Nantillois and south of the wooded area known as the Bois des Ogons while waiting for ammunition for the supporting artillery to be brought up so another attack could be made against the wooded area known as the Bois de Fays. On October 4 the attack was begun again. Even though he was hit in the chest by shrapnel from German artillery, Brekke traversed the battlefield for three days from October 2-5 carrying his messages, continuing his missions as a runner, and presenting an example of courage to his fellow soldiers. He also received a Citation Star.

Because of his wounds, Brekke returned to the United States several months ahead of his Regiment as a Private of Infantry in the Camp Dodge Detachment of the 145th Infantry 37th Division destined for quick discharge aboard the British troopship *RMS Aquitania* (sister ship of the *Lusitania*) leaving Brest, France March 23, 1919 and arriving at New York City March 30, 1919. He was discharged on April 11, 1919.

The 1920 Census showed Brekke living with his parents and family in Fergus Falls, Minnesota and working as a laborer in an iron factory. In 1928 he married Ruby Williamson and settled in Cut Bank, Montana where he worked as a carpenter and rancher. He and his wife had two daughters and a son. The 1930 Census showed him living as a patient in the United States Veterans Hospital in Clark County, Montana and indicated he was a veteran of the World War. After 1932 he was awarded the Purple Heart for the wounds he received in World War

One. Olaf Brekke died at the age of 57 on April 27, 1957 and is buried in Crown Hill Cemetery, Cut Bank, Glacier County, Montana.

George N. Brigham

Corporal, United States Army
Army Service Number: 558268
Company I 47th Infantry Regiment 4th Division
Date of Action: August 10, 1918
Location: At St. Thibaut, France
War Department General Orders 15 (1919)
Medal # 1077

CITATION

Accompanied by another soldier Corporal Brigham penetrated the enemy's lines and patrolled a sector from the north bank of the River Vesle to the town of Bazoche. These two men entered an enemy dugout and killed two Germans, at the same time locating a machine gun emplacement. Corporal Brigham, though wounded, completed his mission before obtaining first aid.

DETAILS

George Newton Brigham was born in Ellington, Tolland County, Connecticut on February 13, 1896, the son of Frank and Minnie Brigham. Before entering the service, he was employed as a clerk in Rockville, Connecticut. He enlisted as a Private of Infantry in the Regular Army at Fort Slocum, New York on June 6, 1917 and was assigned to Company I 47th Infantry at the Camp at Syracuse, New York on June 12, 1917. On August 10, 1917 he was promoted to Private 1st Class. He was promoted to Corporal on October 15, 1917 and that same month went with his Regiment to Camp Greene, North Carolina where they became part of the 4th Division when it formed in December of 1917.

Brigham sailed to France as a Corporal in Company I 47th Infantry 4th Division aboard the Italian troopship *Caserta* on May 10, 1918. As his emergency contact, he listed his mother Minnie in Rockville, Connecticut. He served in the Aisne-Marne Offensive.

Though cited in different orders Brigham and Private Jacob Kreis also of Company I 3rd Battalion 47th Infantry both were awarded the Distinguished Service Cross for actions they performed together on August 10, 1918 in the Vesle Sector. After an unsuccessful attack against the village of Bazoches on August 9 the Battalion pulled back across the Vesle River near the village of St. Thibaut and dug in. During the day of August 10 aerial reconnaissance indicated the enemy may have evacuated Bazoches. Regimental headquarters directed 3rd Battalion to send out five patrols on the night of August 10 to confirm if that was the case. Each patrol consisted of a non-commissioned officer and a private. The men in the patrols were equipped with pistols, gas masks and canteens only. They were to cross the river at five different points and scout toward the main road (Route Nationale 31) which was north of Bazoches. Four of the patrols could not cross the river due to enemy defenses at their assigned crossing points.

Brigham and Kreis were the only patrol to make it across the river and were on the extreme right of the 47th Infantry's area of responsibility and into the area of the 59th Infantry. They made it to about 300 yards south of Route Nationale 31 and were prevented from going further by heavy enemy traffic on the road and enemy troop movements in and around Bazoches. On their return to the river while moving through some woods Brigham and Kreis encountered a German dugout and fought with the inhabitants. Brigham was shot through the neck during the engagement. He and Kreis made their way into the lines of the 59th Infantry where Brigham received medical treatment for his wound and remained at the aid station. Kreis reported back to their Company with the information they had obtained during their patrol. The details Kreis reported about the enemy's troop dispositions, machine gun emplacements and movements were dispatched to their Battalion and Regimental Headquarters, the 59th Infantry on their right and the French Army unit on their left. The next morning the enemy attacked 3rd Battalion, who were ready thanks to Brigham and Kreis' patrol and therefore were able to repulse the attack.

On August 12, 1918, Brigham was admitted to Base Hospital #27 at Angers, France. He was released from the hospital in time to rejoin his unit and serve in the Toulon Sector, the St. Mihiel Offensive and the Meuse-Argonne Offensive. Brigham was promoted to Sergeant on November 15, 1918.

He served on occupation duty in Germany with the 4th Division and returned to the United States with his Company aboard the troopship *U.S.S. Mobile* (seized German liner *Cleveland*) leaving Brest, France July 16, 1919 and arriving at Hoboken, New Jersey July 27, 1919. He was discharged at Camp Upton, New York on August 1, 1919.

Brigham returned to Ellington, Connecticut where he resumed work as a clerk and soon was married to Adelaphene Busher. The 1930 Census showed him living in Ellington with his wife, son and daughter where he was employed as a dealer at a lumber yard and indicated he was a veteran of the World War. George N. Brigham died at the age of 83 on March 13, 1979 and is buried in Grove Hill Cemetery, Rockville, Tolland County, Connecticut.

George Brown

Private, United States Army
Army Service Number: 560709
Headquarters Company 59th Infantry Regiment 4th Division
Date of Action: October 4 - 5, 1918
Location: Near Bois de Fays, France
War Department General Orders 71 (1919)
Medal # No assigned number discovered

CITATION

As a battalion runner Private Brown repeatedly exposed himself to intense artillery and machine gun fire, crossing open spaces in view of the enemy to deliver important messages. He aided largely in maintaining liaison and his courage was an inspiration to those near him.

DETAILS

George Hunt Brown was born in Glen Easton, Marshall County, West Virginia on May 12, 1901, the son of Charles and Virginia (Jennie) Brown. At sixteen years old he lied about his age and enlisted in the Regular Army on July 20, 1917. Brown sailed to England as a Private in Headquarters Company 59th Infantry 4th Division aboard the British troopship *Megantic* on May 3, 1918. As his emergency contact, he listed his uncle John Hunt in Moundsville, West Virginia. From England he and his Company were sent over to France. He served in the Aisne-Marne Offensive, the Vesle Sector, the Toulon Sector and the St. Mihiel Offensive.

Brown was awarded the Distinguished Service Cross for his

actions in the Meuse-Argonne Offensive. During the afternoon of October 4, 1918, 3rd Battalion 59th Infantry was sent into the wooded area known as the Bois de Fays to assist the 58th Infantry which had gained a foothold in the woods. For two days Brown carried messages to that Battalion from the Headquarters of the 59th Infantry which was established in another wooded area to the southeast known as the Bois de Brieulles. To do so he had to repeatedly cross large open areas between the two woods all the while subjected to enemy artillery and machine gun fire. His actions helped to maintain coordination of the two units' activities and his bravery inspired his fellow soldiers. At the time he was only seventeen years old.

He served on occupation duty in Germany with the 4th Division and returned to the United States as a Private in Headquarters Company 59th Infantry aboard the troopship *U.S.S. Mount Vernon* (captured German liner *Kronprinzessin Cecilie*) leaving Brest, France July 24, 1919 and arriving at Hoboken, New Jersey August 1, 1919. He was discharged on August 7, 1919.

He married Clara Holsey on December 26, 1925. The 1930 Census showed Brown living with his wife, five children and his brother in Moundsville, West Virginia where he was employed as a coal miner and indicated he was a veteran of the World War. The 1940 Census showed him still living in Moundsville with his wife, four sons and two daughters where he was still employed as a coal miner. When he registered for the draft in 1942, he indicated he worked for the Valley Coal Company. George H. Brown died at the age of 72 on June 1, 1973 and is buried in Riverview Cemetery, Moundsville, Marshall County, West Virginia.

Raymond Buma

Corporal, United States Army
Army Service Number: 556232
Machine Gun Company 39th Infantry Regiment 4th Division
Date of Action: September 26, 1918
Location: Near Cuisy, France
War Department General Orders 44 (1919)
Medal # 6769

CITATION

After all his squad members had become casualties, Corporal Buma alone continued to operate his gun and after his ammunition was exhausted, he ran from shell hole to shell hole picking up ammunition and carrying it back to his gun, resumed fire on the enemy that was very instrumental in the success of the attack. He was killed in action shortly afterwards.

DETAILS

Raymond (Rienk) Buma was born in Friesland, Netherlands on November 15, 1896, the son of Minne (Sijtzes) and Thersea (TietJe) Buma. His father immigrated to the United States in 1907 and in 1908 was able to bring the rest of his family including Raymond to live with him in Massachusetts. Before the war Buma was a machinist at the Whitin Machine Works in Whitinsville, Massachusetts. He entered the Army on April 16, 1917 and listed his home of record as Whitinsville. He sailed to France as a Private 1st Class in Machine Gun Company 39th Infantry 4th Division aboard the Italian troopship *Duca D'Aosta* on May 10, 1918. As his emergency contact, he listed his father Minne in Whitinsville, Massachusetts. At some time in France, he was promoted to Corporal. He served in the Aisne-Marne Offensive, the Vesle Sector, the Toulon Sector and the St. Mihiel Offensive.

Buma was awarded the Distinguished Service Cross for his actions in the Meuse-Argonne Offensive. On September 26, 1918 the first day of the Meuse-Argonne Offensive his Regiment fought its way through the village of Cuisy and took the village of Septsarges, making possible the capture of the village of Montfaucon by another unit. During the advance after his squad all became casualties, he single-handedly operated his crew served machine gun and laid down suppressive fire against the enemy. He dashed out alone under enemy fire to gather more ammunition for his gun and then remained at his position covering his unit's attack. Buma's actions on September 26 helped the 39th Infantry to reach their objective in the attack, the ridge one kilometer north of Septsarges. He was killed the next day by German artillery fire.

Note: An article published in the NCO Journal in 2018 states that during his award action Buma operated an M1917 Browning machine gun but is most likely incorrect. It appears that the Browning M1917 was not issued to the 4th Division in time for use in World War One and was only used in combat by American Divisions that deployed to France after June 1918. It is known that the 12th Machine Gun Battalion of the 4th Division was not issued the Browning until after the Armistice and it is quite likely the other machine gun units of the Division were not issued that weapon until hostilities ceased as well.

The Browning M1917 was not used in combat until the Meuse-Argonne Offensive which started on September 26, 1918. The machine gun units of the 4th Division were issued their French Model 1914 Hotchkiss guns on June 20, 1918 and trained with those guns until they went into their first action on July 18. The Division was in almost continual front-line action from mid-July to mid-October 1918. An examination of the operations of the Division from July 18 to September 26 indicates the longest stretch of rest and/or training time away from the front during that period was an eleven-day interval between its service in the Vesle and Toulon Sectors. It would have been impossible therefore to pull out any of the machine gun units from the Division, re-equip them and adequately train them on the Browning in time for the Meuse-Argonne Offensive. It is almost certain that Buma used the machine gun issued to all machine gun Companies of the 4th Division in June 1918 which was the French Model 1914 Hotchkiss.

Buma was buried in grave # 5 in the battlefield cemetery # 179 near the village of Nantillois.

His Distinguished Service Cross was presented to his father Minne Buma.

On June 5, 1919 Raymond Buma was reinterred in Plot # 2 Section 31 Grave # 82 in the Argonne American Cemetery # 1232. On July 25, 1921, his body was disinterred for preparation and shipment. His remains were returned to the United States aboard the United States Army Transport *U.S.A.T. Wheaton* leaving Antwerp, Belgium September 20, 1921 and arriving at Hoboken, New Jersey October 3, 1921. He was one of 2,624 American dead brought home aboard the *Wheaton* on that journey. Raymond Buma was laid to his final rest in Pine Grove Cemetery, Northbridge, Worcester County, Massachusetts. Buma Square in Northbridge, Massachusetts is named for him.

William J. Cahill

Private, United States Army
Army Service Number: 1684623
Company D 59th Infantry Regiment 4th Division
Date of Action: October 3 and October 9, 1918
Location: Near Bois de la Côte Lémont, France and in Bois de Fays, France
War Department General Orders 46 (1919)
Medal # 3313

CITATION

On October 3, while acting in the capacity of company runner Private Cahill carried messages to two platoons of his company through a heavy fire of machine guns and snipers. He successfully delivered the messages after crawling for a distance of 400 yards. On October 9, in company with one other runner he delivered messages to a platoon which was engaged in combat liaison duty in the Bois de Fays, passing through a severe artillery fire while in the execution of this mission.

DETAILS

William James Cahill was born in Dublin, Ireland on July 27, 1885. He immigrated to the United States at the age of sixteen, arriving at Boston, Massachusetts in July 1901. At the age of twenty-one he became a naturalized American citizen on September 21, 1906 at Manchester, New Hampshire. When he registered for the draft in 1917, he gave his birth year as 1888. Before entering the Army, he worked as a laborer for the J.H. Mendall Company in Manchester, New Hampshire. On his registration form he indicated he was married. He also stated that he had prior military service of seven and a half months in the British Army, though details of that service were not given. Upon entering the U.S. Army, he listed his home of record as Manchester, New Hampshire. He entered the Army on March 28, 1918.

Cahill sailed to England as a Private in Company M 59th Infantry 4th Division aboard the armed British troopship *RMS Olympic* on May 5, 1918. As his emergency contact, he listed his wife Agnes in Lowell, Massachusetts. From England he and his Company were sent over to France where at some time he was transferred to Company D. He served in the Aisne-Marne Offensive, the Vesle Sector, the Toulon Sector and the St. Mihiel Offensive.

He was awarded the Distinguished Service Cross for his actions in the Meuse-Argonne Offensive on two separate dates while serving as a runner for his Company. The first date was October 3, 1918. On that date Cahill's Regiment was defending the right flank of the 4th Division near the wooded area known as the Bois de la Côte Lémont with the U.S. 80th Division to their right. He carried messages to the platoons of his Company that were vital in informing them of the movements of both Divisions, crawling under fire for 400 yards to do so.

The second date in his Citation was October 9, 1918. On that day Cahill's Regiment had been pulled back from the front lines for a short rest but he and another runner were sent with messages to a platoon of the Company which was stationed in the wooded area known as the Bois de Fays where they were maintaining a liaison with the 58th Infantry. The runners had to venture through heavy enemy artillery fire to deliver their messages.

Cahill was also awarded the French Croix De Guerre. He served on occupation duty in Germany with the 4th Division and returned to the United States as a Private 1st Class in Company D 59th Infantry aboard the troopship *U.S.S. Mount Vernon* (captured German liner *Kronprinzessin Cecilie*) leaving Brest, France July 24, 1919 and arriving at Hoboken, New Jersey August 1, 1919. He was discharged on August 6, 1919.

William J. Cahill died at the age of 71 on December 15, 1956 and is buried in Saint Patrick Cemetery, Lowell, Middlesex County, Massachusetts.

Alberis Callewaert

Private, United States Army
Army Service Number: 2101504
Headquarters Company 58th Infantry Regiment 4th Division
Date of Action: July 18, August 4 and September 28, 1918
Location: Near Chézy, France; near Les Pres Farm, France and near Bois de Fays, France
War Department General Orders 64 (1919)
Medal # 7198

CITATION

Facing heavy fire Private Callewaert carried ammunition from regimental headquarters to the companies of the assaulting battalion, returning with prisoners. In a later engagement he carried and laid wire while under heavy fire from snipers, machine guns, and artillery, thus maintaining telephonic communication with the front-line companies. Subsequently, while endeavoring to establish telephonic communication, he was killed while carrying wire across ground swept by machine guns and artillery.

DETAILS

Alberis Charlie Callewaert was born in Belgium on November 24, 1889. At the time he entered the service he listed his occupation as Hotel Keeper and his home of record as St. Paul, Minnesota. He sailed to England as a Private in Headquarters Company 58th Infantry 4th Division aboard the British troopship *RMS Themistocles* on May 11, 1918. As his emergency contact, he listed his father Constant in Detroit, Michigan. From England he and his Company were sent over to France.

He was awarded the Distinguished Service Cross for his actions on three separate dates. The first date mentioned in his Citation of July 18, 1918 was the first day of the Aisne-Marne Offensive. Callewaert was a member of the Signal Platoon of Regimental Headquarters Company 58th Infantry and on that day he and the members of his platoon were assigned as ammunition carriers for the Infantry. He was cited for his bravery in carrying ammunition under fire to 2nd Battalion 58th Infantry as it attacked from the village of Chézy-en-Orxois toward the village of Chevillon. He returned to his Company from this detail bringing enemy prisoners he had captured.

The second date mentioned in his Citation of August 4, 1918 was also during the Aisne-Marne Offensive. On that date Regimental Headquarters of the 58th Infantry was established at Les Pres Farm northeast of Chery-Chartreuve as the Regiment attacked northward toward the Vesle River. Callewaert ventured out onto the battlefield under heavy fire to carry and lay wire for the field telephones connecting Regimental Headquarters with the rifle Companies.

Callewaert served in the Vesle Sector, the Toulon Sector and the St. Mihiel Offensive.

The third date mentioned in his Citation of September 28, 1918 was the third day of the Meuse-Argonne Offensive. During that time his Regiment was engaged in the attack against the wooded areas known as the Bois de Fays and the Bois des Ogons. Though his Citation gives the impression he was killed on September 28 Callewaert was actually killed on October 6, 1918. On that date the 58th Infantry had occupied much of the Bois de Fays but had halted its advance and was holding and improving its positions in the Bois de Fays while the Germans sent several small counterattacks against them.

Callewaert was killed while again moving across the battleground setting up field phone communications.

He was buried in Plot # 1 Grave # 18 in the American battlefield cemetery # 700 at the village of Brieulles-sur-Meuse.

His Distinguished Service Cross was presented to his father Constant Callewaert.

On May 14, 1919 Callewaert was reinterred in Plot # 1 Section 66 Grave # 8 in the Argonne American Cemetery # 1232. Alberis C. Callewaert was reinterred and laid to his final rest on October 26, 1921 in Plot H Row 21 Grave 4 in the Meuse-Argonne American Cemetery and Memorial, Romagne-sous Montfaucon, Departement de la Meuse, Lorraine, France.

Willis M. Campbell

Sergeant, United States Army
Army Service Number: 2108668
Company B 59th Infantry Regiment 4th Division
Date of Action: September 29, 1918
Location: Near Brieulles-sur-Meuse, France
War Department General Orders 50 (1919)
Medal # No assigned number discovered

CITATION

Sergeant Campbell made his way forward in the face of annihilating fire to the aid of a wounded comrade who was lying exposed to this great hazard and carried him across an open field to safety.

DETAILS

Willis Moser Campbell was born in New Castle, Lawrence County, Pennsylvania on July 26, 1895 the son of Thomas and Hannah Campbell. He was drafted into the Army as a Private on October 6, 1917 at New Castle, Pennsylvania. Campbell did his initial training at Camp Sherman, Ohio and completed further training with the 348th Infantry at Camp Pike, Arkansas. He was promoted to Private 1st Class on February 7, 1918. He was assigned to Company B 59th Infantry in the 4th Division at Camp Greene, North Carolina on April 14, 1918. Campbell sailed to England as a Private with Company B 59th Infantry 4th Division aboard the British troopship *RMS Megantic* on May 3, 1918. As his emergency contact, he listed his father Thomas in New Castle, Pennsylvania. From England he and his Company were sent over to France. Campbell served in the Aisne-Marne Offensive, the Vesle Sector, the Toulon Sector and the St. Mihiel Offensive. On September 2, 1918 he was promoted to Sergeant.

He was awarded the Distinguished Service Cross for his actions in the Meuse-Argonne Offensive on September 29, 1918. His Company was part of the 4th Division's 8th Brigade attack that cleared the wooded area known as the Bois de Brieulles but was stopped cold by heavy enemy fire before reaching the next wooded area known as the Bois de Fays southwest of the village of Brieulles-sur-Meuse. In the no man's land between the two wooded areas Campbell ventured out across the battlefield under extremely heavy artillery and machine gun fire, lifted up a wounded fellow soldier and carried him out of the line of fire.

Campbell served on occupation duty in Germany with the 4th Division and returned to the United States as a Sergeant in Company B 59th Infantry and part of Company K Third Army Composite Regiment aboard the troopship *U.S.S. Leviathan* (seized German liner *Vaterland*) leaving Brest, France September 1, 1919 and arriving at Hoboken, New Jersey September 8, 1919.

(See the entry for Joseph Bassi for an explanation of the Third Army Composite Regiment.)

Campbell was discharged at Camp Meade, Maryland on September 25, 1919.

At some time after that date he married Helen Jones. The 1930 Census showed him living with his wife and two sons in New Castle, Lawrence County, Pennsylvania where he was employed as a brick layer in construction work and indicated he was a veteran of the World War. The 1940 Census showed him and his family living in Clariton, Allegheny County, Pennsylvania where he was employed in Maintenance at a steel mill. Willis M. Campbell died at the age of 88 at Mercer, Pennsylvania on October 18, 1983.

Michael D. Belis

Charles F. Carbaugh

Private, United States Army
Army Service Number: 558049
Company F 47th Infantry Regiment 4th Division
Date of Action: August 9, 1918
Location: Southeast of Bazoches, France
War Department General Orders 21 (1919)
Medal # 1073

CITATION

Private Carbaugh was sent as a runner to direct a platoon of his company to assemble and return to its position. He displayed unusual leadership in performing his mission by himself, taking command of the disorganized unit, getting it well in hand, and leading it back under a hostile shelling without losses and without confusion.

DETAILS

Charles Frederick "Fred" Carbaugh was born in Stephens City, Frederick County, Virginia on November 29, 1893, the son of James and Minerva Carbaugh. Before entering the Army, he was a clerk with Westinghouse Electric Company in Pittsburg, Pennsylvania. He entered the Army on February 20, 1918. Carbaugh sailed to France as a Private in Company F 47th Infantry 4th Division aboard the troopship *U.S.S. Princess Matoika* (seized German liner *Prinzess Alice*) on May 10, 1918. As his emergency contact, he listed his father James in Stephens City, Virginia. Carbaugh served in the Aisne-Marne Offensive.

He was awarded the Distinguished Service Cross for his actions in the Vesle Sector. On August 9, 1918 most of Carbaugh's Battalion had crossed the Vesle River and was scattered in the area between the river and the village of Bazoches. German counterattacks had caused a platoon from Carbaugh's Company to retreat. He was sent as a runner to instruct that platoon to rejoin the rest of the Company in the line. He accomplished his mission by actually taking command of the leaderless and disordered unit and directing it back into position while under enemy artillery fire the entire time. Even though he was a Private at the time, his actions displayed exemplary leadership ability under heavy fire.

At some time after his award action, he was promoted to Sergeant. He was also awarded a Citation Star. Carbaugh served in the Toulon Sector, the St. Mihiel Offensive, and the Meuse-Argonne Offensive. He served with the 4th Division on occupation duty in Germany and returned to the United States as a Sergeant in Company F 47th Infantry aboard the troopship *U.S.S Mobile* (seized German liner *Cleveland*) leaving Brest,

France July 16, 1919 and arriving at Hoboken, New Jersey July 27, 1919. He was discharged on August 2, 1919.

Carbaugh returned to Frederick County, Virginia where he worked as a painter. On October 28, 1919 he married Virginia Lemley and together they eventually had seven children. The 1920 Census showed Carbaugh and his wife living in Opequon, Frederick County, Virginia where he was employed as a painter. The 1930 Census showed him living with his wife, two daughters and three sons in Strasburg, Shenandoah County, Virginia where he was employed as a Janitor at a public school and indicated he was a veteran of the World War. The 1940 Census showed him living with his wife, three daughters and four sons in Strasburg and indicated his employment as a painter at the Viscose Plant. When he registered for the draft in 1942, he indicated he was still living in Strasburg and worked for the American Viscose Corporation in Front Royal, Warren County, Virginia. Charles F. Carbaugh died of congestive heart failure at the age of 51 on July 4, 1945 and is buried in Riverview Cemetery, Strasburg, Shenandoah County, Virginia.

Marion H. Cardwell

First Lieutenant, United States Army
Headquarters Company 58th Infantry Regiment 4th Division
Date of Action: July 18, 1918
Location: Near Chevillon, France
War Department General Orders 10 (1920)
Medal # 5689

CITATION

After his company had failed in an attack on Hill 208 with the loss of two officers and sixty-five men, Lieutenant Cardwell reorganized the remaining men of his organization and personally led them in a second attack on the same objective. The advance was made in the face of heavy machine gun and trench mortar fire but due to the example and individual bravery of Lieutenant Cardwell the objective was taken and held.

DETAILS

Marion Herman Cardwell was born in Princeton, Kentucky on October 7, 1895, the son of John and Tille Cardwell. He entered Officers Training Camp at Fort Benjamin Harrison, Indiana on May 15, 1917. At some time between June 1917 and April 1918 he married Elizabeth Burns. He was commissioned a 2nd Lieutenant in the Officers Reserve Corps on August 15, 1917 and joined the 58th Infantry as it was forming up at Gettysburg, Pennsylvania. On October 26, 1917 he was offered and accepted a commission as a 2nd Lieutenant of Infantry in the Regular Army. Cardwell sailed to France as a 2nd Lieutenant of the 58th Infantry and part of the advance detachment of the 4th Division aboard the troop transport *U.S.S. Finland* on April 30, 1918. As his emergency contact, he listed his wife Elizabeth in Louisville, Kentucky. In France on June 12, 1918, he was promoted to the temporary wartime rank of 1st Lieutenant. He was assigned to Regimental Headquarters Company.

Cardwell was awarded the Distinguished Service Cross for his actions on July 18, 1918, the first day of the Aisne-Marne Offensive. For the attack that day, the Regimental Machine Gun Company and elements of the Regimental Headquarters Company along with Cardwell were attached to 2nd Battalion which itself was attached to and under direct command of the French Army's 13th Group, Chasseurs à Pied, 164th Division. The French and American units captured the village of Chevillon but had difficulty in seizing the high ground east of the village. It was in this area that Cardwell reorganized his depleted Company and led it in attacking and securing its objective while under heavy enemy fire, after a previous attack had failed. (The "Hill 208" mentioned in his Citation could not be found on 4th Division or 58th Infantry maps of the operation and is most likely a misidentification of Hill 172, the high ground to the east of Chevillon against which 2nd Battalion 58th Infantry spent most of July 18 attacking.) Cardwell's individual bravery in the attack so inspiring his men is credited with being the reason the hill was captured.

Cardwell served in the Vesle Sector, the Toulon Sector, the St. Mihiel Offensive, and the Meuse-Argonne Offensive. He was promoted to the wartime temporary rank of Captain of Infantry on October 28, 1918. He was also awarded a Citation Star and the French Croix De Guerre with gilt star.

Cardwell served with the 4th Division in Germany on occupation duty and returned to the United States as a Captain and the Regimental Intelligence Officer of the 58th Infantry aboard the troopship *U.S.S. Mount Vernon* (captured German liner *Kronprinzessin Cecilie*) leaving Brest, France July 24, 1919 and arriving at Hoboken, New Jersey August 1, 1919. He resigned from the Army on September 15, 1919.

The city directory for Louisville, Kentucky for 1921 showed Cardwell living in Louisville and indicated he was employed as the manager of Exide Battery Service Station. The 1930 Census showed him living with his wife Elizabeth and two daughters in Louisville, Kentucky where he was employed as a stockbroker and indicated he was a veteran of the World War. The city directories for Louisville, Kentucky showed Cardwell living in Louisville from 1925-1960 and indicated he was employed by J.J.B. Hilliard & Son (financial management firm) during all those years. Marion H. Cardwell died at the age of 78 on July 1, 1974 and is buried in Cave Hill Cemetery, Louisville, Jefferson County, Kentucky.

Michael D. Belis

James B. Carpenter
Private, United States Army
Army Service Number: 558226
Company H 47th Infantry Regiment 4th Division
Date of Action: August 9, 1918
Location: Near Bazoches, France
War Department General Orders 21 (1919)
Medal # 6358

CITATION

Private Carpenter responded to a call for volunteers to destroy a hostile machine gun, the approach to which was covered by fire from three other machine guns. With seven other soldiers he went forward and skillfully and boldly accomplished the mission. This courageous soldier has since been killed in action.

DETAILS

James Brad Carpenter was born in Cameron, Le Flore County, Oklahoma on October 11, 1895. When he registered for the draft in June 1917, he indicated he was employed as a Rig Builder for the Wallace Ingersoll Company in Avant, Oklahoma. Carpenter sailed to France as a Private in Company H 47th Infantry 4th Division aboard the troopship *U.S.S. Princess Matoika* (seized German liner *Prinzess Alice*) on May 10, 1918. As his emergency contact, he listed his sister Mrs. Lena Woods in Barber, Arkansas. Carpenter served in the Aisne-Marne Offensive.

He was awarded the Distinguished Service Cross for his actions in the Vesle Sector. After crossing the Vesle River on

August 7, 1918 and attempting an unsuccessful attack against the village of Bazoches, 2nd Battalion 47th Infantry including Carpenter and Company H dug in and held a line south of the village all the next day. Machine gun and sniper fire from Bazoches and the area in front of the village was causing casualties among the Americans so scouting parties were sent out to locate enemy machine gun nests and snipers and destroy them. On August 9, Carpenter volunteered for one of those missions and was successful in attacking and destroying a German machine gun while under heavy fire from several enemy machine guns. The next day, August 10, German aircraft dropped 40 to 50 bombs on his Battalion and followed the aerial attack with a heavy artillery barrage. At some time during that air attack and barrage, Carpenter was killed.

His body was not recovered. He is listed with the American Battle Monuments Commission as Missing In Action. The name of James B. Carpenter is inscribed on the Tablets of the Missing at Oise-Aisne American Cemetery and Memorial, Fere-en-Tardenois, Departement de l'Aisne, Picardie, France.

His Distinguished Service Cross was presented to his sister Mrs. Lena Woods.

Joseph H. Carvo

Private First Class, United States Army
Army Service Number: 2268262
Company I 47th Infantry Regiment 4th Division
Date of Action: July 29 - 30, 1918
Location: Near Sergy, France
War Department General Orders 71 (1919)
Medal # 2293

CITATION

Acting as a runner, Private First Class Carvo carried messages repeatedly over open ground swept by terrific machine gun fire, aiding materially in the maintenance of liaison.

DETAILS

Joseph Harry Carvo was born in Crookston, Polk County, Minnesota on October 8, 1893, the son of Edward and Adaline Carvo. Before entering the Army, he was a farmer in Washington State. He entered the Army on October 4, 1917 and listed his home of record as Yakima, Washington. He sailed to France as a Private 1st Class in Company I 47th Infantry 4th Division aboard the troop transport *U.S.S. Louisville* on May 19, 1918. As his emergency contact, he listed his father Edward in Wapato, Washington.

Carvo was awarded the Distinguished Service Cross for his actions in the Aisne-Marne Offensive. On July 28, 1918, 1st and 3rd Battalions of the 47th Infantry were attached to the 42nd Division to join in the advance against the village of Sergy. The village had been taken and lost several times. On

July 29, Carvo's Company I spearheaded the attack along with Company L as part of 3rd Battalion 47th Infantry. The Battalion came under German artillery fire the entire day and lost three successive Battalion Commanders wounded and a fourth killed in the space of a few hours. The following morning, July 30 the advance was resumed. Carvo carried messages across the battlefield during two days of the attack amidst heavy machine gun fire from the enemy.

A week after his award actions, Carvo was awarded a Citation Star for his continuing bravery in the Vesle Sector on August 8 near the village of St. Thibaut on the south side of the Vesle River. The Citation read: "On August 8, while under heavy fire this soldier showed exceptional bravery in carrying messages from command post, Company I to the command post of battalion commander. He was under machine gun and shell fire the greater part of the time but completed all missions in a speedy and satisfactory manner."[6] During the action he was wounded in a gas attack and was hospitalized for three weeks after suffering injuries from mustard and chlorine gas. He returned to duty and took part in the Meuse-Argonne Offensive.

Carvo served on occupation duty in Germany with the 4th Division and returned to the United States as a member of a Casual Detachment sent home for demobilization and discharge aboard the transport *U.S.S. Minnesotan* leaving Brest, France July 23, 1919 and arriving at Philadelphia, Pennsylvania on August 3, 1919. He was discharged on August 5, 1919.

He married Stella Faulkner on April 3, 1921 and they eventually had four children. The 1930 Census showed Carvo living in Yakima, Washington with his wife, two sons and two daughters where he was employed as a short order cook and indicated he was a veteran of the World War. After 1932, Carvo was awarded the Purple Heart for the wounds he suffered from gas in WW1 and his Citation Star was converted to the Silver Star Medal.

Joseph H. Carvo died at the age of 45 on August 9, 1939 and is buried in Calvary Cemetery, Yakima, Yakima County, Washington. The U.S. Army named one of ten target ranges at its Yakima Firing Center, Washington in his honor.

Claude E. Cherry

Sergeant, United States Army
Army Service Number: 55908
Company B 11th Machine Gun Battalion 4th Division
Date of Action: August 7, 1918
Location: Near St. Thibaut, France
War Department General Orders 15 (1919)
Medal # No assigned number discovered

CITATION

Sergeant Cherry commanded the third platoon of his company during the engagement near St. Thibaut. On August 7, 1918 he crossed the Vesle River and took up a position in front of his own infantry on terrain constantly swept by heavy artillery, machine gun, and sniper fire and directed his guns so skillfully as to silence a machine gun nest and make possible the Infantry advance. His conspicuous courage was an inspiration to his men. This gallant soldier was killed on August 9, 1918 by a fragment from an aerial bomb.

DETAILS

Claude Earnest Cherry was born in Carthage, Hancock County, Illinois on April 23, 1885, the son of Herman and Sarah Cherry. Before entering the Army, he worked at William E. Pratt Malleable Iron Company in Joliet, Illinois. He enlisted in the Regular Army on June 30, 1916, trained at Jefferson Barracks, Missouri and served on Mexican Border Service at Camp Eagle Pass, Texas. He was later assigned to the 11th Machine Gun Battalion. Cherry sailed to France as a Sergeant in the

11th Machine Gun Battalion and part of the advance detachment of the 4th Division aboard the troop transport *U.S.S. Finland* on April 30, 1918. As his emergency contact, he listed his mother Sarah in Joliet, Illinois. Cherry served in the Aisne-Marne Offensive.

He was awarded the Distinguished Service Cross for his actions in the Vesle Sector. On August 7, 1918 his Company B and Company D also of the 11th Machine Gun Battalion were in support of the 47th Infantry's attack against the village of Bazoches directly across the Vesle River from the village of St. Thibaut. In the absence of an officer, Cherry commanded a platoon of four machine guns during the action. He moved his guns across the river under fire and established an exposed forward position from which they could cover the infantry's advance. Because of his expert direction of the fire of the guns of his platoon, a German machine gun nest was eliminated, greatly aiding in the American attack.

Enemy resistance at the Vesle was formidable and for several days the 47th Infantry pressed the attack but was stopped from capturing and holding Bazoches. On August 9, German aircraft strafed and bombed the American front lines causing numerous casualties, including Cherry who was killed in the bombing.

Claude E. Cherry was buried in the battlefield cemetery # C-92 near the village of Bazoches.

His Distinguished Service Cross was presented to his mother, Mrs. Sarah Demarest.

Claude E. Cherry was reinterred and laid to his final rest on August 21, 1925 in Plot B Row 12 Grave 32 in the Aisne-Marne American Cemetery and Memorial, Belleau, Departement de l'Aisne, Picardie, France.

In 1930, his mother Mrs. Sarah Elizabeth Demarest declined the offer to visit her son's grave in France under the U.S. Government program of Gold Star Pilgrimages. See the entry for Edward S. Blackman for an explanation of the Gold Star Pilgrimages.

Michael D. Belis

Arthur I. Clark

Sergeant, United States Army
Army Service Number: 2258790
Company C 39th Infantry Regiment 4th Division
Date of Action: September 26, 1918
Location: Near Esnes, France
War Department General Orders 98 (1919)
Medal # 1780

CITATION

Sergeant Clark was in command of one platoon of his company which was being held up by intense enemy machine gun fire. Accompanied by two other soldiers he voluntarily made an attack on one of the machine gun nests under heavy fire, firing a rifle grenade into it and forcing its surrender. He then advanced on another machine gun nest and captured it, taking seven prisoners from both nests. His platoon having been forced to fall back by machine gun fire from the rear, he reorganized it and led it in a successful attack on 75 of the enemy whom he discovered nearby.

DETAILS

Arthur Ira Clark was born north of Osceola in Clarke County, Iowa on September 17, 1894, the son of George and Barbara Clark. At the time he entered the service he listed his home of record as Helena, Montana where he worked as a ranch hand. He was drafted into the Army as a Private on October 4, 1917 at Lewistown, Montana. He trained with Company A 362nd Infantry at Camp Lewis, Washington until March 24, 1918 when he was transferred to Company C 39th Infantry in the 4th Division at Camp Greene, North Carolina. He was promoted to Private 1st Class on April 4, 1918. Clark sailed to France as a Private 1st Class in Company C 39th Infantry 4th Division aboard the Italian troopship *Dante Alegheiri* on May 10, 1918. As his emergency contact, he listed his mother in Charles City, Iowa.

Clark served in the Aisne-Marne Offensive and was promoted to Corporal on July 28, 1918. He served in the Vesle Sector and was promoted to Sergeant on August 29, 1918. He served in the Toulon Sector and the St. Mihiel Offensive.

Clark was awarded the Distinguished Service Cross for his actions on September 26, 1918, the first day of the Meuse-Argonne Offensive in the attack from north of the village of Esnes toward the village of Cuisy, the village of Septsarges and Montfaucon Hill. He led attacks on two German machine gun nests, eliminating both nests and capturing seven prisoners. He then gathered his platoon which had fallen back and led it in an attack on a group of the enemy which outnumbered his small command. He was also wounded at some time during the Meuse-Argonne Offensive.

In the official history of the 39th Infantry in World War One, published in 1919, Clark was one of 16 men in the 39th

Infantry chosen by their fellow soldiers as the bravest men in the Regiment. The entry for Clark in that history was:

"Sergeant Arthur I. Clark came over as a 'buck,' having left his Montana homestead in 1917. It was in the 39th's baptism of fire that he earned his first step up. The lack of fear which he displayed later in the Argonne offensive marked him as a scrapper with few peers. 'Courage and leadership' stands out on his D.S.C., which with a wound chevron is fitting testimony of what he did in the big fight."[7]

Clark was also awarded the French Croix De Guerre. He served on occupation duty in Germany with the 4th Division and returned to the United States with his Regiment aboard the troopship *U.S.S. Leviathan* (seized German liner *Vaterland*) leaving Brest, France July 30, 1919 and arriving at Hoboken, New Jersey on August 6, 1919. He was discharged on August 9, 1919.

In 1920, Clark married Daisy Willey. The 1930 Census showed Clark, his wife Daisy, their son and two daughters living with his wife's parents in Iowa where he worked as a laborer at a packing house and indicated he was a veteran of the World War. In 1940, Clark married for a second time to Helen Torrence and they lived in Greenfield, Iowa where he worked on his own farm. Arthur I. Clark died at the age of 68 on August 18, 1963 and is buried in Sunset Memorial Gardens, Des Moines, Polk County, Iowa.

Joseph T. Clement

Major, United States Army
39th Infantry Regiment 4th Division
Date of Action: July 18, 1918
Location: Near La Ferte-Milon, France
War Department General Orders 43 (1922)
Medal # 7989

CITATION

Major Clement reported for duty from the hospital as his regiment was preparing to advance to the attack. Soon thereafter, when it was of urgent necessity that important orders of the regimental commander be carried to the officer commanding the assault battalion, he volunteered for the dangerous mission. Accompanied by two men, he worked his way forward through heavy artillery shell fire, located the assault units, and delivered the attack order. Endeavoring to locate the commander of the assault battalion, he fearlessly went into a heavy artillery barrage and continued this important and hazardous task until he was wounded by shell fire.

DETAILS

Joseph Taylor Clement was born in Charleston, South Carolina on August 28, 1885, the son of Moultrie and Lucia Clement. He was an Apprentice Seaman 3rd Class in the South Carolina Naval Militia August 28, 1903 to June 30, 1904. He was in the South Carolina National Guard from July 1, 1904 to September 29, 1908 first as a Private in Troop E 1st South Carolina Cavalry and then as a Sergeant in Company B 3rd South Carolina Infantry. He attended The Citadel (South Carolina Military Academy) and was commissioned a 2nd Lieutenant in the 4th Field Artillery (Regular Army) on September 25, 1908. He was transferred to the 9th Infantry on January 27, 1911 and on July 1, 1916 was commissioned a 1st Lieutenant of Infantry. He was promoted to Captain on May 15, 1917.

Clement was assigned to the 39th Infantry at Camp Greene, North Carolina on January 18, 1918. He sailed to France as a Captain in the 39th Infantry with Headquarters 7th Infantry Brigade 4th Division aboard the troopship *U.S.S. Princess Matoika* (seized German liner *Prinzess Alice*) on May 10, 1918. As his emergency contact, he listed his wife in care of T.P. Carothers, American Bank Building, Newport, Kentucky. On June 7, 1918 Clement was promoted to the wartime temporary rank of Major.

He was awarded the Distinguished Service Cross for his actions on July 18, 1918, the first day of the Aisne-Marne Offensive. For the operation the 39th Infantry was attached to and under the command of the French 33rd Division. The direction of the attack was northeast from the village of La Ferte-Milon toward the Ourcq River with the ultimate objective being the village of Noroy-sur-Ourcq. The avenue of advance was south of the Buisson de Cresnes which was a large thicket commanding the high ground in the area. As the French moved slowly forward, their progress was hindered by machine gun fire from the Buisson de Cresnes and the 39th Infantry was ordered to attack the wooded area. The French employed the three Battalions of the 39th Infantry separately in the attack with their individual objectives changed several times during the day.

Communication from Regimental Headquarters to the Battalions was vital therefore in the coordination of all efforts. Clement took two enlisted men and ventured out across the battlefield to carry orders to one of the assaulting Battalions. Though he could have delegated the job to a subordinate, he knew that being delivered by such a senior officer as himself the orders would be received with an elevated level of importance. While the German Infantry positions were at first caught by surprise in the offensive, their artillery responded to it with effectiveness throughout the day and blanketed the area with an almost non-stop barrage. Even though Clement was fresh out of the hospital and was still weakened by illness, he braved the heavy enemy artillery fire, delivered the order to the various Companies in the assault Battalion and moved through the exploding shells searching for the Battalion commander until struck down and disabled by serious shrapnel wounds.

Clement was also awarded a Citation Star. (After 1932 he could have applied to have that Citation Star converted to a Silver Star Medal. The Army Registers indicate he did not have that done.) He received the French Croix De Guerre with silver star. He was one of 20 soldiers of the 4th Division and one of only 399 soldiers of the American Expeditionary Forces to be awarded the Italian War Merit Cross (Croce al Merito di Guerra). His wounds were severe enough to keep him out of action for the rest of the war. He returned to the United States in a detachment of sick and wounded aboard the troop transport *U.S.S. Mercury* (seized German liner *Barbarossa*) leaving France in late September 1918 and arriving at Newport News, Virginia on October 4, 1918. He remained in the Army.

Once recovered from his wounds, Clement served in Hawaii for a couple of years. He was promoted to the permanent rank of Major on July 1, 1920 and was retired as a Major on January 4, 1921 with a disability from wounds received in action. He divorced his wife Caroline in 1929 and in 1930 married Vilma Wallace. A city directory for 1931 showed Clement and Vilma living in Washington, D.C. where he was employed as a special representative for RCA (Radio Corporation of America). After 1932, he was awarded the Purple Heart for the wounds he received in World War One. He was recalled to active duty on April 1, 1941, promoted to the temporary rank of Lieutenant Colonel on February 1, 1942 and the temporary rank of Colonel on December 11, 1943. He was released from active duty on March 23, 1944. In 1947 his retirement was increased to the rank of Colonel. Joseph T. Clement died of cancer at the age of 78 on November 16, 1963 and is buried in Arlington national Cemetery.

Herbert Arnold Cohn

Second Lieutenant, United States Army
Company A 39th Infantry Regiment 4th Division
Date of Action: September 26, 1918
Location: East of Montfaucon, France
War Department General Orders 3 (1921)
Medal # 6384

CITATION

Lieutenant Cohn led the assault wave forward through heavy machine gun fire in an attack against a strongly defended enemy position. He was forced to pass through barbed-wire entanglements before entering the enemy trenches. While exposing himself to intense machine gun fire he was badly wounded but his command, inspired by his gallant example, gained and held the objective sought.

DETAILS

Herbert Arnold Cohn was born in Rochester, New York on April 3, 1893, the son of Henry and Madeleine Cohn. He graduated from Harvard in 1914 and worked for American Oil Company in New York City. When he registered for the draft in May 1917, he indicated his occupation as Student at the Officers Training Camp at Plattsburg Barracks, New York. He was commissioned a 2nd Lieutenant of Infantry in the Officers Reserve Corps on August 29, 1917 and assigned to Company A 39th Infantry at the camp at Syracuse, New York. He went with the 39th Infantry to Camp Greene, North Carolina where it soon became a part of the 4th Division. At some time before

Michael D. Belis

going overseas he was transferred to the 10th Machine Gun Battalion also in the 4th Division. Cohn sailed to France as a 2nd Lieutenant in Headquarters Detachment 10th Machine Gun Battalion 4th Division aboard the French troopship *Rochambeau* on May 7, 1918. As his emergency contact, he listed his mother Madeleine in New York City.

Cohn was transferred back to Company A 39th Infantry at some time after reaching France and before his Regiment went into action. He served in the Aisne-Marne Offensive where he was wounded on an unknown date during the time period July 18-21, 1918. He was wounded a second time, again date unknown between August 1-12, 1918.

He was awarded the Distinguished Service Cross for his actions in the Meuse-Argonne Offensive. On September 26, 1918, the attack was led by 3rd Battalion 39th Infantry with Cohn's 1st Battalion 39th Infantry in support. As 3rd Battalion moved against the village of Cuisy 1st Battalion got lost in the dense fog, veered to the left and came up against Montfaucon Hill east of the village of Montfaucon. Cohn's platoon ran straight into some of the most formidable defenses manned by the Germans in the area, trenches on high ground protected by dense barbed wire entanglements and covered by scores of machine guns. Leading his platoon from the front, Cohn charged into the barbed wire and was severely wounded. He was a shining example to his men who, inspired by his bravery pushed on and captured the enemy trench.

Cohn was also awarded a Citation Star. Because of his wounds he was returned to the United States several months ahead of his Regiment as a 2nd Lieutenant in Company A 39th Infantry and part of a detachment of sick and wounded in Officers Detachment Base Hospital #8 aboard the troopship *U.S.S.*

Madawaska (seized German liner *König Wilhelm II*) leaving St. Nazaire, France December 27, 1918 and arriving at Newport News, Virginia January 9, 1919. He was discharged on April 2, 1919 with a 60% disability rating.

On August 11, 1919 Cohn married Florence Gutman. On January 13, 1921 he was awarded the Conspicuous Service Cross (#289) by the State of New York for his service in World War One. Cohn and his wife had a son born in July 1921 who died in March 1922. They had another son born in 1924 and a daughter born in 1929. The 1930 Census showed Cohn living with his wife, son and daughter in New York City where he was employed as a Broker in a Bond House and indicated he was a veteran of the World War. The 1940 Census showed Cohn and his family living in Hempstead, Nassau County, New York where he was employed as an Investigator at a Securities Exchange. In 1942 his second son died in the famous Cocoanut Grove fire in Boston, Masachusetts. Herbert Arnold Cohn died at the age of 54 on February 4, 1948.

James Conway

Private, United States Army
Army Service Number: 2721014
Company C 58th Infantry Regiment 4th Division
Date of Action: September 29, 1918
Location: Near Nantillois, France
War Department General Orders 66 (1919)
Medal # 3770

CITATION

Private Conway, a company runner, repeatedly volunteered for the most dangerous missions, carrying messages through enemy machine gun and shell fire on numerous occasions. Several days later when his ear drum was broken by concussion from a bursting shell, he refused to go to the rear for treatment but remained on duty until his company was relieved.

DETAILS

James Conway was born in Cromane, Killorglin, County Kerry, Ireland on July 10, 1891. Before entering the Army, he worked as a laborer at the Newburyport Gas & Electric Company in Newburyport, Massachusetts. He entered the Army on April 29, 1918. About one month before he shipped out to Europe, he became a naturalized American citizen. On his naturalization confirmation form dated June 25, 1918 his occupation was listed as Soldier (with a further explanation of Teamster).

Conway sailed to England as a Private in Company A 304th Infantry 76th Division aboard the British transport *Talthybius* on July 8, 1918. As his emergency contact, he listed his sister Nellie in New Bedford, Massachusetts. From England the 304th Infantry was sent over to France where its personnel were used as replacements for front line units. Conway was assigned to Company C 58th Infantry sometime after July 27, 1918.

He was awarded the Distinguished Service Cross for his actions in the Meuse-Argonne Offensive. On September 29, 1918 his Battalion attacked toward the wooded area known as the Bois des Ogons, north of the village of Nantillois. German resistance was intense and for the next four days the attack was halted. The enemy pounded the 58th Infantry with small arms, artillery, and aerial attacks. Conway performed the dangerous job of runner with exceptional bravery and continued to carry messages even though injured by concussion, until his Company was relieved on October 2.

He served with the 4th Division on occupation duty in Germany and returned to the United States with Company C 58th Infantry aboard the troopship *U.S.S. Mount Vernon* (captured German liner *Kronprinzessin Cecilie*) leaving Brest, France July 24, 1919 and arriving at Hoboken, New Jersey August 1, 1919. He was discharged on August 7, 1919.

His address listed with the Veterans Bureau in the early 1920's was Bellevue Hospital in New York City. James Conway died at the age of 43 on November 16, 1934 and is buried in St. Mary's Cemetery, Newburyport, Essex County, Massachusetts. His headstone application with the War Department was signed by his sister Nellie Conway.

Michael D. Belis

Frank B. Cook Jr.
Second Lieutenant, United States Army
Company D 4th Engineer Regiment 4th Division
Date of Action: August 11, 1918
Location: Near Ville Savoye, France
War Department General Orders 143 (1918)
Medal # 1088

CITATION
Lieutenant Cook directed the construction of an artillery bridge on the Vesle River under constant machine gun and shell fire and set a splendid example to the members of his command by his disregard of danger. On the morning of August 11, he was wounded while personally looking after the safety of an out-guard during a heavy enemy bombardment.

DETAILS
Frank Bigelow Cook Jr. was born in California on October 5, 1889, the son of Frank B. and Maude Cook. He was commissioned a 2nd Lieutenant in the Officers Reserve Corps on September 2, 1917. He is listed in the official history of the 4th Engineers in World War One as being a 2nd Lieutenant with Company A 4th Engineers at least by January 1, 1918. He was transferred to Company D and sailed to France as a 2nd Lieutenant of the Engineer Reserve Corps in Company D 4th Engineers 4th Division aboard the troopship *U.S.S. Martha Washington* (seized Austrian liner) on April 30, 1918. As his emergency contact, he listed his father Frank in Oakland, California. Cook served in the Aisne-Marne Offensive.

He was awarded the Distinguished Service Cross for his actions in the Vesle Sector. The 4th Division had to cross the Vesle River and a number of footbridges and several bridges large enough to move 75mm artillery guns across the river were constructed by the 4th Engineers, most of the work being done while under fire from the enemy. As these bridges were destroyed by German artillery, the Engineers rebuilt them time and again. Cook directed the successful building of one such artillery bridge and set an example for his men of courage under fire while doing so. After his work with the bridge, Cook went into the village of Ville Savoye to see about some Engineers who were in a forward position and under fire. He was caught in an enemy artillery barrage and was severely wounded and unable to move.

Two of his enlisted men from Company D charged into the barrage to get him. Corporal Tom Barto was killed during the rescue, but Private 1st Class Gilbert Wilcox pulled Cook out of the village and got him to safety. Both Barto and Wilcox were awarded the Distinguished Service Cross for their bravery. Cook recovered from his wounds and was transferred to Company C. He was promoted to 1st Lieutenant on October 12, 1918. He was also awarded the French Croix De Guerre. He served on occupation duty in Germany with the 4th Division and returned to the United States with Company C 4th Engineers aboard the troopship *U.S.S. Von Steuben* (seized German liner *Kronprinz Wilhelm*) leaving Brest, France July 21, 1919 and arriving at Hoboken, New Jersey July 29, 1919. He was discharged on August 9, 1919.

The 1920 Census showed Cook living with his parents in Oakland, Alameda County, California where he was employed as a civil engineer for the State of California. The 1930 Census showed him still living with his parents in Oakland and employed as a civil engineer and indicated he was a veteran of the World War. His father died in 1935. Voter registration records for November 1944 showed him still living at the same family address of his parents' home in Oakland and employed as a civil engineer. Frank B. Cook Jr. died at the age of 76 on October 16, 1965.

Fred E. Cullen
Corporal, United States Army
Note: Fred E. Cullen was actually a Private at the time of his award actions.
Army Service Number: 567377
Company D 12th Machine Gun Battalion 4th Division
Date of Action: August 7, 1918
Location: Near Bazoches, France
War Department General Orders 27 (1920)
Medal # 6001

CITATION
When two American soldiers attempted to cross an open space and were fired upon by an enemy machine gun, one of the soldiers fell wounded and the enemy concentrated their fire upon the fallen man. Under enemy machine gun fire, Corporal Cullen rushed forward 75 yards and in spite of the heavy fire carried his wounded comrade to a place of safety.

DETAILS
Frederick E. Cullen was born in Mottville hamlet, Skaneateles, Onondaga County, New York on January 13, 1898, the son of Frederick J. and Sarah Cullen. He enlisted as a Private in the Regular Army in Company D 39th Infantry at Camp Syracuse, New York on October 2, 1917. On November 26, 1917 at Camp Greene, North Carolina he was transferred to the 12th Machine Gun Battalion. He sailed to England as a Private in Company D 12th Machine Gun Battalion 4th Division aboard the British troopship *RMS Aquitania* on May 7, 1918. As his

emergency contact, he listed his mother Sara in Skaneateles, New York. From England, Cullen and his Battalion were sent over to France where in June they were issued their French Model 1914 Hotchkiss machine guns. Cullen served in the Aisne-Marne Offensive.

He was awarded the Distinguished Service Cross for his actions in the Vesle Sector on August 7, 1918. On that date Cullen's Company was supporting 1st Battalion 59th Infantry as it relieved 1st Battalion 58th Infantry near the Rouen-Rheims Road north of the Vesle River east of the village of Bazoches. Cullen charged out into deadly German machine gun fire, picked up a wounded fellow soldier, and carried him to safety.

Cullen was also awarded a Citation Star. He was promoted to Private 1st Class on August 31, 1918. He served in the Toulon Sector and was promoted to Corporal on September 13, 1918. He served in the St. Mihiel Offensive and the Meuse-Argonne Offensive. Cullen served on occupation duty in Germany with the 4th Division and was reduced to the rank of Private on July 3, 1919. He returned to the United States with Company D 12th Machine Gun Battalion aboard the troopship *U.S.S. Von Steuben* (seized German liner *Kronprinz Wilhelm*) leaving Brest, France July 21, 1919 and arriving at Hoboken, New Jersey July 29, 1919. He was discharged on August 4, 1919.

The 1920 Census showed Cullen living with his parents in Skaneateles, New York where he was employed as a laborer in a felt mill. He married Hulda Nejberg on June 12, 1922. The 1930 Census showed Cullen living in Skaneateles, New York with his wife, daughter, and son where he was employed as a general laborer and indicated he was a veteran of the World War. On July 7, 1931, he was awarded the Conspicuous Service Cross (#2129) by the State of New York for his actions in

World War One. After 1932, his Citation Star was converted to the Silver Star Medal. In 1949 he moved to California. He returned to New York in 1962 and lived in Auburn, Cayuga County. He retired as a painting contractor. Fred E. Cullen died at the age of 65 when he fell down the cellar stairs in his home on May 23, 1963 and fractured his skull. He is buried in Mottville Cemetery, Mottville, Onondaga County, New York. His grave marker indicates he was back in uniform during World War Two, but no details of that service could be found.

Michael D. Belis

Walter Currie
Private, United States Army
Army Service Number: 567917
Company A 59th Infantry Regiment 4th Division
Date of Action: August 7 and September 29, 1918
Location: Near Ville Savoye, France and near Brieulles, France
War Department General Orders 98 (1919)
Medal # 6121

CITATION

On the Vesle River, when his company was in need of ammunition and after several men had been killed in the attempt to secure it, Private Currie volunteered and went for ammunition across an open field swept by machine gun fire. He successfully returned with the ammunition thereby greatly assisting his company to hold its position. He was severely wounded near Brieulles-sur-Meuse while making a gallant stand against the enemy with twelve other men, the only survivors of his platoon.

DETAILS

Walter Currie was born in Detroit, Michigan on July 29, 1895, the son of Arthur and Mary Currie. When he registered for the draft in June 1917, he indicated he was employed as a stagehand at the Cadillac Theatre in Detroit, Michigan. He entered the Army on July 28, 1917 at Camp Crane, Pennsylvania. Currie sailed to England as a Private in Company A 59th Infantry 4th Division aboard the British troopship *Megantic* on May 3, 1918. As his emergency contact, he listed his mother Mary in

Detroit, Michigan. From England he and his Regiment were sent over to France. He served in the Aisne-Marne Offensive.

He was awarded the Distinguished Service Cross for his actions on two separate dates. The first date mentioned in his Citation occurred in the Vesle Sector when the 8th Brigade of the 4th Division was attacking across the Vesle River north of the village of Ville Savoye. The 1st Battalion of the 58th Infantry had advanced and secured positions along the road east of the village of Bazoches and during the day of August 7, 1918 Currie's 1st Battalion of the 59th Infantry carried out the process of relieving the 58th Infantry in place. Several German counterattacks were repulsed as Currie's Company and one other Company from his Battalion along with Company D of the 12th Machine Gun Battalion relieved the exhausted 58th Infantry. During the fighting, Currie ventured across open ground under heavy machine gun fire to obtain and return with badly needed ammunition for his Company even though several soldiers had already been killed in the attempt.

Currie then served in the Toulon Sector and the St. Mihiel Offensive.

The second date mentioned in Currie's Citation occurred on September 29, 1918 in the Meuse-Argonne Offensive during the 4th Division's attack toward the village of Brieulles-sur-Meuse on the Meuse River. In that attack the enemy resisted with heavy artillery and counterattacks. On September 29, Currie's Battalion was spearheading the advance on the right of the 4th Division's attacking line. For the next three days all positions were then held and consolidated. During an enemy counterattack Currie was wounded and one of only thirteen men left out of his original platoon of fifty yet they still held their position.

Due to the severity of his wounds, Currie returned home several months ahead of his Regiment. He was assigned to the 338th Infantry which was scheduled to be demobilized in April 1919. He returned to the United States with Company G 338th Infantry aboard the troopship *U.S.S. Leviathan* (seized German liner *Vaterland*) leaving Brest, France March 26, 1919 and arriving at Hoboken, New Jersey April 2, 1919. He was discharged on April 11, 1919.

The 1920 Census showed him living with his parents in Detroit, Michigan where he was employed as a theater usher. About 1922 he married his wife, Norma. The 1930 Census showed Currie living with his wife, daughter and son in Minden, Sanilac County, Michigan where he was employed as a manager of a drug store and indicated he was a veteran of the World War. When he registered for the draft in 1942, he indicated he was living in Detroit, Michigan and was unemployed. Walter Currie died at the age of 69 on October 2, 1964.

Earl W. Curtis

Private, United States Army
Army Service Number: 2101765
Company B 59th Infantry Regiment 4th Division
Date of Action: September 29, 1918
Location: Near Brieulles-sur-Meuse, France
War Department General Orders 46 (1919)
Medal # 3312

CITATION

Advancing alone across open territory and exposed to extremely heavy machine gun fire, Private Curtiss rescued a fellow soldier who was lying wounded beyond the front line. He accomplished his mission even after being painfully wounded in the head during his return.

DETAILS

Earl Washington Curtis was born in Houston, Texas County, Missouri on January 11, 1895. Before entering the Army, he was living in Virden, Illinois and was employed as a coal miner. He entered the Army on September 18, 1917. Curtis sailed to England as a Private in Company B 59th Infantry 4th Division aboard the British troopship *Megantic* on May 3, 1918. As his emergency contact, he listed Harry B. Fisher, a friend in Yukon, Missouri. From England he and his Regiment were sent over to France. He served in the Aisne-Marne Offensive, the Vesle Sector, the Toulon Sector, and the St. Mihiel Offensive.

He was awarded the Distinguished Service Cross for his actions in the Meuse-Argonne Offensive. On September 29,

1918 in an area south of the village of Brieulles-sur-Meuse, his Company, as part of the 4th Division's 8th Brigade was in the advance that cleared the wooded area known as the Bois de Brieulles. The attack was stopped by heavy enemy fire before reaching the next wooded area known as the Bois de Fays. In the no man's land between the two areas, Curtis ventured out across the battlefield under heavy fire, took hold of a wounded fellow soldier and brought him back to safety. Curtis was wounded himself in the performance of this act. For his bravery he was also awarded the French Croix De Guerre.

He served on occupation duty in Germany with the 4th Division and returned to the United States as a Private in Company B 59th Infantry and part of Company K Third Army Composite Regiment aboard the troopship *U.S.S. Leviathan* (seized German liner *Vaterland*) leaving Brest, France September 1, 1919 and arriving at Hoboken, New Jersey September 8, 1919.

(See the entry for Joseph Bassi for an explanation of the Third Army Composite Regiment.)

He was discharged on September 27, 1919.

Curtis returned to Missouri where he became a farmer. He married Georgia Hern on July 17, 1922. The 1930 Census showed Curtis living with his wife, four sons and daughter in Piney, Texas County, Missouri where he was employed as a farmer and indicated he was a veteran of the World War. The 1940 Census showed him living with his wife, four sons and two daughters in Piney where he was still a farmer. Curtis and his wife had 6 children, 14 grandchildren and 23 great-grandchildren. Earl W. Curtis died at the age of 92 on February 14, 1987 and is buried in Ozark Cemetery, Houston, Texas County, Missouri.

Clinton Day

Private First Class, United States Army
Army Service Number: 2264227
Company C 58th Infantry Regiment 4th Division
Date of Action: August 7, 1918
Location: Near St. Thibaut, France
War Department General Orders 66 (1919)
Medal # 3751

CITATION

Private Day repeatedly volunteered and carried messages from his company in the front line across an open field swept by enemy machine gun and sniper fire to the battalion command post. He also voluntarily made trips across this dangerous area for the purpose of filling canteens for wounded soldiers and securing stretchers.

DETAILS

Clinton Day was born in Fillmore, Millard County, Utah on October 1, 1891, the son of James and Emma Day. Before entering the Army, he was a farmer. He was drafted into the Army as a Private on October 3, 1917 and trained with Company L 362nd Infantry. Day was assigned to Company C 58th Infantry on February 26, 1918 and sailed to England as a Private in Company C 58th Infantry aboard the British transport *City of Brisbane* on May 7, 1918. As his emergency contact, he listed his mother Emma in Fillmore, Utah. From England he and his Company were sent over to France. He served in the Aisne-Marne Offensive.

He was awarded the Distinguished Service Cross for his actions in the Vesle Sector. Though his Citation gives the location as being near the village of St. Thibaut, the actual location was northeast of St. Thibaut and across the Vesle River from and nearer to the village of Ville Savoye. On August 7, 1918, Day and his Battalion were in defensive positions along the Rouen-Rheims Road and repulsed two German counterattacks early in the morning and a third later that night. Day's bravery in continually crossing the battlefield to carry messages kept his Company and Battalion Headquarters in constant contact, ensuring the successful coordination of their activities. In addition, Day assisted in caring for the wounded by leaving positions of safety and venturing out under fire to get water and stretchers for them.

He served in the Toulon Sector, the St. Mihiel Offensive, and the Meuse-Argonne Offensive. On November 18, 1918, he was promoted to Private 1st Class. Day served on occupation duty in Germany with the 4th Division and returned to the United States with Company C 58th Infantry aboard the troopship *U.S.S. Mount Vernon* (captured German liner *Kronprinzessin Cecilie*) leaving Brest, France July 24, 1919 and arriving at Hoboken, New Jersey on August 1, 1919. He was discharged on August 11, 1919.

The 1920 Census showed Day living with his parents in Fillmore, Utah where he was employed as a farm laborer. The 1940 Census showed him living with his wife, two daughters and two sons in Fillmore where he was employed as a Laborer for the WPA (Works Progress Administration) in road construction. He was married three times, his first two wives dying. He had eight grandchildren. Clinton Day died at the age of 81 on May 30, 1973 and is buried in Fillmore Cemetery, Fillmore, Millard County, Utah.

Michael D. Belis

Charles E. Deleuw

First Lieutenant, United States Army
4th Engineer Regiment 4th Division
Date of Action: August 11, 1918
Location: Near Ville Savoye, France
War Department General Orders 138 (1918)
Medal # 6982

CITATION

Lieutenant Deleuw was in command of a detachment of engineers engaged in constructing an artillery bridge across the river Vesle under constant fire from machine guns and bombardment by both high explosive and gas shells. Although he was suffering from the effects of gas, this officer remained in charge of the party, directing the work and furnishing his men a splendid example of courage under fire and disregard for personal safety.

DETAILS

Charles Edmund DeLeuw was born in Jacksonville, Morgan County, Illinois on July 3, 1891, the son of Oscar and Mary Deleuw. He received degrees from the University of Illinois in 1912 and 1916. From June to November 1916, he served as a Private First Class and then Corporal in the Illinois National Guard on Mexican Border Service. He married Martha Guthrie in August 1917. DeLeuw was commissioned a 1st Lieutenant in the Officers Reserve Corps on August 15, 1917. He was with the 4th Engineers at least by January 1, 1918. He sailed to France as a 1st Lieutenant with Headquarters Detachment 4th Engineers 4th Division aboard the troopship *U.S.S. Mar-*

tha Washington (seized Austrian liner) on April 30, 1918. As his emergency contact, he listed his wife Martha in Riverside, Illinois.

After reaching France he was assigned to Company A 4th Engineers, for a while commanded Headquarters Detachment 4th Engineers, and then was Topographic Officer of the Regiment. He served in the Aisne-Marne Offensive.

Deleuw was awarded the Distinguished Service Cross for his actions in the Vesle Sector during the efforts of the 4th Engineers to build and maintain bridges across the Vesle River. On August 11, 1918, he was tasked with building and emplacing a bridge large enough for the artillery units to bring their 75mm guns across the river near the village of Ville Savoye. He commanded a detachment of enlisted men from Company D 4th Engineers and though affected by poison gas from German artillery, he supervised the successful construction of the bridge while under fire, inspiring his men with his courage. Four of those enlisted men were also awarded the Distinguished Service Cross for their actions during the building of that artillery bridge that day, Private Charles Glenn, Private First Class Arthur J. Goetsch, Private Cornelius J. O'Brien, and Sergeant William J. Wood.

Deleuw was awarded the Knight's Cross of the Order of the Belgian Crown (Chevalier De L'Ordre de la Couronne.) He was the only soldier in the 4th Division to be given this Belgian decoration during World War One and one of only 143 soldiers of the American Expeditionary Forces to receive this award.

He served in the Toulon Sector and the St. Mihiel Offensive. Deleuw was seriously wounded by artillery fire on September 29, 1918 during the Meuse-Argonne Offensive. He and 1st Lieutenant Phillip W. Stafford, the Assistant Regimental

Supply Officer for the 4th Engineers were standing and talking in front of a field hospital near Cuisy when a German shell landed about thirty yards away, killing Stafford and severely wounding Deleuw. At some time after being wounded, he was promoted to Captain. Because of his wounds, Deleuw returned to the United States several months ahead of his Regiment as a Captain of Engineers and part of a casual detachment of officers aboard the transport *U.S.S. Wilhelmina* leaving Bordeaux, France January 5, 1919 and arriving at Hoboken, New Jersey on January 19, 1919. He was discharged on January 22, 1919.

The 1920 Census showed Deleuw, his wife and daughter living in Riverside, Cook County, Illinois where he worked as a Civil Engineer. His wife died in 1920. He was a founding partner of DeLeuw Cather and Associates, one of the largest engineering firms in the United States. He was married at least two more times. He worked as an engineer for the State of Illinois beginning in 1933 and in 1938 he became Chief Subway Engineer for the City of Chicago. When he registered for the draft in 1942, he indicated he was living in Highland Park, Lake County, Illinois and was self-employed at Charles-Deleuw & Company Consulting Engineers in Chicago. Charles Deleuw died at the age of 79 on October 28, 1970 and is buried in Wheaton Cemetery, Wheaton, DuPage County, Illinois.

Walter H. Detrow

Private, United States Army
Army Service Number: 2110211
Company B 47th Infantry Regiment 4th Division
Date of Action: August 1, 1918
Location: Near Sergy, France
War Department General Orders 44 (1919)
Medal # 7175

CITATION

After all the officers and noncommissioned officers of his platoon had been lost, Private Detrow assumed command of the platoon, successfully leading it from its critical situation to the objective through a terrific machine gun and shell fire. He performed this gallant act without any previous instructions or orders and acted entirely upon his own initiative.

DETAILS

Walter Harrison Detrow was born in Mahoning County, Ohio on May 17, 1893, the son of John and Barbara Detrow. Before entering the Army, he was a farmer in Washingtonville, Ohio. He was drafted into the Army as a Private on September 20, 1917. Detrow was assigned to Company B 47th Infantry on April 12, 1918. He sailed to France as a Private in Company B 47th Infantry 4th Division aboard the troopship *U.S.S. Princess Matoika* (seized German liner *Prinzess Alice*) on May 10, 1918. As his emergency contact, he listed his sister Mrs. D.F. Wentz in New Springfield, Ohio.

Detrow was awarded the Distinguished Service Cross for his actions in the Aisne-Marne Offensive. He and his Company were part of 1st Battalion 47th Infantry and along with 3rd Battalion 47th Infantry fought for four days to take and hold the village of Sergy. Both Battalions were attached to units of the 42nd Division for the attack. 1st Battalion 47th Infantry was attached to the 167th Infantry Regiment. By the time their part in the battle was over, the two Battalions of the 47th Infantry had lost 27 officers and 462 enlisted men killed, wounded, or gassed and 6 enlisted men missing in action. In Detrow's platoon, all officers and non-commissioned officers were either killed or wounded during the attack, leaving his platoon leaderless. Exhibiting the tenacity of the American soldier Detrow, even though serving at the lowest enlisted rank of Private, assumed command of the remnants of his platoon and led it through heavy fire onto the Battalion objective, the high ground overlooking Sergy.

For his bravery and leadership ability, Detrow was promoted to Corporal on August 14, 1918. He was promoted to Sergeant on August 25, 1918. He was also awarded a Citation Star. He was one of three soldiers of the 4th Division who received the French Military Medal (Médaille Militaire) during World War One and one of only 304 soldiers of the American Expeditionary Forces to be given that award. He also received the French Croix De Guerre with palm.

He served in the Vesle Sector, the Toulon Sector, and in the St. Mihiel Offensive. During the Meuse-Argonne Offensive beginning on October 4, 1918, 7th Brigade including Detrow and his Regiment were being held as the Division Reserve in the area near the village of Brieulles-sur-Meuse at the Meuse River. The American attack had stalled and both Americans and Germans relegated themselves to static defensive lines and patrolling. In this environment, Detrow was killed in action on October 7, 1918.

Detrow was buried in Plot 2 Grave # 32 in the American battlefield cemetery # 989 at the village of Brieulles-sur-Meuse.

His Distinguished Service Cross was presented to his mother Mrs. J. W. Detrow.

On June 6, 1919, Walter H. Detrow was reinterred in Plot # 1 Section 50 Grave # 30 in the Argonne American Cemetery # 1232. He was reinterred and laid to his final rest on November 23, 1921 in Plot B Row 38 Grave 25, in the Meuse-Argonne American Cemetery, Romagne-sous-Montfaucon, Department de la Meuse, Lorraine, France.

Michael D. Belis

Albert Dietz
Sergeant, United States Army
Army Service Number: 562072
Company I 59th Infantry Regiment 4th Division
Date of Action: October 6, 1918
Location: In the Bois de Fays, France
War Department General Orders 35 (1919)
Medal # 2933

CITATION
When one of Sergeant Dietz's men was wounded and his clothing and bandolier of ammunition caught fire, he cried for help and Sergeant Dietz left a place of safety and regardless of his personal safety went through intense machine gun fire and rescued him.

DETAILS

Albert Dietz was born in Wheatland, Indiana on December 9, 1897, the son of Albert and Mary Dietz. In his registration for the draft in 1917 he indicated that he was married and working as a laborer for William Newcomb in Vincennes, Indiana. He also stated that he had prior military service as a Private in the Army Coast Artillery Corps and had served in the Hawaiian Islands. Dietz was drafted into the Army on March 8, 1918 and was assigned to the 59th Infantry at Camp Greene, North Carolina where he took his initial training. He sailed to Europe as a Private assigned to Company I 59th Infantry and part of Casual Detachment No. 1 of the 4th Division aboard the troopship *U.S.S. Louisville* on May 19, 1918. As his emergency contact, he listed his wife Bertha in Vincennes, Indiana.

He served in the Aisne-Marne Offensive and the Vesle Sector. Dietz was promoted to Corporal on August 27, 1918. He served in the Toulon Sector and the St. Mihiel Offensive and was promoted to Sergeant on September 14, 1918.

He was awarded the Distinguished Service Cross for his actions in the Meuse-Argonne Offensive. After several days of attacks against the German defenses in the wooded area known as the Bois de Fays, by the night of October 4-5, 1918 the 58th Infantry, the 10th Machine Gun Battalion and Dietz' 3rd Battalion of the 59th Infantry had established a significant foothold in the woods. On October 6, the enemy mounted several counterattacks in the Bois de Fays and during one of those attacks Dietz left his place of cover and ventured through heavy machine gun fire to rescue one of his men who was wounded, unable to move and whose smoldering clothing and ammunition bandoleer threatened to cause him further injury. Dietz brought the man to safety while under intense fire from the Germans.

Dietz was one of 20 soldiers of the 4th Division and one of only 399 soldiers of the American Expeditionary Forces to be awarded the Italian War Merit Cross (Croce al Merito di Guerra). He was promoted to First Sergeant on October 16, 1918.

Dietz returned to the United States a couple of months ahead of his Regiment as a Sergeant of Company I 59th Infantry assigned to the Saint Aignan Casual Company No. 4993 of Special Discharges aboard the French liner *La Savoie* leaving Le Harve, France May 31, 1919 and arriving at New York City June 9, 1919. He was discharged on June 14, 1919.

A 1921-1922 city directory for Vincennes, Indiana showed Dietz and his wife Bertha living there where he was employed as a miner. Sometime after 1922 he enlisted in the Army. In 1932 he divorced Bertha and married Florence Hartman. His marriage license from 1932 gave his occupation as U.S. Army. Albert Dietz died at the age of 69 on February 1, 1967 and is buried in Golden Gate National Cemetery, San Bruno, San Mateo County, California. His grave marker indicates he served in uniform again during World War Two and attained the rank of Chief Warrant Officer.

Lester C. Dill

Private, United States Army
Army Service Number: 557748
Company B 47th Infantry Regiment 4th Division
Date of Action: August 1, 1918
Location: At Sergy, France
War Department General Orders 145 (1918)
Medal # 7911

CITATION

After being wounded twice while he was carrying a message, Private Dill bandaged his wounds under fire and delivered his message.

DETAILS

Lester Calvin Dill was born in Plainfield, Cumberland County, Pennsylvania on December 25, 1901, the son of Lewis and Musie Dill. At fifteen he lied about his age and enlisted as a Private in the Regular Army at Fort Slocum, New York on June 2, 1917 and was assigned to Company B 47th Infantry. At the time he entered the Army, he listed his home of record as Towanda, Pennsylvania. On August 9, 1917, he was assigned to the rank/position of Cook but reverted back to the rank of Private at some time before shipping overseas. He sailed to France as a Private in Company B 47th Infantry 4th Division aboard the troopship *U.S.S. Princess Matoika* (seized German liner *Prinzess Alice*) on May 10, 1918. As his emergency contact, he listed his mother Musie in Towanda, Pennsylvania.

Dill was awarded the Distinguished Service Cross for his

actions in the Aisne-Marne Offensive. He and his Company were part of 1st Battalion 47th Infantry and along with 3rd Battalion 47th Infantry fought for four days to take and hold the village of Sergy. By the time they occupied the town and the high ground around it, the two Battalions had lost 27 officers and 462 enlisted men killed, wounded or gassed and 6 enlisted men missing in action. On July 30, 1918, while performing the dangerous job of runner, Dill was hit twice by machine gun bullets. Instead of seeking medical care he bandaged his wounds himself and continued to carry out his duties as a runner, delivering his messages that day and the next. At the time he was sixteen years old.

Dill served in the Toulon Sector and the St. Mihiel Offensive. He returned to the United States several months ahead of his Regiment in order to receive an early discharge. Dill sailed as a member of a detachment of sick and wounded diptheria carriers from Base Hospital #118 aboard the transport *U.S.S. Santa Paula* leaving St. Nazaire, France April 8, 1919 and arriving at Hoboken, New Jersey April 20, 1919. After a short stay at Debarkation Hospital No. 1 for diptheria quarantine, he was discharged at Camp Dix, New Jersey on May 20, 1919.

The 1920 Census showed Dill living with his parents in Towanda, Bradford County, Pennsylvania and working as a laborer. The 1930 Census showed him still living with his parents but in Sheshequin, Bradford County, Pennsylvania, working as a laborer driving a farm truck and indicated he was a veteran of the World War. The 1940 Census showed him as a soldier in the U.S. Army living at Fort Ontario, New York. His rank is only given as "N.C.O." and the unit is not specified in the Census. He eventually attained the rank of Major and served during World War Two and Korea. His date of retirement at the rank of Major is listed in the Army Registers as August 31, 1954. Lester C. Dill died at the age of 89 on March 24, 1990 and is buried in Riverside Cemetery, Oswego, Oswego County, New York.

Michael D. Belis

Joseph Dilworth
Private, United States Army
Army Service Number: 573543
Company A 39th Infantry Regiment 4th Division
Date of Action: September 26, 1918
Location: Near Montfaucon Hill, France
War Department General Orders 89 (1919)
Medal # 7380

CITATION
After his squad leader had become a casualty, Private Dilworth assumed command and led his men against machine gun nests, materially assisting in the capture of two guns and prisoners. He was killed in the performance of duty.

DETAILS
Joseph Francis Dilworth was born in New York City, New York on September 11, 1890, the son of John and Mary Dilworth. In 1913 he was married to Helga Johnson, and they had one son. He served a two-and-a-half-year enlistment in the United States Navy aboard the gunboat/dispatch vessel *U.S.S. Dolphin*. Before entering the U.S. Army, he served as a Corporal on Mexican Border Service with Company G 1st Connecticut Infantry (National Guard.) When he registered for the draft in 1917, he listed his occupation as a helper in the printing room at Cheney Brothers Silk Company in South Manchester, Connecticut.

Dilworth enlisted as a Private of Infantry in the Regular Army on March 12, 1918 at Camp Greene, North Carolina. He was assigned to Company A 39th Infantry 4th Division stationed at Camp Greene. He sailed to France as a Private in Company A 39th Infantry 4th Division aboard the Italian troopship *Dante Alegheiri* on May 10, 1918. As his emergency contact, he listed his wife Helga in Manchester, Connecticut. He served in the Aisne-Marne Offensive, the Vesle Sector, the Toulon Sector and the St. Mihiel Offensive. In a letter home after his first battle, he wrote that combat was similar to going into a football game, that he felt nervous for the first few minutes but after that he didn't mind.

Dilworth was awarded the Distinguished Service Cross for his actions in the Meuse-Argonne Offensive. On the morning of September 26, 1918, the first day of the Offensive, dense fog covered the battlefield. During the advance, Dilworth's Battalion was disoriented in the fog and inadvertently moved to the left, coming up against the strong German defenses at Montfaucon Hill. After his Sergeant had become a casualty, Dilworth, though only a Private took command of his squad and led the squad in an attack against machine gun emplacements defending the hill. Under his direction the squad captured two machine guns and took several prisoners. Dilworth was killed during the action.

Dilworth was buried in Row B Grave 43 in the American battlefield cemetery # 840 at the village of Cuisy.

His Distinguished Service Cross was presented to his widow Mrs. Helga Dilworth.

On May 8, 1919 Joseph F. Dilworth was reinterred in Plot 3 Section 42 Grave 108 in the Argonne American Cemetery # 1232. He was reinterred and laid to his final rest on November 29, 1921 in Plot E Row 15 Grave 35 in the Meuse-Argonne American Cemetery, Romagne-sous-Montfaucon, France.

In 1920, Manchester Memorial Hospital at Manchester,

Connecticut was opened and dedicated to the men and women of Manchester who served in the war 1914-1918. In 1933 a bronze plaque mounted on a boulder was installed in front of the hospital. The plaque features the names of the forty-five residents of Manchester killed in World War One. Joseph F. Dilworth is one of the names inscribed on the plaque.

Frank J. Downs

Sergeant, United States Army
Army Service Number: 559681
Company B 58th Infantry Regiment 4th Division
Date of Action: July 19, 1918
Location: Near Courchamps, France
War Department General Orders 27 (1920)
Medal # 5869

CITATION

During an enemy counterattack, when the units on his right and left were falling back and his Company was taking heavy casualties, Sergeant Downs (at the time a Corporal) promptly grasping the situation, fearlessly led his platoon forward 400 yards, broke up the enemy attack, and established a new line of resistance. His heroic conduct enabled adjoining units to advance.

DETAILS

Frank Joseph Downs was born in Sunbury, Northumberland County, Pennsylvania on March 6, 1891, the son of Frank W. and Agnes Downs. In his registration for the draft in 1917, he listed his home of record as Pottsville, Schuylkill County, Pennsylvania, and his occupation as railroader. He enlisted in the Regular Army on July 22, 1917 as a Private in Co B 58th Infantry as it was forming up at Gettysburg, Pennsylvania. He was promoted to Private 1st Class on November 26, 1917 and to Corporal on December 11, 1917.

Downs sailed to England as a Corporal in Company B 58th Infantry 4th Division aboard the armed British troopship *Moldavia* on May 6, 1918. As his emergency contact, he listed his father Frank in Pottsville, Pennsylvania. On May 23, 1918, as the ship was within sight of land and about to arrive in England the *Moldavia* was torpedoed by the German submarine SM UB-57 and sank with the loss of 56 men, 55 of which were Downs' fellow soldiers from Company B 58th Infantry. The other loss was a soldier from Company A 58th Infantry which had also sailed aboard the *Moldavia*. Downs was rescued by one of the escorting British destroyers and brought to England. From England he and his Company were sent over to France.

Downs was awarded the Distinguished Service Cross for his actions in the Aisne-Marne Offensive. On July 19, 1918 Downs' Battalion of the 58th Infantry was supporting the attack of a Battalion from the French 133rd Infantry Regiment. The attack was stopped about 800 meters east of the village of Courchamps and met by German counterattacks which began to push the Allied forces back. Downs as a Corporal assumed command of his platoon after its leaders had become casualties, led the platoon forward in an advance against an enemy counterattack, and halted the enemy attack thereby allowing the rest of his Company and other units to cease their retreat and hold their positions. His actions allowed supporting units to join up and the Allied advance to continue. He also received a Citation Star.

Downs was promoted to Sergeant on July 27, 1918. He served in the Vesle Sector, the Toulon Sector, and the St. Mihiel Offensive. In the Meuse-Argonne Offensive, he was wounded by a machine gun bullet to his right shoulder on September 30, 1918. After recovering from his wound, he served with the 4th Division on occupation duty in Germany. He returned to the United States with his Company aboard the troopship *U.S.S. Mount Vernon* (captured German liner *Kronprinzessin Cecilie*) leaving Brest, France July 24, 1919 and arriving at Hoboken, New Jersey August 1, 1919. He was discharged at Camp Dix, New Jersey on August 7, 1919.

The 1920 Census showed him living with his parents in Pottsville, Pennsylvania where he was employed as a brakeman for the railroad. The 1930 and 1940 Census both showed him at the same address but unemployed and marked as disabled in war. Frank J. Downs died of Acute Myocardial Failure at the age of 57 on May 19, 1948 and is buried in Pomfret Manor Cemetery, Sunbury, Northumberland County, Pennsylvania.

Michael D. Belis

Thomas D. Drake
Sergeant, United States Army
Army Service Number: 560684
Company A 59th Infantry Regiment 4th Division
Date of Action: July 19, 1918 and September 29, 1918
Location: Near Courchamps, France and near Brieulles, France
War Department General Orders 10 (1920)
Medal # 5670

CITATION
After having successfully led his platoon to its objective in the attack of 19 July, Sergeant Drake gathered together groups of other companies of the battalion which had become disorganized due to heavy losses and established under intense fire a line of defense which was held until the unit was relieved. On September 29 he was wounded in the hand but refused to go to the rear and continued to perform his duties, frequently exposing himself to heavy machine-gun fire in order to control his command.

DETAILS
Thomas Davison Drake was born in Tunnelton, Preston County, West Virginia on October 24, 1900, the son of Eli and Isadora Drake. On April 23, 1917, thirteen days after the United States declared war on Germany Drake, at the age of sixteen lied about his age and enlisted as a Private in Company A 4th Infantry (Regular Army) at Gettysburg, Pennsylvania. When elements of the 4th Infantry were used to create the 59th Infantry, Drake thus became one of the first members of the 59th

Infantry. He went with the Regiment to Camp Greene, North Carolina where it became part of the 4th Division. Before May 1918, he had been promoted to Private 1st Class, Corporal, and Sergeant. Drake sailed to France as a Sergeant in Company A 59th Infantry and part of the advance detachment of the 4th Division aboard the troopship *U.S.S. Great Northern* on May 2, 1918. As his emergency contact, he listed his mother in Lumberport, West Virginia.

He was awarded the Distinguished Service Cross for his actions on two separate dates. The first date was in the Aisne-Marne Offensive. His 1st Battalion of the 59th Infantry was attached to and under direct control by the French 133rd Infantry 164th Division for the Offensive and on the first day had been held in reserve. On July 19, 1918, the second day of the Offensive 1st Battalion 59th Infantry attacked from a point east of the village of Courchamps. The Americans struggled through enemy artillery, machine gun, and rifle fire and advanced to about 300 meters east of the Courchamps-Priez road. The Battalion Commander was wounded, and seven other officers were either killed or wounded. Casualties were heavy among the enlisted men and the attack was stopped. Drake exposed himself to heavy enemy fire to gather together survivors from several companies, withdrew them to the road, organized them into a single force and established a defensive position along the road which they held until relieved by a Battalion of the French 152nd Infantry Regiment. He was at the time only seventeen years old.

Drake served in the Vesle Sector, the Toulon Sector, and the St. Mihiel Offensive.

The second date in his Citation was in the Meuse-Argonne Offensive. On September 29, 1918 1st Battalion 59th Infantry was leading the right side of the Division attack near the wooded area known as the Bois de Brieulles. It reached the German trenches on the northern slope of Hill 281 where it was halted by heavy enemy fire. Though he was wounded, Drake remained with his men as they dug in opposite the enemy positions and directed them in defending their line until relieved. He would not become eighteen years old for another month.

At some time after the Meuse-Argonne Offensive Drake was promoted to 1st Sergeant. He received two Citation Stars. He served on occupation duty in Germany with the 4th Division and returned to the United States as a 1st Sergeant in Company A 59th Infantry and part of Company "K" Third Army Composite Regiment aboard the troopship U.S.S. Leviathan (seized German liner Vaterland) leaving Brest, France September 1, 1919 and arriving at Hoboken, New Jersey September 8, 1919.

(See the entry for Joseph Bassi for an explanation of the Third Army Composite Regiment.)

Drake was discharged on September 27, 1919.

On November 12, 1920 Drake enlisted in the Regular Army as a Sergeant and from 1920 to 1923 served in Companies E and C 40th Infantry and Company C 10th Infantry on duty in the United States and Hawaii. On October 16, 1923 he was commissioned a 2nd Lieutenant of Infantry with date of rank back to July 2, 1923. He served as an officer in the United States and in the Panama Canal Zone before World War Two, being promoted to 1st Lieutenant on September 20, 1928, Captain on August 1, 1935 and Major on July 2, 1940. After 1932 he was awarded the Purple Heart for wounds he received in World War One and his two Citation Stars were converted to Silver Star Medals. He was married in 1937. He was promoted to the temporary wartime ranks of Lieutenant Colonel on December 24, 1941 and Colonel on December 30, 1942.

In World War Two, Drake commanded the 168th Infantry Regiment 34th Infantry Division in North Africa. In the Kasserine Pass/Faid Pass/Sidi Bou Zid areas Drake and his Regiment fought a last-ditch rear-guard action which allowed a large part of the American forces to withdraw and prevented the Kasserine Pass battle from becoming a larger disaster than it already was. Drake was captured and taken prisoner by the Germans in that battle on February 16, 1943. For his courage and leadership during that action he was awarded an oak leaf cluster to his Distinguished Service Cross in Headquarters U.S. Army-Mediterranean Theater of Operations General Orders No. 32 (1945).

In captivity, Drake was the senior officer and POW Commander at Oflag 64 in Altburgund, Germany until his repatriation because of stomach ulcers on July 26, 1944. His further awards for his World War Two service included two more Silver Star Medals, Legion of Merit, Bronze Star Medal, Air Medal and two more Purple Hearts. He retired from the Army as a Colonel with a disability in the line of duty on April 30, 1947. Thomas D. Drake died at the age of 69 on April 8, 1970 and is buried in Greenwood Memorial Park, San Diego, San Diego County, California.

Michael D. Belis

Charles T. Dunbar

Corporal, United States Army
Army Service Number: 569264
Company F 4th Engineer Regiment 4th Division
Date of Action: August 5, 1918
Location: West of Fismes, France
War Department General Orders 145 (1918)
Medal # 1080

CITATION

Corporal Dunbar was a member of a small detachment of Engineers which went out in advance of the front line of the Infantry through an enemy barrage from 77-millimeter and 1-pounder guns to construct a footbridge over the River Vesle. As soon as their operations were discovered, machine gun fire was opened upon them. Undaunted, the party continued at work, removing the German wire entanglements and successfully completing a bridge which was of great value in subsequent operations.

DETAILS

Charles Thomas Dunbar was born in Jumping Branch, Summers County, West Virginia on January 28, 1891, the son of William and Sarah Dunbar. At the time he entered the Army he listed his home of record as Sidney, Nebraska. He entered the Army on June 3, 1917. Dunbar sailed to France as a Corporal in Company F 4th Engineers 4th Division aboard the troopship *U.S.S. Martha Washington* (seized Austrian liner) on April 30, 1918. As his emergency contact, he listed his father William in Bellepoint, West Virginia.

Dunbar was awarded the Distinguished Service Cross for his actions in the Aisne-Marne Offensive. After several days of trying to cross the Vesle River under fire and only managing to put across small numbers of men, on August 5, 1918 the 4th Division made a determined effort to get across the river. The 4th Engineers, including Dunbar, worked under fire incessantly to build and emplace footbridges which the Infantry could utilize to make a mad dash across the river. Dunbar was part of a detachment of Engineers who were working along the river near the village of Ville Savoye west of the village of Fismes. On August 5, that detachment particularly distinguished themselves by remaining at their work under the heaviest of fire and completing a bridge for the Infantry.

(See the entry for William B. Beach for an overview of the kinds of activities carried out by the soldiers of the 4th Engineers involved in building the bridges across the Vesle River.)

Dunbar was one of three soldiers in the 4th Division to be awarded the Belgian Military Decoration (Décoration Militaire) during World War One. All three soldiers were Corporals from Company F 4th Engineers, and all received the Belgian Military Decoration and the Distinguished Service Cross for their actions at the Vesle River on August 5, 1918. Dunbar was one of only 20 soldiers of the American Expeditionary Forces to receive the Belgian Military Decoration.

He was one of 20 soldiers of the 4th Division and one of only 399 soldiers of the American Expeditionary Forces to be awarded the Italian War Merit Cross (Croce al Merito di Guerra).

At some time after the date of his award action, Dunbar was promoted to Sergeant. He served in the Vesle Sector, the Toulon Sector, the St. Mihiel Offensive, and the Meuse-Argonne

Offensive. He served on occupation duty in Germany with the 4th Division and returned to the United States with Company F 4th Engineers aboard the troopship *U.S.S. Von Steuben* (seized German liner *Kronprinz Wilhelm*) leaving Brest, France July 21, 1919 and arriving at Hoboken, New Jersey July 29, 1919. He was discharged on August 2, 1919.

The 1920 Census showed him living with his brother's family in La Grande, Oregon and working as a carpenter. In either 1921 or 1922, he married Ada Luton. The 1930 Census showed him living in La Grande with his wife, son and daughter and working as a brakeman for a railroad and indicated he was a veteran of the World War. When he registered for the draft in 1942, he was still living in La Grande with Ada and indicated he was employed by the Union Pacific Railroad. Charles T. Dunbar died at the age of 65 on April 5, 1956 and is buried in Grandview Cemetery, La Grande, Union County, Oregon.

Charles B. Duncan

Captain, United States Army
Commanding Officer Battery F 77th Field Artillery Regiment 4th Division
Date of Action: September 29, 1918
Location: Near Bois de Septsarges, France
War Department General Orders 71 (1919)
Medal # 7473

CITATION

When an enemy shell landed in the ammunition dump of his battery, Captain Duncan jumped in among the burning shells and succeeded in getting the fuses away and extinguishing the fire. Later he was mortally wounded by enemy shell fire.

DETAILS

Charles Benjamin Duncan was born in New York City, New York on December 1, 1893, the son of Charles B. and Anne Duncan. When he was four years old his father died and when he was six his mother died, and he went to live with his grandmother. At the time he entered the Army, he listed his home of record as Nashville, Tennessee. At the age of 18, Duncan entered the United States Military Academy at West Point on June 14, 1912. From the beginning he had trouble at the Academy with his mathematics studies. On June 28, 1913, he was turned back a class because of deficiency in his mathematics studies. He was discharged from the Academy on January 15, 1915 for that same reason of deficiency in mathematics.

Duncan then attended the University of Virginia in 1916. In 1917 he was commissioned a 1st Lieutenant in the Cavalry branch of the Regular Army and assigned to the 2nd Cavalry. In June of 1917, the 2nd Cavalry was divided into three units which were renamed the 2nd, the 18th and the 19th Cavalry. Duncan was part of the unit which became the 19th Cavalry. On July 20, 1917, the 19th Cavalry was re-designated as the 19th Cavalry Provisional Artillery. Duncan was given several months of artillery training and became commanding officer of Battery F. In November 1917, the 19th Cavalry Provisional Artillery was officially reflagged as the 77th Field Artillery. Duncan thus was one of the first members of the 77th Field Artillery upon its creation in 1917. In December 1917, he was promoted to the wartime temporary rank of Captain of Cavalry and assigned to the 77th Field Artillery with date of rank back to August 5, 1917.

In the fall of 1917, the 77th was moved to Camp Greene, North Carolina where it became part of the 4th Division when the Division was formed. Duncan sailed to England with the 77th Field Artillery Regiment as Captain and Commanding Officer of Battery F aboard the Australian troopship *Hororata* on May 19, 1918. As his emergency contact, he listed his uncle Roper Eastman in Jacksonville, Florida. From England he and his Regiment were sent over to France where they received their French 75mm guns in early June. Duncan served in the Vesle Sector and the St. Mihiel Offensive. (The 77th Field Artillery did not participate in the Aisne-Marne Offensive or the Toulon Sector.)

He was awarded the Distinguished Service Cross for his actions in the Meuse-Argonne Offensive. On September 29, 1918, an enemy artillery round landed in the ammunition area of his battery and set the area on fire. Ignoring the danger to himself, Duncan ran into the flames and put out the fire, thereby preventing the ammunition from exploding and undoubtedly saved the lives of his men. He was wounded that same day by German artillery and died of his wounds the next day.

The official history of the 77th Field Artillery describes Duncan's death:

"CHARLES B. DUNCAN, Captain of Cavalry, Commanding Battery F

On the evening of September 29, 1918, Captain Charles B. Duncan was mortally wounded near his battery positions in the Bois-de-Septsarges. During a heavy attack of high explosive shells, Captain Duncan was hit in the leg. The large femoral vein was severed. At once Lieutenant Simmons and a medical man applied a tourniquet in an endeavor to stop him from bleeding to death. He was immediately taken to the first aid station in an unconscious condition. While there he regained consciousness. He made light of his injury and spoke confidently of early return to his fighting battery. He was removed to Field Hospital No. 19 at Cuisy. There he slowly weakened and in the early morning of September 30th he breathed his last. He was buried in the military cemetery northeast of the village church of Cuisy. In the death of Captain Duncan, the Regiment lost a much beloved and devoted officer and friend. For an extraordinary act of gallantry in extinguishing a fire near one of his ammunition dumps, thereby saving the lives of many of his men this gallant officer was posthumously awarded the Distinguished Service Cross." [8]

Duncan was buried in Row A Grave 1 in the Amerian battlefield cemetery # 840 at the village of Cuisy.

His Distinguished Service Cross was presented to his grandmother, Mrs. William Duncan.

Charles B. Duncan was the only soldier from the 77th Field Artillery to receive the Distinguished Service Cross in World War One.

On May 8, 1919, Charles B. Duncan was reinterred in Plot 1 Section 45 Grave 6 in the Argonne American Cemetery # 1232. He was laid to his final rest on December 6, 1921 in Plot E Row 8 Grave 39 in the Meuse-Argonne American Cemetery and Memorial, Romagne-sous Montfaucon, Departement de la Meuse, Lorraine, France.

At the University of Virginia on the south side of the Rotunda, his name is inscribed on the bronze plaque listing the names of students and graduates of the University who were killed in World War One.

Peter W. Ebbert

Captain, United States Army
Headquarters Company 2nd Battalion 58th Infantry Regiment 4th Division
Date of Action: August 7, 1918
Location: At Ville Savoye, France
War Department General Orders 27 (1919)
Medal # No assigned number discovered

CITATION

Lieutenant Ebbert, acting as battalion supply officer, conducted numerous details of food and ammunition through the heavy enemy artillery barrage. Later in the day he volunteered for observation duty and was posted in a prominent tower where he was killed by a direct artillery hit.

Note: The heading of the Citation for the Distinguished Service Cross for Ebbert gives his rank as Captain. The text of Ebbert's Citation gives his rank as Lieutenant at the time of his award action. The Official Army Register indicates he held the temporary rank of Captain at the time of his award action.

Michael D. Belis

DETAILS

Peter William Ebbert was born in New York City on August 3, 1895, the son of Peter and Julia Ebbert. He enlisted in Company L 5th Infantry New Jersey National Guard on June 26, 1916 and served on active duty with that organization from June 26 to November 4, 1916, and again from March 25 to August 14, 1917, rising from Private to Corporal to Sergeant. His duty with the New Jersey National Guard included several months in Arizona on Mexican Border Service. On August 15, 1917, he was commissioned a 2nd Lieutenant of Infantry in the Regular U.S. Army with date of rank back to August 7, 1917 and was immediately promoted to 1st Lieutenant. At the time he entered the Army he listed his home of record as Glen Rock, New Jersey. He married his wife Marion in November 1917.

Ebbert sailed to France as a 1st Lieutenant of the 39th Infantry and a member of the advance detachment of the 4th Division aboard the troop transport *U.S.S. Kroonland* on April 30, 1918. As his emergency contact, he listed his wife in Glen Rock, New Jersey. At some time after reaching France, he was transferred to the 58th Infantry still in the 4th Division. He was promoted to the wartime temporary rank of Captain on June 12, 1918.

Ebbert served in the Aisne-Marne Offensive. He was awarded a Citation Star as Supply Officer of 2nd Battalion 58th Infantry for leading details of much needed food and ammunition to the frontline troops while he and those details were under fire during enemy barrages on July 28, 1918.

He was awarded the Distinguished Service Cross for his actions in the Vesle Sector. On August 7, 1918, Ebbert and 2nd Lieutenant Lowell H. Riley also of the 58th Infantry were killed in action together at the village of Ville Savoye near the Vesle River and both were awarded the Distinguished Service Cross. The following passage from the official history of the 4th Division in World War One describes the event:

> "In Villesavoye [sic] there stands the ruin of an old church. On the corner of this church at one time there was built a tower. On August 4th, the town was captured and occupied by the 58th Infantry. It was necessary that the enemy's movements across the Vesle be observed. The one position from which these observations could be made was the tower of the old church, and it was by far the most outstanding target for the Germans. First Lieutenant Peter W. Ebbert and Second Lieutenant Lowell H. Riley, 58th Infantry, volunteered to establish an observation station in this tower. For two days they occupied it and sent in very valuable information. During the whole of the time, they were subjected to heavy artillery fire. On the evening of August 7th, an enemy shell crashed into the tower and both officers were instantly killed." [9] (The passage gives Ebbert's rank incorrectly. He held the temporary rank of Captain at the time.)

Ebbert was buried in the battlefield cemetery # C-64 near the village of Ville Savoye.

His Distinguished Service Cross was presented to his widow, Mrs. Peter W. Ebbert.

Peter W. Ebbert was reinterred in Plot 1 Section G Grave # 18 in the American Cemetery # 617 at Fismes. He was reinterred and laid to his final rest on August 21, 1922 in Plot C Row 5 Grave 9 in the Oise-Aisne American Cemetery, Seringes-et- Nesles, France.

On July 6, 1932 Mrs. Julia A. Ebbert the mother of Peter W.

Ebbert sailed aboard the *S.S. President Harding* and visited her son's grave in France under the Gold Star Pilgrimage program. See the entry for Edward S. Blackman for an explanation of the Gold Star Pilgrimages.

Peter W. Ebbert never saw his daughter Caroline who was born after he deployed to France.

Andrew Edmiston

Second Lieutenant, United States Army
Company A 39th Infantry Regiment 4th Division
Date of Action: August 2, 1918
Location: In the Forêt de Fére, France
War Department General Orders 9 (1923)
Medal # 1589

CITATION

While in command of a platoon of his regiment it was severely bombed by enemy planes and under intense artillery fire. Serious disorganization followed the bombing. Taking steps for the safety of his unit, he went forward under heavy artillery fire and assisted in the reorganization of units and, though badly wounded, insisted upon the evacuation of all other wounded before accepting first aid for himself. His splendid example of courage and devotion inspired every man of his command.

DETAILS

Andrew Edmiston was born in Weston, West Virginia on November 13, 1892, the son of Matthew and Ella Edmiston. He attended the Kentucky Military Institute at Lyndon, Kentucky and then the University of West Virginia. On his registration for the draft in June 1917 he gave his name as Andrew Edmiston Jr. and indicated he was a farmer. He attended Officers Training Camp at Fort Benjamin Harrison, Indiana, was commissioned a 2nd Lieutenant in the Officers Reserve Corps on August 27, 1917 and assigned to the 39th Infantry. Edmiston sailed to France as a 2nd Lieutenant in Company A 39th Infantry 4th Division aboard the Italian troopship *Dante Alegheiri* on May 10, 1918. As his emergency contact, he listed his uncle Andrew in Weston, West Virginia.

Edmiston was awarded the Distinguished Service Cross for his actions in the Aisne-Marne Offensive. At the end of July 1918, the 39th Infantry moved into the wooded area known as the Forêt de Fére southwest of the village of Sergy. Enemy artillery pounded the area causing considerable casualties in the Regiment. On the night of August 1, 1918, the 39th Infantry took up positions in the Forêt de Fére in preparation for the next day's advance. At about 8 o'clock that night, a German aircraft discovered the American formations in the woods. The 1st Battalion which included Edmiston's Company A was in line in a column of twos, and the enemy aircraft flew down that line dropping strings of small bombs.

Every Company in the Battalion was hit by the German

bombs. As the aircraft exited the area, enemy artillery and machine guns added to the turmoil, resulting in total casualties in the Battalion for the night of 27 killed and 94 wounded. Edmiston moved about in the open under fire during the night and into the next morning, gathering disorganized men together into a cohesive unit and seeing to the care of the wounded. Though wounded himself, he supervised the removal of the casualties to the forward aid stations and did not seek aid for himself until all enlisted men were cared for.

His further front line service could not be determined but he apparently was returned to duty and wounded again. He sailed home several months ahead of his Regiment as a 2nd Lieutenant of Company A 39th Infantry and part of a detachment of Convalescent Casual Officers aboard the troop transport *U.S.S. Pastores* leaving Bordeaux, France January 31, 1919 and arriving at Newport News, Virginia February 14, 1919. In the passenger list he was indicated as having a gunshot wound to the left arm. He was discharged on August 11, 1919.

The 1920 Census showed Edmiston living in Hacker's Creek, West Virginia where he was employed as a Real Estate Agent and Farmer. He was married in April 1920. In 1922 he moved to Weston, West Virginia where he was the mayor from 1924-1926. The 1930 Census showed him living with his wife and daughter in Weston, West Virgnia where he was employed as a newspaper editor and indicated he was a veteran of the World War. After 1932, he was awarded the Purple Heart with oak leaf cluster for the wounds he received in World War One.

In 1932 he was elected as a Democrat to the U.S. House of Representatives where he served until January 1943. After 1939, he was awarded the State of West Virginia Distinguished Service Medal. He was State director of War Manpower for

Michael D. Belis

West Virginia from June 1943 to June 1945. His wife Merle died in 1950 and he remarried in 1953. Andrew Edmiston died at the age of 73 on August 28, 1966 and is buried in Machpelah Cemetery, Weston, Lewis County, West Virginia.

Emil J. Eklund

First Lieutenant, United States Army
Company E 58th Infantry Regiment 4th Division
Date of Action: October 5, 1918
Location: Near Bois de Fays, France
War Department General Orders 16 (1920)
Medal # 5728

CITATION

The courageous conduct of Lieutenant Eklund while in command of the left flank platoon of the division during a strong enemy counterattack, exposed to heavy machine gun and rifle fire, had a great morale effect on his men. The attack was repulsed after a hand-to-hand encounter in which many casualties were suffered.

Note: Though his Citation gives his rank as First Lieutenant Eklund was a Second Lieutenant at the time of the award action.

Michael D. Belis

DETAILS

Emil Julius Eklund was born in Lead, South Dakota on December 12, 1888, the son of Charles and Cora Eklund. He enlisted as a Private in the Regular Army at Fort George Wright, Washington on June 12, 1917 and was assigned to Company D 4th Engineers. He was promoted to Private 1st Class on August 1, 1917, to Corporal on August 7, 1917 and to Sergeant on September 1, 1917. He attended Officers Training Camp at Fort Oglethorpe, Georgia in January 1918 but did not graduate. Eklund sailed to France as a Sergeant in Company D 4th Engineers 4th Division aboard the transport *U.S.S. Martha Washington* (seized Austrian liner) on April 30, 1918. As his emergency contact, he listed his father Charles in Los Angeles, California. He was commissioned a 2nd Lieutenant in France on June 1, 1918 and transferred to Company E 58th Infantry also in the 4th Division. He served in the Aisne-Marne Offensive, the Vesle Sector, the Toulon Sector and the St. Mihiel Offensive.

He was awarded the Distinguished Service Cross for his actions in the Meuse-Argonne Offensive. On October 5, 1918, Eklund and his platoon of about fifty men protected the left flank of the entire 4th Division in the wooded area known as the Bois de Fays. To his left was the American 80th Division. The enemy attempted to exploit the weakness in communication between the two American Divisions and made three counterattacks on this flank that day. Eklund commanded his platoon in the defense against these attacks, one of which consisted of at least 300 Germans assaulting his position. This attack ended with the enemy inside the American line resulting in face-to-face combat with bayonets and rifle butts. The Los Angeles Examiner reported that Eklund "exposed himself to intense rifle fire during an attack" and that his "bravery so stimulated his men that they repulsed the enemy in a fierce hand-to-hand encounter."[10]

Eklund was promoted to 1st Lieutenant on November 14, 1918. He served on occupation duty in Germany with the 4th Division and returned to the United States as a 1st Lieutenant in Company E 58th Infantry aboard the troopship *U.S.S. Mount Vernon* (captured German liner *Kronprinzessin Cecilie*) leaving Brest, France July 24, 1919 and arriving at Hoboken, New Jersey on August 1, 1919. He was discharged on August 16, 1919 at San Francisco, California.

The 1920 Census showed him living with his parents in Los Angeles, California where he was employed as an electrician. He married sometime after January 1920, had a daughter in 1921 and a son in 1924. His wife Hallie was a schoolteacher in California and had served in Europe as a Y.M.C.A. canteen service worker during the war, returning from France aboard the troopship *U.S.S. America* (seized German liner *Amerika*) leaving Brest, France September 6, 1919 and arriving at Hoboken, New Jersey September 15, 1919.

In those days there was no USO and organizations such as the Y.M.C.A., the Salvation Army, the Knights of Columbus, the Jewish Welfare Board, and a few others did the work of seeing to the welfare of the soldiers that the USO was later created to do. It is possible that Eklund met his future wife while they were both overseas, but that could not be confirmed. Eklund's parents were living in Los Angeles, California and after the war Eklund went to live with his parents so he may have met Hallie in California after he moved there in 1919.

On March 15, 1929 Eklund was admitted to the U.S. National Home for Disabled Volunteer Soldiers, Pacific Branch in Sawtelle, Los Angeles County, California, having been diagnosed with Obstructive Pulmonary Tuberculosis. Emil J. Eklund died two years later at the age of 42 on June 18, 1931.

Edwin A. Elliott

Sergeant, United States Army
Army Service Number: 556067
Company E 39th Infantry Regiment 4th Division
Date of Action: October 12, 1918
Location: North of Nantillois, France
War Department General Orders 15 (1921)
Medal # No assigned number discovered

CITATION

An ammunition detail having failed several times to carry ammunition over a barraged zone, Sergeant Elliott voluntarily gathered and conducted an ammunition detail over 3 kilometers under extremely heavy artillery and machine-gun fire to the front line. In advance of his men, he dragged a full box of Springfield ammunition for over a kilometer and distributed it to the front line. Later, he volunteered to carry, and carried, a message back to regimental headquarters.

DETAILS

Edwin Alexander Elliott was born in Troy, Bell County, Texas on August 18, 1891, the son of Owen and Mary Elliott. His father Owen was a physician in Bosqueville, McLennan County, Texas. By the time he was eighteen Elliott was working in a grocery store in Bosqueville. When he registered for the draft in June 1917, he indicated he was a student and secretary for the Y.M.C.A. He was drafted into the Army as a Private at Camp Greene, North Carolina on January 10, 1918. At the time of induction his home of residence was listed as Elm Mott, McLennan County, Texas. He was assigned to Headquarters Company 39th Infantry 4th Division at Camp Greene. Elliott was promoted to Private 1st Class on March 18, 1918.

He sailed to France as a Private 1st Class in Headquarters Company 39th Infantry 4th Division aboard the Italian troopship *Duca D'Aosta* on May 10, 1918. As his emergency contact, he listed his father Dr. O.C. Elliott in Elm Mott, Texas. Elliott served in the Aisne-Marne Offensive and the Vesle Sector. On August 27, 1918 he was reduced to the rank of Private. He was transferred to Company E 39th Infantry on August 28, 1918. Elliott served in the Toulon Sector and the St. Mihiel Offensive. He was promoted to Sergeant on September 21, 1918.

He was awarded the Distinguished Service Cross for his actions in the Meuse-Argonne Offensive. On October 12, 1918, the 39th Infantry occupied a line in the northern edge of the wooded area known as the Bois de Forêt, north/northeast of the village of Nantillois and very near the village of Brieulles-sur-Meuse. At 12:30 p. m. a heavy artillery concentration lasting for two hours fell on the front line in Bois de Forêt. This resulted in heavy casualties among the Americans, and a withdrawal from the northern edge of the wood. After the firing had ceased, the old positions were reoccupied.

During the barrage, a detail of men carrying ammunition failed numerous times to reach the soldiers in the front lines. Without being ordered to do so, Elliott assembled several soldiers together, had them gather crates of ammunition, and led them through the artillery fire, bringing that ammunition to the front. Forging ahead of his little group, he personally dragged a heavy case of ammunition for over a kilometer. After reaching the front lines with his much-needed cargo, Elliott volunteered to carry a message back to Headquarters, and re-crossed the fire swept battleground again.

Six days later, on October 18, his Regiment was pulled out of the front lines. On October 22, 1918 Elliott was discharged as an enlisted man in order to accept a commission. On October 23, 1918 he was commissioned a 1st Lieutenant and Chaplain of Infantry. He was assigned to Headquarters Base Section # 5, American Expeditionary Forces at Brest, France. He returned to the United States aboard the troopship *U.S.S. Prinz Friedrich Wilhelm* (surrendered German liner) leaving Brest, France August 14, 1919 and arriving at Hoboken, New Jersey August 23, 1919. Elliott was discharged on August 28, 1919.

He returned to Texas where he married Ora Leveridge on December 25, 1919. In 1925 he and his wife had a daughter. The 1925 National Guard Register indicated he was a Chaplain with Headquarters Company 144th Infantry Regiment, Texas National Guard in Fort Worth, Texas. The 1930 Census showed Elliott and his wife and daughter living in Austin, Travis County, Texas where he was employed as a teacher at a university. He became head of the Economics Department at Texas Christian University. In 1935 he and his wife had a second daughter. The

1940 Census showed Elliott and his wife and daughters living in Fort Worth, Tarrant County, Texas where he was employed as Regional Director of the National Labor Relations Board.

He was Regional Director of the National Labor Relations Board, Region 16, based in Fort Worth, Texas from 1934 until his retirement in 1961. At some time after 1969 he moved to Danbury, Connecticut. Edwin A. Elliott died at the age of 94 on August 5, 1986 and is buried in Christian Church Cemetery, Danbury, Fairfield County, Connecticut. His personal papers are held in the Special Collections, University of Texas at Arlington Libraries.

Harold W. Enright

Private, United States Army
Army Service Number: 2658588
Company I 47th Infantry Regiment 4th Division
Date of Action: September 28, 1918
Location: Near Bois de Brieulles, France
War Department General Orders 46 (1919)
Medal # 2028

CITATION

Private Enright charged an enemy machine gun nest which was inflicting heavy losses upon our troops and delayed the advance. He wounded the gunner and captured the gun thereby enabling our advance to continue.

DETAILS

Harold William Enright was born in Rochester, Monroe County, New York on August 23, 1895. He entered the Army on May 25, 1918, and listed his home of record as Warren, Illinois. Enright sailed to England as a Private and replacement in the Camp Gordon July Automatic Replacement Draft of Infantry aboard the Canadian transport *Northumberland* on July 22, 1918. As his emergency contact, he listed his uncle, J.F. Enright in Warren, Illinois. From England he was sent over to France. After further training in France, he was assigned to Company I 47th Infantry 4th Division.

Enright was awarded the Distinguished Service Cross for his actions in the Meuse-Argonne Offensive. On September 28, 1918 the 47th Infantry attacked through the wooded area known as the Bois de Brieulles in an attempt to secure the next wooded area, the Bois de Fays. Enright single handedly attacked and eliminated a German machine gun nest that was holding up the advance. At some time after the action, he was promoted to Private 1st Class. He was also awarded a Citation Star.

He was awarded the French Croix De Guerre. He was one of 20 soldiers of the 4th Division and one of only 399 soldiers of the American Expeditionary Forces to be awarded the Italian War Merit Cross (Croce al Merito di Guerra). The Italian War Merit Cross was presented to him on June 11, 1919 in a 4th Division awards ceremony at Remagen, Germany.

Enright served on occupation duty in Germany with the 4th Division and returned to the United States with Company I 47th Infantry aboard the *U.S.S. Mobile* (seized German liner *Cleveland*) leaving Brest, France July 16, 1919 and arriving at Hoboken, New Jersey July 27, 1919. He was discharged on August 4, 1919.

He married Grace Akins in 1919. The 1920 Census showed Enright and Grace living with Grace's parents in Warren, Illinois where Enright worked as a salesman for a meat market. The 1930 Census showed Enright and his wife and son living with his wife's parents in Warren, Illinois where Enright was employed as an insurance salesman and indicated he was a veteran of the World War. When he registered for the draft in 1942, he indicated he and Grace were still living in Warren, Illinois and he was employed by Bankers Life Company in Des Moines, Iowa. Harold W. Enright died at the age of 61 on July 7, 1957 and is buried in Elmwood Cemetery, Warren, Jo Daviess County, Illinois.

Charles H. Epler

Private, United States Army
Army Service Number: 560697
Company A 59th Infantry Regiment 4th Division
Date of Action: August 8, 1918
Location: Near Ville Savoye, France
War Department General Orders 95 (1919)
Medal # 2274

CITATION

After several unsuccessful attempts to silence an enemy machine gun nest had been made, Private Epler and another soldier volunteered to put the nest out of action. His companion was killed but Private Epler succeeded in throwing grenades into the nest, setting fire to the ammunition boxes with which it was surrounded killing several of the crew and stopping the fire of the gun.

DETAILS

Charles H. Epler was born in Dayton, Ohio January 24, 1887, the son of Martin and Mary Epler. Before entering the Army, he worked as a laborer for The Favorite Stove and Range Company in Piqua, Ohio. He enlisted as a Private in the Regular Army at Columbus Barracks, Ohio on July 29, 1917 and was assigned to Company A 59th Infantry. He was promoted to Private 1st Class on November 10, 1917 and to Corporal on March 21, 1918. Epler sailed to England as a Corporal in Company A 59th Infantry 4th Division aboard the British troopship *Megantic* on May 3, 1918. As his emergency contact, he listed his mother Mary in Dayton, Ohio. From England he and his Company were sent over to France. Epler served in the Aisne-Marne Offensive. He was reduced to the rank of Private on July 31, 1918.

He was awarded the Distinguished Service Cross for his actions in the Vesle Sector on August 8, 1918. On that day his Regiment was situated near the Rouen-Rheims Road across the Vesle River from the village of Ville Savoye. Patrols were sent out to determine the enemy's main defensive positions. In the afternoon the Germans blasted the 59th Infantry with an artillery barrage lasting for about three hours. The Regiment also received heavy fire from machine guns and snipers hidden in the wooded area around the Château du Diable situated on their right flank. Epler and another soldier attacked an enemy machine gun position after several attacks against it by others had failed. The other soldier was killed but Epler continued the attack alone and put the German machine gun out of action using hand grenades.

Epler served in the Toulon Sector and the St. Mihiel Offensive. He was promoted to Private 1st Class on September 21, 1918. He served in the Meuse-Argonne Offensive and was promoted to Corporal on November 7, 1918. Epler was also awarded a Citation Star and the French Croix De Guerre with gilt star.

He served on occupation duty in Germany with the 4th Division. On February 1, 1919 he was reduced to the rank of Private. He returned to the United States with his Company aboard the troopship *U.S.S. Mount Vernon* (seized German liner *Kronprinzessin Cecilie*) leaving Brest, France July 24, 1919 and arriving at Hoboken, New Jersey August 1, 1919. He was discharged on August 8, 1919.

City directories for Dayton, Ohio showed him living there in 1933 and 1934 where he was employed as a molder. Charles H. Epler died at the age of 47 when struck by an automobile at Dayton, Ohio on November 12, 1934. He is buried in Dayton National Cemetery, Dayton, Montgomery County, Ohio.

Michael D. Belis

Etienne Escudier
First Lieutenant, Army of France
79th Infantry Regiment Army of France
Attached to 59th Infantry Regiment 4th Division
Date of Action: September 29, 1918
Location: In the Bois de Brieulles, France
War Department General Orders 71 (1919)
Medal # 3593

CITATION
Though he was not required to do so by the duties of his position, Lieutenant Escudier volunteered to ascertain the source of an extremely heavy artillery fire which was being directed upon the American Infantry. In accomplishing this mission, he exposed himself to heavy shell and machine gun fire for three hours and secured accurate information, displaying absolute fearlessness and indifference to his own personal safety.

DETAILS
Etienne Escudier was one of a number of French Army officers who served as advisors to the U.S. 4th Division during World War One. He had been assigned as a Liaison Officer to the 59th Infantry in June 1918 and served with them in the Aisne-Marne Offensive, the Vesle Sector, the Toulon Sector and the St. Mihiel Offensive. He spoke French, English and German and formed a close friendship with Colonel Frederic M. Wise, the U.S. Marine Corps Colonel who took command of the 59th Infantry in early September 1918.

Escudier was awarded the Distinguished Service Cross for

his actions in the Meuse-Argonne Offensive on September 29, 1918. On that day, the 8th Brigade of the 4th Division, including the 59th Infantry Regiment. spearheaded the Division's attack out of the wooded area known as the Bois de Brieulles toward the next wooded area known as the Bois de Fays. German resistance was fierce, with intense artillery barrages and it was during this phase of the battle that Escudier earned his award. Though as an advisor and not required to actively participate in combat in the front lines, he volunteered for a dangerous scouting job. For three hours Escudier moved alone about the battlefield in the forward area under heavy fire until he determined the positions of the enemy artillery and recorded their locations. His report allowed the American artillery batteries to plot accurate counterfire.

Escudier served on occupation duty in Germany with the 4th Division. On April 28, 1919 the two Infantry Brigades of the 4th Division were assembled in formation on the parade ground at Remagen, Germany for an award ceremony. Major General Mark L. Hersey, commanding the 4th Division decorated officers and soldiers of the Division with the Distinguished Service Cross and the French Croix De Guerre. On that day one of the officers on whose tunic Hersey pinned the Distinguished Service Cross was 1st Lieutenant Etienne Escudier.

As his emergency contact, Escudier listed his father, Paul Escudier, 20 Rue Moncey, Paris, France.

Michael D. Belis

Charles H. Evans
Private, United States Army
Army Service Number: 574149
Company B 39th Infantry Regiment 4th Division
Date of Action: September 26, 1918
Location: Near the Bois de Brieulles, France
War Department General Orders 98 (1919)
Medal # 7788

CITATION
When his company was held up by heavy enemy machine gun fire, Private Evans and two other soldiers advanced in the face of intense fire and captured the enemy machine gun nest from which the fire had been coming, killing two of the enemy and capturing three prisoners with their machine gun.

DETAILS
Charles Hurlbut Evans was born in Jamaica, Guthrie County, Iowa on April 5, 1893, the son of Wallace and Martha Evans. At the time he registered for the draft in 1917 he was living in Weede, Dawson County, Montana where he worked as a farmer. He was drafted into the Army as a Private on March 25, 1918 and assigned to the 39th Infantry at Camp Greene, North Carolina. Evans was promoted to Private First Class on May 2, 1918. He sailed to France as a Private 1st Class in Company B 39th Infantry 4th Division aboard the Italian troopship *Dante Alegheiri* on May 10, 1918. As his emergency contact, he listed his father Wallace in Panora, Iowa. Evans served in the Aisne-Marne Offensive, the Vesle Sector, the Toulon Sector, and the St. Mihiel Offensive.

He was awarded the Distinguished Service Cross for his actions on the first day of the Meuse-Argonne Offensive. The actual location of his award action was about two to three miles southwest of the Bois de Brieulles. On September 26, 1918, Evans and his Company as part of 1st Battalion 39th Infantry attacked Montfaucon Hill northwest of the village of Cuisy. The high ground against which they attacked was defended with barbed wire obstacles and a large number of machine gun nests. Evans led two fellow soldiers and together they attacked and captured a German machine gun nest which had been holding up the advance of his Company. The fighting at Montfaucon Hill continued from mid-morning until about three in the afternoon when the Battalion broke off its attack after suffering heavy casualties.

Evans also received a Citation Star and on October 18, 1918 he was promoted to Corporal. He was awarded the French Croix De Guerre. Evans served on occupation duty in Germany with the 4th Division and returned to the United States as a Corporal in Company B 39th Infantry and part of Company I Third Army Composite Regiment aboard the troopship *U.S.S. Leviathan* (seized German liner *Vaterland*) leaving Brest, France September 1, 1919 and arriving at Hoboken, New Jersey September 8, 1919.

(See the entry for Joseph Bassi for an explanation of the Third Army Composite Regiment.)

Evans was discharged on September 24, 1919.

The 1920 Census showed Evans living with his parents in Panora, Guthrie County, Iowa where he was employed as a laborer for the railroad. His father died in 1921 and sometime after that he moved with his mother to Seattle, Washington. The 1930 Census showed Evans living with his mother, his sister, and members of his sister's family in Seattle where Evans worked as a laborer in a can factory and indicated he was a veteran of the World War. The 1940 Census showed him living with his mother in Seattle and still employed as a laborer in a can factory. Charles H. Evans died at the age of 68 in Marin County, California on March 27, 1962.

Daris V. Ford

Private, United States Army
Army Service Number: 568604
Company C 4th Engineer Regiment 4th Division
Date of Action: August 6, 1918
Location: On the River Vesle, east of St. Thibaut, France
War Department General Orders 78 (1919)
Medal # No assigned number discovered

CITATION

While his company was advancing ahead of the Infantry toward the Vesle River to put in foot bridges, Private Ford, acting as liaison messenger, displayed undaunted courage and utter disregard for his personal safety, by time after time carrying messages through a terrific enemy barrage and heavy machine gun fire, each time successfully accomplishing his mission.

DETAILS

Daris Vernon Ford was born in Gilkey Township, Rutherford County, North Carolina on January 26, 1889, the son of George and Sarah Ford. In his registration for the draft in 1917 he indicated he was married and employed as a Locomotive Fireman for the Union Pacific Railroad in North Platte, Nebraska. Ford entered the Army on October 26, 1917 at Fort Logan, Colorado. He sailed to France as a Private in Company C 4th Engineers 4th Division aboard the troopship *U.S.S. Martha Washington* (seized Austrian liner) on April 30, 1918. As his emergency contact, he listed his wife Gertrude in North Platte, Nebraska. He served in the Aisne-Marne Offensive.

Ford was awarded the Distinguished Service Cross for his actions in the Vesle Sector. On August 6, 1918 the Infantry elements of the 4th Division attacked toward the Vesle River while the Engineers hurriedly worked to build and put in place temporary foot bridges and artillery bridges on which the Infantry could use to run across the river. Ford ventured out into German artillery and machine gun fire to relay communications between the Engineers and the Infantry during the advance. He also received a Citation Star.

He was one of 20 soldiers of the 4th Division and one of only 399 soldiers of the American Expeditionary Forces to be awarded the Italian War Merit Cross (Croce al Merito di Guerra).

Ford served in the Toulon Sector, the St. Mihiel Offensive, and the Meuse-Argonne Offensive. He served on occupation duty with the 4th Division in Germany and returned to the United States as a Staff Sergeant in Company C 4th Engineers aboard the troopship *U.S.S. Von Steuben* (seized German liner *Kronprinz Wilhelm*) leaving Brest, France July 21, 1919 and arriving at Hoboken, New Jersey July 29, 1919. He was discharged on August 4, 1919 at Camp Dodge, Iowa.

In 1920 his ten-year-old son Jay died. The city directory of North Platte, Nebraska for 1921 showed Ford and his wife Gertrude living in North Platte where he had returned to his job as a Fireman with the Union Pacific Railroad. The 1930 Census showed him living away from his wife in Cedar Bluff, Alabama where he was employed as a bridge painter and indicated he was a veteran of the World War. He was admitted to the U.S. National Homes for Disabled Volunteer Soldiers, Bath Branch, Bath, New York from May to September 1932. The 1940 Census showed Ford and his wife Minnie (Gertrude) living with her brother's family in Jacksonville, Florida and indicated Ford was employed as a painter for the railroad. Daris V. Ford died at the age of 70 on May 24, 1959 in Thomas County, Georgia.

Michael D. Belis

Pietro Formica

Private First Class, United States Army
Army Service Number: 567932
Company A 59th Infantry Regiment 4th Division
Date of Action: August 8, 1918
Location: Near Ville Savoye, France
War Department General Orders 95 (1919)
Medal # 2265

CITATION

After several other soldiers had been killed in attempting to carry a message across an open field, under intense enemy fire Private Formica volunteered for this perilous mission and successfully accomplished it. He continued to display marked courage in carrying messages under fire until he was wounded the next day.

DETAILS

Pietro E. Formica was born in Torino (Turin), Italy on June 26, 1890 the son of Giuseppe and Dominica Formica. He immigrated to the United States at the age of 23, arriving at the port of New York on February 3, 1913. Before entering the Army, he was an iron worker at Imperial Ship Building Corporation in Detroit, Michigan. He enlisted as a Private in the Regular Army at Columbus Barracks, Ohio on July 20, 1917. On August 2, 1917, Formica was assigned to Company A 59th Infantry at Gettysburg, Pennsylvania and moved with it to Camp Greene, North Carolina on October 17, 1917. He temporarily served with Company A 10th Machine Gun Battalion from November 20, 1917 to April 15, 1918 when he returned to Company A 59th Infantry. Formica sailed to England as a Private in Company A 59th Infantry 4th Division aboard the British troopship *Megantic* on May 3, 1918. As his emergency contact, he listed his mother Dominica in Torino, Italy. From England he and his Company were sent over to France. He served in the Aisne-Marne Offensive.

Formica was awarded the Distinguished Service Cross for his actions in the Vesle Sector on August 8, 1918. On that day his Company was near the Rouen-Rheims Road across the Vesle River from the village of Ville Savoye and sending out patrols to determine the enemy's positions. German artillery fire was especially intense that day and the 59th Infantry was also taking heavy fire from the wooded area around the Château du Diable on their right flank. Even though several soldiers had been killed attempting to carry a vital message across an open area, Formica volunteered for the job and accomplished it under heavy enemy fire.

Formica served in the Toulon Sector and the St. Mihiel Offensive and was promoted to Private 1st Class on September 20, 1918. His Citation states that he was wounded on August 9 but he was actually wounded on September 29, 1918 during the Meuse-Argonne Offensive. On that day 8th Brigade which included his Regiment spearheaded the attack against the wooded area known as the Bois de Fays. Formica took a bullet through the lung in the advance. He was held at the forward aid station for a couple of days until he was stabilized and then sent to the rear. He was in Base Hospital No. 36 from October 1-26 and then in Base Hospital No. 43 from October 27-November 14, 1918. He returned to his Company and served with the 4th Division on occupation duty in Germany.

Formica was presented with his Distinguished Service Cross by Major General Mark L. Hersey, Commanding General of the 4th Division in an awards ceremony at Remagen, Germany on June 8, 1919. He also was awarded a Citation Star. He received the Verdun Medal from the French Government. Formica returned to the United States with his Company aboard the troopship U.S.S. Mount Vernon (seized German liner Kronprinzessin Cecilie) leaving Brest, France July 24, 1919 and arriving at Hoboken, New Jersey August 1, 1919. He was discharged at Camp Grant, Illinois on August 12, 1919.

An Act of Congress in May 1918 allowed for any alien who served in the military service of the United States during World War One to file a petition for naturalization and have several requirements pursuant to the process waived. Tens of thousands such veterans applied for citizenship under the provisions of this Act and Formica was one of them. He became a naturalized American citizen on February 10, 1920.

Formica's first job after return to civilian life was as an elevator operator. He married Filomena "Nancy" Gatti in September 1921 and they eventually had four children. They lived in Bridgeport, Connecticut for 42 years where Formica worked as a technician at A and C Dental Laboratory until retirement. After 1932 he was awarded the Purple Heart for the wound he received in World War One and his Citation Star was converted to the Silver Star Medal. He was a member of the Legion of Valor. Pietro E. Formica died at the age of 71 on July 12, 1961 and is buried in Saint Michaels Cemetery, Stratford, Fairfield County, Connecticut.

Ernest Fosnes

Corporal, United States Army
Army Service Number: 567092
Company A 59th Infantry Regiment 4th Division
Date of Action: July 19 and August 8, 1918
Location: Not given
Note: The locations of his award actions were near Courchamps and near Ville Savoye.
War Department General Orders 5 (1920)
Medal # 6019

CITATION

On July 19, 1918 Corporal Fosnes exposed himself to intense machine gun and artillery fire to assist in the reorganization of his company which had become temporarily disorganized due to heavy losses. On August 8, 1918 when his platoon had become separated from the company during the attack he exposed himself to direct machine gun fire in order to encourage the members of his platoon in their task. He was mortally wounded a short time afterwards.

DETAILS

Ernest Fosnes was born in Montevideo, Chippewa County, Minnesota on September 18, 1890, the son of Christopher (Kristoffer) and Sarah Fosnes. His registration for the draft in 1917 indicated that he had prior military service of three years and ten months as a Seaman in the United States Navy. Before entering the Army, he was employed as a farm hand in Baiting Hollow, Suffolk County, New York. He was drafted into the

Michael D. Belis

Army as a Private on February 15, 1918. He was promoted to Corporal on March 28, 1918. Fosnes sailed to England as a Corporal in Company A 59th Infantry 4th Division aboard the British troopship *Megantic* on May 3, 1918. As his emergency contact, he listed his father Christopher in Montevideo, Minnesota. From England he and his Company were sent over to France.

Fosnes was awarded the Distinguished Service Cross for his actions on two separate dates. The first date mentioned in his Citation was July 19, 1918, the second day of the Aisne-Marne Offensive. Attached to and under direct command of the French 164th Division the 59th Infantry led the advance that day in the darkness of early morning along a line east of the village of Courchamps. German artillery, machine gun and small arms fire took a heavy toll on Fosnes' 1st Battalion. The Battalion Commander was wounded, and seven other officers of the Battalion were either killed or wounded. The 59th Infantry attack was stopped by the ferocity of the enemy resistance and, taking heavy casualties, 1st Battalion pulled back to the Courchamps-Priez road. Moving about in the open under enemy fire, Fosnes helped to gather the disheartened and disorganized men of his Company and directed them in establishing defensive positions along the road.

The second date mentioned in his Citation was August 8, 1918 in the Vesle Sector. On that date Fosnes' Company was located near the Rouen-Rheims Road across the Vesle River from the village of Ville Savoye. In the advance, his platoon became separated from the rest of the Company and Fosnes braved heavy fire to get them up and moving to rejoin the Company in the attack. The next day, August 9, 1918 the 59th Infantry was met with strong resistance from the Germans in the wooded area around the Château du Diable. Several attacks against the entrenched enemy in those woods were launched without success. Fosnes died of wounds received in action during one of those attacks on August 9, 1918.

He was buried in Grave # 45 in the American battlefield cemetery # 412 at the village of Villers-sur- Fère.

His Distinguished Service Cross was presented to his father Christopher Fosnes.

On June 3, 1919 Ernest Fosnes was reinterred in Plot 3 Section K Grave 106 in the American Cemetery # 608 at Sering-es-et-Nesles. On May 2, 1921 his remains were moved to Plot 3 Section L Grave 121 in the same cemetery. At some time after 1921, the graves in the cemetery were re-numbered and his grave became Block B Row 4 Grave 45. On February 24, 1928, the graves were rearranged and Ernest Fosnes' final resting place became Plot B Row 4 Grave 22 in the Oise-Aisne American Cemetery, Seringes-et-Nesles, France.

On the day Fosnes died and in the area in which he died, an incident occurred in his Regiment that would in the future be tied with the 4th Division as part of the history and lineage of the 59th Infantry Regiment. During the day of August 9, 1918, a squad of five machine gunners dressed in olive drab uniforms came from the direction of the Château du Diable and approached a platoon from the 59th Infantry. The platoon Sergeant was one of the oldest and most experienced soldiers in the Regiment. When told by one of his men that the approaching soldiers were Americans, the Sergeant replied, "They have no business coming from that direction. Let 'em have it."[11] His platoon opened fire and killed them. The dead men were later inspected and confirmed to be German soldiers in American uniforms.

Michael D. Belis

The 59th Infantry was inactivated in 1922 before the Army approved a Distinctive Unit Insignia for the Regiment. Though there was a 59th Armored Infantry during World War Two, it was not a "straight-leg" regular Infantry Regiment. The 59th Infantry as a regular Infantry Regiment did not exist again after World War One until it was reorganized on March 1, 1952. When the Distinctive Unit Insignia for the 59th Infantry Regiment was approved by the Department of the Army on December 23, 1952, it was emblazoned with the newly approved official motto for the 59th Infantry of "Let 'Em Have It."

Earl R. Fretz

First Lieutenant, United States Army
Note: Earl R. Fretz was actually a Second Lieutenant at the time of his award action and was not promoted to First Lieutenant until later.
12th Machine Gun Battalion 4th Division
Date of Action: July 18, 1918
Location: Near Courchamps, France
War Department General Orders 5 (1920)
Medal # 6321

CITATION

After all the officers of Company E 59th Infantry had become casualties, Lieutenant Fretz voluntarily assumed command of the Infantry company in addition to his machine gun platoon and personally led it forward to its objective. The gallantry displayed by this officer while exposed to heavy machine gun and artillery fire was an important factor in the success of the advance.

Michael D. Belis

DETAILS

Earl Russell Moyer Fretz was born in Fretz Valley Pines, Ottsville, Bucks County, Pennsylvania on July 2, 1895, the son of Henry and Amanda Fretz. He received a Bachelor of Philosophy degree from Brown University in 1916 and was enrolled as a first-year law student at Harvard University when the United States declared war against Germany. He dropped out of school, enlisted in the Army in May 1917, and after graduating from Officers Training Camp at Plattsburg Barracks, New York was commissioned a 2nd Lieutenant in the Officers Reserve Corps on August 15, 1917. He was assigned to the 39th Infantry at the Camp at Syracuse, New York and moved with them to Camp Greene, North Carolina as they became part of the 4th Division.

In December 1917, he was transferred to the 11th Machine Gun Battalion, also in the 4th Division at Camp Greene and given a commission as a 2nd Lieutenant of Infantry in the Regular Army. He had married Gertrude Tuttle in the summer of 1917 and the day before he shipped overseas, their son was born. Fretz sailed to France with Company B 11th Machine Gun Battalion aboard the troopship *U.S.S. Rijndam* (former Dutch liner) on May 10, 1918. As his emergency contact, he listed his wife Gertrude in Dorchester, Massachusetts. In France in June 1918, he was transferred to the 12th Machine Gun Battalion, still in the 4th Division.

Fretz was awarded the Distinguished Service Cross for his actions during the Asine-Marne Offensive. On July 18, 1918, the first day of the Offensive, he was advancing with his machine gun platoon in support of the 59th Infantry when Company E 59th Infantry which was nearby lost all of its officers wounded. Fretz assumed command of the Infantry Company and brought it together with his machine gun platoon into action in the face of heavy enemy fire. He established both units into position effectively, so they were able to hold against German counterattacks from seven that night until relieved at three in the afternoon the next day. Sergeant William Ryan of Fretz' machine gun platoon remarked that Fretz's "bravery was a marked example of a true soldier and an inspiration to the men under his command." [12]

Fretz was promoted to the temporary rank of 1st Lieutenant in October 1918. He also received a Citation Star. He served in the Vesle Sector, the Toulon Sector, the St. Mihiel Offensive, and the Meuse-Argonne Offensive. Most of the 4th Division, except the artillery, was pulled out of the front lines by October 19, 1918, yet somehow Fretz must have been at the front as he was wounded in the leg by a machine gun bullet on October 21, 1918. He was removed to Hospital #115 in Vichy. While still in the hospital, he developed pneumonia and died on November 7, 1918, four days before the end of hostilities.

Fretz was buried in Grave 104 in American cemetery # 536 at the village of Vichy-Les-Bains.

His Distinguished Service Cross was presented to his widow, Mrs. Gertrude Fretz.

On December 21, 1922, Earl R. Fretz was reinterred and laid to his final rest in Plot B Row 10 Grave 16 in the Aisne-Marne American Cemetery and Memorial, Belleau, Departement de l'Aisne, Picardie, France.

The name of Earl Russell Fretz is one of the names carved in stone on the wall of the Memorial Room in the Memorial Church of Harvard at Harvard University in Cambridge, Massachusetts.

Henry J. Garst

Corporal, United States Army
Note: The Citation for Henry J. Garst gives his rank as Corporal. His Form No. 724-1 A.G.O. Statement of Service card indicates he only ever held the rank/position of Cook.
Army Service Number: 558199
Company H 47th Infantry Regiment 4th Division
Date of Action: August 9, 1918
Location: Near Bazoches, France
War Department General Orders 46 (1919)
Medal # 6684

CITATION

Responding to a call for volunteers to destroy a hostile machine gun, Corporal Garst, with two other soldiers, boldly went forward through machine gun fire and accomplished this mission.

DETAILS

Henry John Garst was born in St. Louis, Missouri on May 13, 1896, the son of George and Theresa Garst. He enlisted in the Regular Army on February 14, 1917 at Jefferson Barracks, Missouri. He was assigned to Company H 9th Infantry and when elements of the 9th Infantry were used to make the 47th Infantry, he was transferred to Company H 47th Infantry on June 1, 1917. Garst sailed to France as a Cook in Company H 47th Infantry 4th Division aboard the troopship *U.S.S. Princess Matoika* (seized German liner *Prinzess Alice*) on May 10, 1918. As his emergency contact, he listed his father George in St. Louis, Missouri. Garst served in the Asine-Marne Offensive.

He was awarded the Distinguished Service Cross for his actions in the Vesle Sector. On August 9, 1918, Garst as part of 2nd Battalion 47th Infantry was dug in with most of his Company on the north side of the Vesle River across from the village of St. Thibaut and coming under fire from machine guns in and around the village of Bazoches. The day was devoted to hunting out enemy machine gun nests and snipers. Garst, along with Sergeant Louis Scionti of Company F and Bugler Richard Marcella also of Company F, volunteered, went forward under fire, and destroyed one of those German machine guns. Scionti and Marcella also received the Distinguished Service Cross for the action.

Garst also received a Citation Star. He was one of 20 soldiers of the 4th Division and one of only 399 soldiers of the American Expeditionary Forces to be awarded the Italian War Merit Cross (Croce al Merito di Guerra).

Garst served in the Toulon Sector, the St. Mihiel Offensive, and the Meuse-Argonne Offensive. He returned to the United States several months ahead of his Regiment as a Cook and part of a detachment of sick and wounded in the St. Aighan Casual Company No. 910 aboard the Armored Cruiser *U.S.S. Pueblo* leaving Brest, France February 9, 1919 and arriving at Hoboken, New Jersey February 22, 1919. He married Mary Connor in 1919. He was transferred to Company M 57th Infantry and given a furlough pending discharge.

The 1920 Census showed Garst and his wife living with her parents in Philadelphia, Pennsylvania where his occupation was listed as Soldier - Home Service. He was discharged on June 4, 1920. Garst and his wife Mary had a son born in September 1920 who died of pneumonia in March 1921. They had a daughter born in 1921 and another daughter born in 1922.

Garst was admitted to Saint Joseph's Sanitarium (Tuberculosis) in Albuquerque, Bernalillo County, New Mexico on August 1, 1926. He died of chronic active pulmonary tuberculosis at the age of 30 on August 31, 1926 at Saint Joseph's Sanitarium. He contracted the disease in France during his service in the war. On his death certificate his occupation was listed as a Policeman in Philadelphia, Pennsylvania. Henry J. Garst is buried in Holy Sepulchre Cemetery, Cheltenham, Montgomery County, Pennsylvania.

Isaac Gataino

Corporal, United States Army
Army Service Number: 558294
Company I 47th Infantry Regiment 4th Division
Date of Action: August 8, 1918
Location: Near St. Thibaut, France
War Department General Orders 35 (1919)
Medal # 1846

CITATION

Corporal Gataino showed exceptional courage and judgment when patrolling the country to the flank of his company under heavy machine gun and artillery fire. He obtained liaison with the flank company and brought back valuable information regarding the river to the front of our lines.

DETAILS

Isaac John Gataino was born in Greece on September 11, 1898. He entered the Army on April 6, 1917 and listed his home of record as Chicago, Illinois. He sailed to France as a Corporal in Company I 47th Infantry 4th Division aboard the troop transport *U.S.S. Louisville* on May 19, 1918. As his emergency contact, he listed his father Joseph in Chicago, Illinois. Gataino served in the Aisne-Marne Offensive.

He was awarded the Distinguished Service Cross for his actions in the Vesle Sector. On August 8, 1918, Gataino's Battalion was dug in along the road east of the village of St. Thibaut. Nearly seven thousand German artillery rounds were fired on St. Thibaut, its vicinity, and on the 47th Infantry that day. Gataino made his way through heavy enemy fire and conducted a one-man patrol to the Vesle River in order to ascertain the disposition of the enemy at the river and determine the conditions at the river relative to planning a crossing. He established contact with the Company on his flank and together with their reports and his own observations he was able to inform his Company Commander of exact conditions and German positions at the river.

Though it did not specify a date, the official history of the 47th Infantry in World War One recorded that Gataino suffered wounds from a gas attack and was later returned to duty. Gataino was also awarded the French Croix De Guerre. At some time after his award action, he was assigned the rank/position of Bugler. He served on occupation duty in Germany with the 4th Division and returned to the United States as a Bugler with Company I 47th Infantry aboard the troopship *U.S.S. Mobile* (seized German liner *Cleveland)* leaving Brest, France July 16, 1919 and arriving at Hoboken, New Jersey July 27, 1919.

Gataino married Julia Tarr on December 27, 1919. He was discharged during the demobilization of the 4th Division on October 28, 1920. In 1935, Gataino and his wife and daughter were living in Chicago, Illinois. By 1938 they had moved to Los Angeles, California. The 1940 Census showed them living in Los Angeles, California where he was employed as the manager of a grocery store. His wife Julia died in 1973. Isaac Gataino died at the age of 76 on September 16, 1974 and is buried in Forest Lawn Memorial Park (Glendale), Los Angeles County, California.

Reuben L. George

Corporal, United States Army
Army Service Number: 2284172
Company A 59th Infantry Regiment 4th Division
Date of Action: September 29, 1918
Location: Near Brieulles, France
War Department General Orders 95 (1919)
Medal # 2257

CITATION

After his platoon had become badly disorganized under heavy fire and all the sergeants had been killed or wounded, Corporal George took charge of the platoon, reorganized it with great courage and initiative and led it on in the attack against hostile machine guns. He was wounded shortly afterwards but he remained throughout the night where he had fallen refusing to be evacuated till all the other wounded had been cared for.

DETAILS

Reuben Lawrence George was born in Hollister, San Benito County, California on January 29, 1893, the son of Edwin and Amelia George. At the time he entered the Army he indicated he was married and listed his home of record as Chualar, Monterey County, California. He indicated his civilian occupation as farm laborer. George entered the Army on September 20, 1917. He sailed to England as a Private in Company A 59th Infantry aboard the British troopship *Megantic* on May 3, 1918. As his emergency contact, he listed his mother Amelia in San Jose, California. From England he and his Regiment were sent over to France. He served in the Aisne-Marne Offensive, the Vesle Sector, the Toulon Sector, and the St. Mihiel Offensive.

George was awarded the Distinguished Service Cross for his actions in the Meuse-Argonne Offensive. On September 29, 1918, the 8th Brigade, including the 59th Infantry, spearheaded the advance after being held as the 4th Division Reserve for the first three days of the Offensive. George's Company as part of 1st Battalion 59th Infantry reached the German trenches on Hill 281 near the village of Brieulles-sur-Meuse and was held up by heavy enemy fire from the nearby wooded area known as the Bois de Fays and by artillery fire from across the Meuse River. After consolidating their position on the northern slope of the hill, they dug in for the night.

During the advance, George, as a Corporal, took command of his platoon after Sergeant Arthur Paulson and Sergeant George Greepa were killed and Sergeant Homer Bleau was wounded (and died the next day). Paulson and Bleau both received the Distinguished Service Cross for their actions that day. George was severely wounded leading the attack and lay on the battlefield overnight exposed and unable to move before being evacuated the next day to the aid station. He also received a Citation Star.

George was in a hospital for over a month. Anxious to get back and join the fighting, he left the hospital before being released. Upon leaving the hospital, it took him a couple of days to get back to his Company and he rejoined them the day after the Armistice was signed. He served on occupation duty in Germany with the 4th Division and returned to the United States as a Corporal in Company A 59th Infantry aboard the troopship *U.S.S. Mount Vernon* (captured German liner *Kronprinzessin Cecilie*) leaving Brest, France July 24, 1919 and arriving at

Hoboken, New Jersey August 1, 1919. He was discharged on September 29, 1919.

George went home to California where he became a grain farmer. At some time between 1920 and 1928, he married his wife, Rhea. Their daughter was born in 1928. The 1930 Census showed him living with his wife, daughter and stepson in Hollister, San Benito County, California where he was employed as a farmer and indicated he was a veteran of the World War. In 1949, he moved to Idaho and bought a ranch. His wife Rhea died in 1966 and in 1970 he married Alice Davis. He was a hands on "do it all himself" rancher until a stroke forced him to quit and sell the ranch. Reuben L. George died at the age of 82 on April 19, 1975 and is buried in Indian Valley Cemetery, Adams County, Idaho.

Charles Glenn

Private, United States Army
Army Service Number: 568781
Company D 4th Engineer Regiment 4th Division
Date of Action: August 11, 1918
Location: Near Ville Savoye, France
War Department General Orders 145 (1918)
Medal # 1091

CITATION

Although his eyes had been burned by gas, Private Glenn volunteered for duty and assisted in the construction of an artillery bridge across the Vesle River under constant machine gun and artillery fire, setting a conspicuous example of personal bravery and devotion to duty.

DETAILS

Charles Glenn was born in Kentucky on April 20, 1879. (His Form No. 724-1 1/2 A.G.O. Statement of Service card indicates his place of birth as Owenorburg, Kentucky.) He enlisted in the Regular Army on June 9, 1917 at Fort George Wright, Washington. He was promoted to Private 1st Class on September 1, 1917 and was assigned the rank/position of Cook on December 5, 1917. He married Martha Morrison on December 10, 1917 in Vancouver, Washington. Glenn sailed to France as a Cook in Company D 4th Engineers, 4th Division aboard the troopship *U.S.S. Martha Washington* (seized Austrian liner) on April 30, 1918. As his emergency contact, he listed his wife Martha in Vancouver, Washington. Glenn's rank/position was changed to Private on July 15, 1918. He served in the Aisne-Marne Offensive.

He was awarded the Distinguished Service Cross for his actions in the Vesle Sector. The last day of battle for the Infantry units of the 4th Division in the Vesle Sector was August 11, 1918 and it was a day of consolidation of territory for both sides. While considerable patrolling was carried out, the 4th Division had ceased its large-scale attacks in anticipation of being relieved in place by the 77th Division later that night. The day settled into being an artillery duel between the Germans and the Americans. Glenn was wounded by gas and despite his wounds he volunteered to help emplace a bridge across the Vesle River near the village of Ville Savoye that day which could bring across the 75mm guns of the 4th Division artillery so as to better fire upon the enemy positions on the north side of the river. He was one of four enlisted men from Company D 4th Engineers who along with their officer were awarded the Distinguished Service Cross for their actions during the building of that artillery bridge across the Vesle River under fire on August 11, 1918.

Glenn was promoted to Private 1st Class on September 1, 1918. He served in the Toulon Sector, the St. Mihiel Offensive, and the Meuse-Argonne Offensive. On November 1, 1918, he was reduced to the rank of Private. He returned to the United States several months ahead of his Regiment as a Private of Company D 4th Engineers and part of the Sick and Wounded Convalescent Detachment No. 196 aboard the troop transport *U.S.S. Manchuria* leaving St. Nazaire, France May 11, 1919 and arriving at Hoboken, New Jersey May 22, 1919. In the passenger list his name carried the notation of his medical condition as "flat foot."

Michael D. Belis

He was discharged on June 7, 1919. The 1920 Census showed Glenn living with his wife Mattie in Vancouver, Washington where he was employed as a carpenter. Charles Glenn died at the age of 64 on December 31, 1943 and is buried in Custer National Cemetery, Crow Agency, Big Horn County, Montana.

Cornelius T. Glynn

Corporal, United States Army
Army Service Number: 1685561
Company K 59th Infantry Regiment 4th Division
Date of Action: October 5 - 6, 1918
Location: Near Bois de Fays, France
War Department General Orders 71 (1919)
Medal # 7475

CITATION

Corporal Glynn showed marked bravery as battalion runner, repeatedly carrying messages through heavy artillery and machine gun fire. He remained on duty night and day, aiding materially in maintaining liaison.

DETAILS

Cornelius Thomas Glynn was born in Hartford, Connecticut on May 10, 1896, the son of Thomas and Bridget Glynn. Before entering the Army, he worked as a Printing Presser at Calhoun Press in Hartford. He was drafted into the Army on March 28, 1918. On May 1, 1918, he was transferred to Company K 59th Infantry. Glynn sailed to England as a Private in Company K 59th Infantry 4th Division aboard the armed British troopship *RMS Olympic* on May 5, 1918. As his emergency contact, he listed his father, Thomas in Hartford, Connecticut. From England he and his Company were sent over to France. Glynn served in the Aisne-Marne Offensive, and the Vesle Sector. He was promoted to Corporal on September 5, 1918. He served in the Toulon Sector and the St. Mihiel Offensive.

Glynn was awarded the Distinguished Service Cross for his actions in the Meuse-Argonne Offensive. On October 4, 1918, the 58th Infantry assaulted and occupied the wooded area known as the Bois de Fays. Glynn's 3rd Battalion 59th Infantry was sent into the Bois de Fays to reinforce the 58th Infantry. For two days, on October 5-6, 1918, Glynn acted as a runner for his Battalion, moving through the woods and crossing the open ground between the Bois de Fays and the wooded area to the southeast known as the Bois de Brieulles where his Regimental Headquarters was located. German machine guns and artillery fire made movement dangerous and deadly. As a Corporal, Glynn could have designated a lower ranking soldier to undertake the high-risk job of runner, yet he took it upon himself to perform the role, sparing his men the added danger. He was also awarded the French Croix De Guerre.

At some time after October 6, 1918, Glynn was promoted to Sergeant. He returned to the United States several months ahead of his Regiment in Le Mans Casual Company No. 1203 aboard the *U.S.S. Frederick*, an older battleship formerly known as the *U.S.S. Maryland*, now being used to transport sick and wounded soldiers. He left Brest, France February 19, 1919 and arrived at Hoboken, New Jersey March 3, 1919. He was discharged on March 13, 1919.

The 1920 Census showed him living with his parents in Hartford where he was employed as a Pressman at a printing shop. The 1930 Census showed him living with his parents and working as a Pressman and indicated that he was a veteran of the World War. When he registered for the draft in 1942, he indicated his place of employment as Hartford High School in Hartford, Connecticut. Cornelius T. Glynn died at the age of

59 on September 2, 1955 and is buried in Northwood Cemetery, Windsor, Hartford County, Connecticut.

Michael D. Belis

Arthur J. Goetsch
Sergeant, United States Army
Army Service Number: 568732
Company D 4th Engineer Regiment 4th Division
Date of Action: August 11, 1918
Location: At Ville Savoye, France
War Department General Orders 147 (1918)
Medal # 1090

CITATION
Although his eyes had been burned by gas, Sergeant (then Private First Class) Goetsch volunteered for duty and assisted in the construction of an artillery bridge across the Vesle River under constant machine gun and artillery fire, setting a conspicuous example of personal bravery and devotion to duty.

DETAILS
Arthur James Goetsch was born in Davenport, Scott County, Iowa on December 17, 1892, the son of Hannis and Agnes Goetsch. At the time he entered the Army he listed his home of record as Walnut, Iowa. He entered the Army on June 1, 1917. Goetsch sailed to France as a Sergeant in Company D 4th Engineers 4th Division aboard the troopship *U.S.S. Martha Washington* (seized Austrian liner) on April 30, 1918. As his emergency contact, he listed his mother Agnes in Walnut, Iowa. Goetsch served in the Aisne-Marne Offensive.

He was awarded the Distinguished Service Cross for his actions in the Vesle Sector. August 11, 1918 was a day of consolidation of territory for both sides. While considerable patrolling was carried out, the 4th Division had ceased its large-scale attacks in anticipation of being relieved in place by the 77th Division later that night. The day settled into being an artillery duel between the Germans and the Americans. Despite his eye wounds from gas, Goetsch helped while under fire to emplace a bridge across the Vesle River that day which could bring across the 75mm guns of the 4th Division artillery so as to better fire upon the enemy positions on the north bank of the river. His courage was an inspiration to his fellow soldiers. He was one of four enlisted men from Company D 4th Engineers who along with their officer were awarded the Distinguished Service Cross for their actions during the building of that artillery bridge across the Vesle River under fire on August 11, 1918.

At some time after August 11, 1918, Goetsch was promoted to Sergeant 1st Class. He was awarded the French Croix De Guerre. He served in the Toulon Sector, the St. Mihiel Offensive, and the Meuse-Argonne Offensive. Goetsch served on occupation duty in Germany with the 4th Division and returned to the United States as a Sergeant 1st Class in Company D 4th Engineers aboard the troopship *U.S.S. Von Steuben* (seized German liner *Kronprinz Wilhelm*) leaving Brest, France July 21, 1919 and arriving at Hoboken, New Jersey July 29, 1919. He was discharged on August 4, 1919.

Goetsch married Della Lehnhardt on July 6, 1920 in Iowa. The 1930 Census showed him living with his wife, son and two daughters in Davenport, Iowa where he was employed as an electrician and indicated he was a veteran of the World War. The 1940 Census showed him and his family still in Davenport where he was then employed as a Firefighter. Arthur J. Goetsch died at the age of 78 on March 25, 1971 and is buried in Davenport Memorial Park, Davenport, Scott County, Iowa.

David S. Grant

Second Lieutenant, United States Army
Company F 39th Infantry Regiment 4th Division
Date of Action: August 5, 1918
Location: At St. Thibaut, France
War Department General Orders 17 (1928)
Medal # No assigned number discovered

CITATION

While leading his platoon in an attack upon the enemy's fortified position, with utter disregard for his own personal safety Lieutenant Grant advanced steadily at the head of his platoon through severe machine gun and artillery fire thereby being an inspiration to his men. When the order was given to continue the advance in small detachments Lieutenant Grant led the first of these against the enemy's fire until he fell mortally wounded. Although he realized the seriousness of his wound he refused to be cared for and directed the disposition of his platoon until he made the supreme sacrifice.

DETAILS

David Swain Grant was born in Asheville, North Carolina on November 13, 1892, the son of James and Louise Grant. He attended North Carolina State University in 1909-1910. During the year 1915 he was in England where he performed volunteer hospital work for the Red Cross to aid the British war effort. After returning to the United States, he enlisted in the North Carolina National Guard on June 1, 1916 with the rank of Sergeant 1st Class. He served in the Medical Detachment of Field Hospital #1 of the North Carolina National Guard commanded by Major F. J. Clemenger on Mexican Border Service. He then served in Company K 1st Infantry North Carolina National Guard until May 1917 when he was discharged.

Grant attended Officers Training Camp in 1917 at Fort Oglethorpe, Georgia, was commissioned a 2nd Lieutenant of Infantry in the Officers Reserve Corps on November 27, 1917, and assigned to the 39th Infantry at Camp Greene, North Carolina. He sailed to France as a 2nd Lieutenant in Company F 39th Infantry 4th Division aboard the Italian troopship *Dante Alegheiri* on May 10, 1918. As his emergency contact, he listed his mother Mrs. T.E. Clayton in Asheville, North Carolina.

Grant was awarded the French Croix De Guerre with silver star for his actions on July 19, 1918, the second day of the Aisne-Marne Offensive in the valley near the Ourcq River when he led his platoon from a supporting position through a heavy barrage to join with elements of the French 20th Infantry in the front line. His act of reinforcing the French allowed them to continue holding their frontline position and eliminated a withdrawal that might have been necessary otherwise.

He was awarded the Distinguished Service Cross for his actions a few weeks later, still in the Aisne-Marne Offensive. By August 5, 1918, the 39th Infantry had occupied the village of St. Thibaut at the Vesle River. Grant's Company F was in position on the eastern edge of the village and on the morning of August 5 was one of four Companies designated to lead the day's attack in an attempt to cross the river.

As soon as the Companies began to move the Germans laid down a heavy artillery barrage which prevented Company F from advancing. By noon the artillery fire had diminished to such an extent that Company F was able to move out from its position in the village but enemy machine gun fire from the village of Bazoches across the river forced Grant and his Company to take cover behind the railroad embankment which ran parallel to the river. Grant was ordered to move his platoon forward in small groups in the belief that less fire would be drawn that way. He led the first group himself and was struck down by machine gun fire as he attempted to cross the railroad tracks. Though seriously wounded, he remained on the field directing his soldiers across the embankment until he succumbed to his wounds and died.

Always leading his men from the front, Grant was an inspiration through his courage and daring. A number of recommendations for awards in the 4th Division were either lost or destroyed during the fighting in the Vesle River area and his Distinguished Service Cross recommendation was apparently one of those, as the orders for his award are dated ten years after the act.

Grant was buried in the American Plot Row A Grave 6 in

the American Cemetery # 847 at Bazoches. He was reinterred in Plot 5 Section F Grave 37 in the American cemetery # 617 at the village of Fismes on May 19, 1919. David S. Grant was reinterred and laid to his final rest on November 12, 1923 in Plot B Row 12 Grave 11 in the Aisne-Marne American Cemetery and Memorial, Belleau, Departement de l'Aisne, Picardie, France.

His Distinguished Service Cross was presented to his mother, Mrs. Thad E. Clayton.

The name of David S. Grant is inscribed on the plaque in the Shrine Room of the Memorial Bell Tower at North Carolina State University.

Frank B. Gresham

Sergeant, United States Army
Army Service Number: 1098437
Company G 39th Infantry Regiment 4th Division
Date of Action: September 26, 1918
Location: Near the Bois de Fays, France
War Department General Orders 46 (1919)
Medal # 3308

CITATION

After his patrol had been twice scattered by machine gun fire, Sergeant Gresham continued his reconnaissance, accompanied by only one other soldier and secured the information for which he had been sent. Upon rejoining his company, he was placed in command of his platoon, whose commander had been wounded and succeeded in reorganizing it under heavy shell fire.

DETAILS

Frank Butler Gresham was born in Edgefield County, South Carolina on October 17, 1882, the son of William and Martha Gresham. By 1902 he was married to Fannie Wilkes and had a daughter. A city directory for Augusta, Georgia in 1903 showed him and Fannie living in that city where he was employed as a carpenter. He enlisted as a Private in the Regular Army on July 6, 1904. Upon entering the Army, he listed his previous occupation as boiler maker. He served in Company 91 of the Coast Artillery Corps and was discharged at Jackson Barracks on July 5, 1907. He enlisted a second time on March 7, 1908 and served three more years in the Coast Artillery Corps, this time in Company 90. He was discharged from this enlistment as a Private on March 6, 1911 at Fort McKinley, Maine. He enlisted again in the Regular Army on August 14, 1917 and was assigned to Company C 56th Infantry at Fort Oglethorpe, Georgia. He was assigned the rank/position of Mechanic on August 17, 1917 and the rank/position of Bugler on April 28, 1918.

He sailed to France as a Bugler with Company C 56th Infantry 7th Division aboard the troopship *U.S.S. Leviathan* (seized German liner *Vaterland*) on August 3, 1918. As his emergency contact, he listed his mother Elizabeth in Augusta, Georgia. After reaching France his rank was changed to Private on August 21, 1918. He was transferred to Company G 39th Infantry 4th Division on August 23, 1918. He served in the Toulon Sector and was promoted to Corporal on September 7, 1918. He served in the St. Mihiel Offensive and was promoted to Sergeant on September 23, 1918.

Gresham was awarded the Distinguished Service Cross for his actions in the Meuse-Argonne Offensive. On September 26, 1918, the first day of the Offensive the 1st and 3rd Battalions of the 39th Infantry attacked from a starting point near the village of Cuisy with their ultimate objective being the wooded area known as the Bois de Fays. Gresham's 2nd Battalion was designated as the reserve Battalion and followed the attacking units at some distance behind. Leading a scouting patrol in advance of his platoon, Gresham showed exceptional bravery in completing his mission with only one other man after the patrol had been separated and scattered by heavy enemy fire. In rejoining his platoon, he assumed command of the platoon after its Lieutenant had been wounded. Gresham directed the platoon in the advance under heavy artillery fire. For his courage and leadership, he was also awarded the French Croix De Guerre with gold star.

Gresham was promoted to Supply Sergeant on November 8, 1918 and then went back to the rank of Sergeant on November 16, 1918. He served on occupation duty in Germany with the 4th Division. On May 20, 1919 he was reduced to the rank of Private. He returned to the United States with his Company aboard the troopship *U.S.S. Leviathan* (seized German liner *Vaterland*) leaving Brest, France July 30, 1919 and arriving at Hoboken, New Jersey August 6, 1919. He was discharged on August 14, 1919.

He enlisted in the Army again on an undetermined date. By August 18, 1920 he was serving with the 5th Military Police Company at Camp Gordon, Georgia. Gresham died while on a leave of absence on July 3, 1921 when he fell from the third story of the Southern Hotel in Savannah Beach, Georgia onto the pavement below. His death certificate indicated he was a Bugler at Fort Screven, Georgia at the time of his death. The certificate also indicated that acute alcoholism contributed to his death. He was 38 years old.

Gresham's death certificate indicated he was widowed. (Though details could not be found, his wife Fannie had apparently died sometime between 1903 and 1910. The 1910 Census showed Gresham living in a boarding house in Maine as a soldier and his marital status checked as single.) His obituary in a newspaper article published in 1921 stated that he was survived by his daughter, his mother, two sisters and six brothers. Frank B. Gresham is buried in Laurel Grove Cemetery (North), Savannah, Chatham County, Georgia.

Glenn M. Grove

Sergeant, United States Army
Army Service Number: 567696
Company D 11th Machine Gun Battalion 4th Division
Date of Action: September 26, 1918
Location: Near Nantillois, France
War Department General Orders 145 (1918)
Medal # 7552

CITATION

Sergeant Grove, with two officers, using captured German Maxim guns pushed forward to a heavily shelled area from which the other troops had withdrawn and by their accurate and effective fire kept groups of the enemy from occupying advantageous positions. When given permission to withdraw, Sergeant Grove declined to do so but maintained fire superiority all afternoon until it became too dark to see. His conspicuous gallantry furnished an inspiration to the other members of the command.

DETAILS

Glenn Miller Grove was born in James Creek, Huntingdon County, Pennsylvania on December 23, 1898, the son of John and Rozena Grove. He enlisted as a Private in the Regular Army in Company B 58th Infantry as it was forming up at Gettysburg, Pennsylvania on July 23, 1917. On August 3, 1917 he was promoted to the rank/position of Mechanic. On September 1, 1917, he was promoted to Private 1st Class. He moved with his Regiment to Camp Greene, North Carolina on November 22, 1917 and was transferred to Company C 10th Machine Gun Battalion as it was forming up and becoming a part of the 4th Division. On March 1, 1918, he was transferred to Company D 11th Machine Gun Battalion, still in the 4th Division.

Grove sailed to France as a Private 1st Class in Company D 11th Machine Gun Battalion 4th Division aboard the troopship *U.S.S. Rijndam* (former Dutch liner) on May 10, 1918. As his emergency contact, he listed his mother Rozena in Tyrone, Pennsylvania. He was promoted to Corporal in France on June 4, 1918. He served in the Aisne-Marne Offensive, the Vesle Sector and the Toulon Sector. On September 13, 1918 he was promoted to Sergeant. He served in the St. Mihiel Offensive.

Grove was awarded the Distinguished Service Cross for his actions in the Meuse-Argonne Offensive. On September 26, 1918, the first day of the Offensive, the 7th Brigade of the 4th Division including the 11th Machine Gun Battalion had reached the Corps objective which was on a line that ran approximately east and west through the village of Nantillois. The Divisions on either side had not reached the Corps objective and the advance of the 4th Division was halted for hours waiting for the flanking units to catch up. When at 5:30 p.m. the other Divisions had still not reached the objective, the order was given for the 4th Division to resume the attack. The halt had given the enemy time to regroup and rebuild his defenses.

After resuming the attack and achieving a minimal gain, without the flanking Divisions there to protect its flanks, the 4th Division pulled back to the line of the Corps objective and dug in for the night. Using captured German Maxim machine guns (most likely MG 08/15 models) Grove, along with 1st Lieutenant Homer S. Jarvis and 1st Lieutenant Earl M. McKinley, both also from 11th Machine Gun Battalion moved to an exposed forward position and covered the withdrawal of the 4th Division Infantry in their area and prevented the enemy from launching any successful counterattack. They maintained their fire upon the enemy until it became too dark to see. Only then did they pull back and join the withdrawn units. Jarvis and McKinley were also awarded the Distinguished Service Cross for the action.

Grove was awarded the French Croix De Guerre with gilt star. After serving in the first phase of the Meuse-Argonne Offensive he was pulled out of the front lines and sent to Officers School in the rear area on October 18, 1918. He finished the course at the school on January 18, 1919. On April 21, 1919, he was discharged as an enlisted man and the next day, April 22, 1919 was commissioned a 2nd Lieutenant of Infantry and was transferred to Company A 324th Service Battalion, Quartermaster Corps which was also in France. He returned to the United States with that unit aboard the transport *U.S.S. Liberator* leaving Brest, France July 5, 1919 and arriving at Hoboken, New Jersey July 18, 1919. He was discharged at Camp Dix, New Jersey on July 24, 1919.

He returned to Pennsylvania and married Carrie Cupper. The 1930 Census showed Grove living with his wife and daughter in Tyrone, Blair County, Pennsylvania where he was employed as a Salesman for a Sporting Goods Store and indicated he was a veteran of the World War. His son was born in 1932. Grove was back in uniform during World War Two and served as a Signal Corps Officer in the continental United States from August 26, 1942 until March 2, 1946, reaching the rank of Captain. Glenn M. Grove died of heart disease at the age of 52 on June 14, 1951 and is buried in Eastlawn Cemetery, Tyrone, Blair County, Pennsylvania. He was a member of the Legion of Valor.

James P. Growdon

Captain, United States Army
Company F 4th Engineer Regiment 4th Division
Date of Action: August 5, 1918
Location: West of Fismes, France
War Department General Orders 21 (1919)
Medal # 1082

CITATION

After reconnoitering a sector of the River Vesle in advance of the front lines of the infantry for the purpose of selecting a site for a footbridge, Captain Growdon went with a small party of engineers through an enemy barrage from 77-millimeter and one-pounder guns and assisted in directing the construction work. As soon as the operations were discovered, machine gun fire was opened upon the party, but they continued at work, removing the German wire entanglements and successfully completing a bridge which was of great value in subsequent operations.

DETAILS

James Paul Growdon was born in Pioneer, Cedar County, Iowa on January 10, 1884. He studied engineering at the University of Nebraska and worked for the Chicago, Burlington and Quincy Railroad before the war. He was married to Hazel Parsons on December 27, 1910. Growdon entered the Army on May 1, 1917 and received a commission in the Engineers Reserve Corps. He was assigned to the 4th Engineers as a Captain and Commanding Officer of Company F on January 1, 1918.

Michael D. Belis

He sailed to France in command of Company F 4th Engineers 4th Division aboard the troopship *U.S.S. Martha Washington* (seized Austrian liner) on April 30, 1918. As his emergency contact, he listed his wife Hazel in McMinnville, Oregon.

Growdon was awarded the Distinguished Service Cross for his actions in the Aisne-Marne Offensive. On August 5, 1918, the 4th Division made a determined effort to cross the Vesle River and the 4th Engineers were instrumental in getting the Division across. Growdon ventured out beyond the front lines alone and scouted the river west of the village of Fismes close to the village of Ville Savoye in order to determine where to emplace a footbridge for the 58th and 59th Infantry to use. On August 5, 1918, he and his detachment particularly distinguished themselves by remaining at their work under the heaviest of fire and completing a bridge for the Infantry.

(See the entry for William B. Beach for an overview of the kinds of activities carried out by the soldiers of the 4th Engineers involved in building the bridges across the Vesle River.)

Growdon served in the Vesle Sector, the Toulon Sector, the St. Mihiel Offensive, and the Meuse-Argonne Offensive. He was awarded the French Croix De Guerre. He received a wartime temporary promotion to Major on November 7, 1918 and was given command of 2nd Battalion 4th Engineers. He served on occupation duty in Germany with the 4th Division and returned to the United States as a Major in Headquarters 4th Engineers aboard the troopship *U.S.S. Von Steuben* (seized German liner *Kronprinz Wilhelm*) leaving Brest, France July 21, 1919 and arriving at Hoboken, New Jersey July 29, 1919. He was discharged from the Army on September 11, 1919.

Growdon became employed by Alcoa Aluminum in 1925, eventually retiring from the company in 1954. In the 1930's he divorced his wife Hazel, married Isabel McCauley, and moved to Pittsburgh, Pennsylvania. He was back in uniform during World War Two, returning to active duty on January 2, 1943. He served overseas from October 31, 1943 to August 4, 1944 and was separated from active duty on September 21, 1944. His obituary mentioned that he saw action in Italy. The Army Registers indicate Growdon was given a retirement as a Colonel on June 29, 1948. James P. Growdon died at the age of 85 on June 23, 1969 and is buried in Calvary Cemetery, Pittsburgh, Allegheny County, Pennsylvania.

Leonard E. Guy

Sergeant, United States Army
Army Service Number: 572657
Company C 58th Infantry 4th Division
Date of Action: September 27, 1918
Location: Near Nantillois, France
War Department General Orders 81 (1919)
Medal # 1327

CITATION

Sergeant Guy displayed exceptional courage in attacking single-handed a machine gun emplacement, capturing the gun and taking as prisoners three machine gunners.

DETAILS

Leonard Evans Guy was born in Anoka, Minnesota on February 18, 1892, the son of Levi and Martha Guy. He enlisted in the United States Marine Corps on March 10, 1913. He served in the Philippines Islands and on Guam and eventually reached the rank of Corporal before receiving a bad conduct discharge in July 1915. When he registered for the draft in June 1917 he was living in Great Falls, Montana and working as a switchman for the Great Northern Railroad.

Guy was drafted into the Army as a Private on March 17, 1918 and assigned to Company C 58th Infantry at Camp Greene, North Carolina. He was promoted to Corporal on April 1, 1918. He sailed to England as a Corporal in Company C 58th Infantry 4th Division aboard the British transport *City of Brisbane* on May 7, 1918. As his emergency contact, he listed his mother Martha in Aladdin, Washington. From England he and his Company were sent over to France. He served in the Aisne-Marne Offensive and the Vesle Sector. Guy was promoted to Sergeant on August 13, 1918. He served in the Toulon Sector and the St. Mihiel Offensive.

Guy was awarded the Distinguished Service Cross for his actions in the Meuse-Argonne Offensive. At the end of the day on September 26, 1918, Guy's Company C 58th Infantry had halted its attack and dug in south of the village of Cuisy. On September 27, 1918, Company C continued its advance through Septsarges and by the end of the day was near the village of Nantillois where it stopped for the night. During this phase of the battle the 7th Brigade had been designated as the spearhead of the attack and Guy's Regiment as part of 8th Brigade was the reserve force which followed approximately two kilometers behind the attacking line. He and his fellow soldiers had to deal with pockets of German resistance bypassed by the leading elements of the Division. Guy attacked and eliminated an enemy machine gun by himself in the advance and took three prisoners during the engagement.

Guy was also awarded a Citation Star. He was one of 20 soldiers of the 4th Division and one of only 399 soldiers of the American Expeditionary Forces to be awarded the Italian War Merit Cross (Croce al Merito di Guerra).

He served on occupation duty in Germany with the 4th Divison. Guy returned to the United States as a Sergeant of Company C 58th Infantry in Casual Company # 2279 aboard the troopship *U.S.S. Prinz Friedrich Wilhelm* (surrendered German liner) leaving Brest, France July 5, 1919 and arriving at Hoboken, New Jersey July 14, 1919. He was discharged on July 23, 1919.

Michael D. Belis

By 1928 he was living in Alameda County, California where he once again was a switchman for a railroad. After 1932 his Citation Star was converted to the Silver Star Medal. The 1940 Census showed him living in Oakland, California with his wife Fiona and still working as a switchman. His 1942 registration for the draft indicated he worked for the Western Pacific Railroad. Leonard E. Guy died at the age of 59 on August 7, 1951 and is buried in Golden Gate National Cemetery, San Bruno, San Mateo County, California.

Harold Lamonte Hall

Note: In at least two of the official Citation listings and a couple of other accounts his name is given as Harold De La Monte Hall. His actual name was Harold Lamonte Hall.
Private, United States Army
Army Service Number: 561842
Company A 59th 4th Division
Date of Action: September 29, 1918
Note: The date of Hall's award actions and death is given in his Citation as September 29, 1918. The Register of Burials of Deceased American Soldiers, 1917 – 1922 from the Office of the Quartermaster General gives his date of death (and therefore the date of his award action) as September 30, 1918.
Location: Near Bois de Brieulles, France
War Department General Orders 37 (1919)
Medal # 6702

CITATION

When his company was in a perilous position Private Hall volunteered and carried a message to Battalion Headquarters, a distance of 1,000 yards under heavy artillery and machine gun fire. On his return journey he was killed.

DETAILS

Harold Lamonte Hall was born in Charleston, Kanawha County, West Virginia on September 19, 1899, the son of Perry and Mary Hall. His father died in 1911 and in February 1917 his mother remarried. Hall entered the Army on April 8, 1917. He sailed to England as a Private in Headquarters Company 59th Infantry 4th Division aboard the British troopship *Megantic* on May 3, 1918. As his emergency contact, he listed his mother Mrs. Mary E. Amberg in Buffalo, West Virginia. From England he and his Company were sent over to France.

Hall served in the Aisne-Marne Offensive, the Vesle Sector, the Toulon Sector and the St. Mihiel Offensive. At some time in France, he was transferred to Company A.

He was awarded the Distinguished Service Cross for his actions in the Meuse-Argonne Offensive. On September 29, 1918, his Brigade spearheaded the 4th Division's advance. First Battalion 59th Infantry which included Hall's Company attacked through the wooded area known as the Bois de Brieulles, occupied the northern slope of Hill 281 and dug in. The next day September 30 the Battalion remained in defensive positions along a line on the northern edge of the Bois de Brieulles. They were receiving enemy machine gun fire from the nearby wooded area known as the Bois de Fays and artillery fire from across the Meuse River when Hall volunteered to deliver a message from his Company Commander to Battalion Headquarters. He successfully accomplished his mission while under heavy enemy fire but was killed as he returned to his Company back through that same deadly fire.

Hall was buried in Plot 1 Grave # 27 in the American battlefield cemetery # 702 at the village of Septsarges.

His Distinguished Service Cross was presented to his mother, Mrs. Mary E. Amberg.

On May 20, 1919 he was reinterred in Plot 3 Section 65 Grave # 131 in the Argonne American Cemetery # 1232. On July 27, 1921 his body was disinterred for preparation and shipment. His remains were returned to the United States aboard the United States Army Transport *U.S.A.T. Cantigny* leaving Antwerp, Belgium September 1, 1921 and arriving at Hoboken, New Jersey September 12, 1921. He was one of 1,199 deceased American soldiers brought home by the *Cantigny* on that voyage. The remains of Harold Lamonte Hall were cremated, and he was laid to his final rest in Hall-Oldaker Cemetery, Putnam County, West Virginia.

Michael D. Belis

Arthur M. Hamilton

Corporal, United States Army
Army Service Number: 560474
Company E 58th Infantry 4th Division
Date of Action: October 6, 1918
Location: Near Brieulles, France
War Department General Orders 53 (1920)
Medal # 6305

CITATION

Corporal Hamilton and a comrade under heavy enemy fire went to the rescue of wounded lying in advance of our lines and returned to our lines with two wounded American soldiers. In accomplishing this mission, they advanced to within 75 yards of the enemy lines over an area which the enemy raked with their fire.

DETAILS

Arthur Marion Hamilton was born in Clearfield, Iowa on October 21, 1886, the son of Marion and Emily Hamilton. In his registration for the draft in February 1918, he listed his home of record as Des Moines, Iowa and his occupation as Farmer. He entered the Army on February 10, 1918. Hamilton sailed to England as a Private of Company C 12th Machine Gun Battalion 58th Infantry 4th Division aboard the British transport *S.S. Rhesus* on May 7, 1918. As his emergency contact, he listed his brother John in Des Moines, Iowa. From England he and his Company were sent over to France. At some time between May 7 and July 18, 1918, he was promoted to Corporal. Hamilton was wounded on July 18, 1918, the first day of the Aisne-Marne Offensive. After he recovered, he returned to duty with his Company.

He was awarded the Distinguished Service Cross for his actions in the Meuse-Argonne Offensive. On October 6, 1918, his Regiment was situated in the wooded area known as the Bois de Fays near the village of Brieulles-sur-Meuse. Hamilton and Private First Class Forrest L. Martz of Company C 12th Machine Gun Battalion rescued two wounded comrades under the most dangerous of conditions, coming as close as 75 yards to the enemy lines. Both received the Distinguished Service Cross for their actions.

(See the entry for Joseph Bassi for a description of the battlefield conditions in the Bois de Fays upon which Hamilton and Martz ventured to get their injured fellow soldiers.)

Hamilton returned to the United States several months ahead of his Regiment as part of a detachment of sick and wounded in Le Mans Casual Company # 1211 aboard the Armored Cruiser *U.S.S. Pueblo* leaving Brest, France February 9, 1919 and arriving at Hoboken, New Jersey February 22, 1919. He was discharged on March 26, 1919.

Michael D. Belis

Hamilton was married to Elsie Winter in Des Moines, Iowa on September 3, 1927. On the marriage record he indicated this was his second marriage. He also indicated on the marriage record that he was at the time employed as a Police Officer in Des Moines, Iowa. The 1930 Census showed Hamilton and Elsie living in Des Moines where he was employed as a Clerk with the City Police Department and indicated he was a veteran of the World War. After 1932 he was awarded the Purple Heart for the wounds he received in World War One. Arthur M. Hamilton died at the age of 52 on September 26, 1939 and is buried in Glendale Cemetery, Des Moines, Polk County, Iowa.

William H. Hammond
First Lieutenant, United States Army
Company I 39th Infantry 4th Division
Date of Action: September 26 - 27, 1918
Location: Near Montfaucon, France
War Department General Orders 95 (1919)
Medal # 7555

CITATION
First Lieutenant Hammond fearlessly led his platoon against a German counterattack and succeeded in breaking it up. Sighting a German patrol taking American prisoners to the rear, he led a combat patrol which routed the Germans and rescued the captured Americans. In the advance in which he took part the next day, he was severely wounded in the chest but refusing first-aid treatment continued to urge his men forward, although unable himself to go.

Michael D. Belis

DETAILS

William Hays Hammond was born in Visalia, Tulare County, California on October 28, 1886, the son of John and Stella Hammond. He attended Stanford University and then the University of California from 1906-1908. He worked as the County Clerk for Tulare County, California. He served in the California National Guard 6th Infantry and then 2nd Infantry from 1906-1909 rising from Private to Sergeant. He married Madge Bliven December 27, 1909. Their son was born in 1915. In March 1917, Hammond again joined the 2nd Infantry of the California National Guard and was assigned to the Machine Gun Company of that Regiment. He was commissioned a 2nd Lieutenant in the Officers Reserve Corps on April 6, 1917. When part of the California National Guard 2nd Infantry was used to create the 159th Infantry, Hammond became a member of the 159th Infantry.

Hammond was promoted to 1st Lieutenant in the 159th Infantry on December 21, 1917. He sailed to England as a 1st Lieutenant in Company L 159th Infantry 40th Division aboard the Australian transport *H.M.A.T. Osterley* on August 8, 1918. As his emergency contact, he listed his father John in Washington, D.C. From England the 159th Infantry was sent over to France and once there its personnel were used as replacements for other units. Hammond was then assigned to Company I 39th Infantry in the 4th Division. He served in the St. Mihiel Offensive.

Hammond was awarded the Distinguished Service Cross for his actions in the Meuse-Argonne Offensive. On September 26, 1918, the 39th Infantry was leading the 4th Division attack on the left flank. Hammond's Company I was in the front of the assault along with Company M. In advancing over a hill near the village of Montfaucon the American advance stalled due to heavy fire from enemy machine guns and minenwerfer (trench mortars) and at that moment the Germans counterattacked. Hammond led a force which consisted of his platoon, another platoon from his Company, and a platoon from Company K and repulsed the enemy attack. Shortly after driving off the enemy, Hammond observed the movements of a German patrol with captured Americans. He led a patrol from his platoon which subdued the Germans, liberated three soldiers from 1st Battalion 39th Infantry who had been captured earlier that day and took fifteen Germans as prisoners. The next day, Hammond's Company was again in the front of the assaulting line and, as he led his platoon forward, he was severely wounded and unable to continue. He remained on the field and urged his soldiers on until he had to be carried from the field.

In addition, he received the French Knight of the Legion of Honor (Chevalier de la Légion d'Honneur) and was twice awarded the French Croix De Guerre with palm. Because of his wounds, he was evacuated back to the United States several months ahead of his Regiment aboard the *U.S.S. Rhode Island* (an older battleship being used as a troop transport for sick and wounded) leaving Brest, France February 12, 1919 and arriving at Newport News, Virginia on February 28, 1919.

He returned to the California National Guard and in March 1919 was assigned to the 32nd Infantry Regiment at Camp Kearney, San Diego, California. Hammond was given a commission in the Regular Army as a Captain on July 1, 1920. He served in the Philippines and in the Panama Canal Zone during the 1920's. He was promoted to Major in 1932 and also that year was awarded the Purple Heart for the wounds he received during World War One. He was promoted to Lieutenant Colonel in 1940 and Colonel in 1942. He retired from the Army in 1946. William H. Hammond died at the age of 80 on October 16, 1967 and is buried in Golden Gate National Cemetery, San Bruno, San Mateo County, California.

James W. Hanbery
First Lieutenant, United States Army
Company L, 59th Infantry Regiment 4th Division
Date of Action: July 19, 1918
Location: At Chateau-Thierry, France
War Department General Orders 31 (1922)
Note: General Orders 31 (1922) rescinded a previous award of the Citation Star issued to James W. Hanbery in General Orders 59 (1921) and upgraded the award to the Distinguished Service Cross.
Medal # 7947

CITATION
For extraordinary heroism in action at Chateau-Thierry, France, 19 July 1918, in command of the attacking unit of the assault company of his battalion. After gaining his objective, in an advance through heavy machine-gun and artillery fire, the battalion on his left having been held up by enemy machine-gun

nests, Lieutenant Hanbery's company and battalion became exposed to grazing and flanking fire which threatened the destruction of the entire battalion. Lieutenant Hanbery reorganized the attacking line and although wounded, led a brilliant and successful attack against the enemy machine-gun nests until again wounded and rendered helpless, when he refused succor in order not to endanger the lives of his men.

DETAILS

James Willis Hanbery was born in Hopkinsville, Kentucky on December 19, 1890, the son of John and Florence Hanbery. At an early age his family moved to Oklahoma. He received a Bachelor of Arts Degree from Phillips University in Enid, Oklahoma in 1913. When the war started, he was a history teacher at the State Manual Training Normal College in Pittsburg, Kansas. Sometime between June 1917 and May 1918 he was married. He had been enlisted in the volunteer force that was to be raised by Theodore Roosevelt to serve in the war in 1917, a proposal which was formally refused by President Woodrow Wilson. With the dissolution of that volunteer force, Hanbery entered Army Officers Training Camp at Fort Sheridan, Illinois and upon graduation received a commission in the Officers Reserve Corps on November 27, 1917.

He sailed to England as a 1st Lieutenant in Company L 59th Infantry 4th Division aboard the armed British troopship *RMS Olympic* on May 5, 1918. As his emergency contact, he listed his wife in Pittsburg, Kansas. From England he and his Company were sent over to France. By July 8, 1918 he was commanding Company L 59th Infantry as a 1st Lieutenant.

Hanbery was awarded the Distinguished Service Cross for

his actions in the Aisne-Marne Offensive. Though his Citation gives the location of Chateau-Thierry, the actual location was about seven miles northwest of there and near the village of Courchamps. For the operation, 3rd Battalion 59th Infantry including Hanbery's Company was attached to the 13th Chasseurs of the 164th French Infantry Division.

At 4:25 on the morning of July 19, 1918, preceded by tanks 3rd Battalion 59th Infantry moved eastward into the wooded area known as the Bois de l'Orme. Hanbery and his men maneuvered through heavy machine gun and artillery fire, advanced through the woods and reached their objective which was the Courchamps-Priez road. German machine gun nests held up the Battalion on his left and began to hinder the progress of his own Battalion. Though shot in the leg during the action, Hanbery led his Company in an attack upon the nests and eliminated a number of them. He continued to press his Company foward as the spearhead of the Battalion's advance and reached a position near the wooded area known as the Bois de Leipzig. Here they were stopped by intense enemy machine gun fire and then bombarded by German artillery which rained high explosive and gas shells upon them, seriously wounding the Battalion Commander, killing or wounding fifteen officers, and causing heavy casualties among the enlisted men. The Battalion was forced to withdraw back near the Bois de l'Orme.

Hanbery was struck down, hit in the head and back by shell fragments and was temporarily unable to move. He ordered his men to leave him and retreat to safety with the Battalion. Once back with the rest of the Battalion, the survivors of his Company reported his actions and lamented they had been ordered to leave their commander behind. They believed he was lost and presumed dead. As his men left the field, Hanbery managed to crawl into a shell hole and lost consciousness. He awoke eight hours later and used his pocketknife to dig the bullet out of his leg. He bound up the wound and began to drag himself over to aid two wounded soldiers lying on the battlefield. A German artillery round killed them before he could get to them. He crawled under fire in the direction of the American lines for over a mile until he was picked up by friendly forces and brought to a field hospital.

Later while recovering in a rear hospital, Hanbery was surprised to read his name in a list of killed in action in a Paris English language newspaper. Another newspaper in Enid, Oklahoma posted that he had been killed at Chateau-Thierry and nearly the entire town turned out for a parade and memorial service for what they thought was their first resident lost in the war. From the hospital, Hanbery was able to get a cablegram sent to his family notifying them he was very much alive. He left France in late October 1918 sailing in a detachment of sick and wounded aboard the troop transport *U.S.S. Kroonland* and arrived at Newport News, Virginia November 1, 1918. He was discharged on March 19, 1919.

The 1920 Census showed him living with his wife Millie in Omaha, Nebraska where he was employed as a reporter for a newspaper. Sometime after 1920 he was either divorced or widowed. He moved to Long Beach, California in 1924 and married Beatrice Dail on June 24, 1929. The 1930 Census showed him living with his wife Beatrice and her mother in Long Beach, California where he was employed as a manager in the insurance industry and indicated he was a veteran of the World War. Hanbery became prominent in Long Beach business, veterans, and civic organizations. After 1932, he was awarded the Purple Heart for wounds he received in World

War One. He was active in the Fourth Division Association. James W. Hanbery died at the age of 69 on June 29, 1960 and is buried in Forest Lawn Memorial Park (formerly Sunnyside Mausoleum & Memorial Garden), Long Beach, Los Angeles County, California.

Michael D. Belis

Mathias Willoughby Haney

First Lieutenant, United States Army
Company A 39th 4th Division
Date of Action: September 26 - 28, 1918
Location: Near Montfaucon Hill, France
War Department General Orders 46 (1919)
Medal # 6986

CITATION

Captain Haney, then a Lieutenant, displayed exceptional skill in extricating his company from a perilous position into which it had moved because of a dense fog and in so doing captured prisoners whose number exceeded that of his own command. Taking command of his battalion next day at a critical time, he succeeded in stopping a threatened retreat and under heavy machine gun and shell fire re-established the line. On September 28, near Septsarges Captain Haney led his battalion forward through heavy fire, advancing his line one kilometer and holding it against counterattacks until he was relieved.

DETAILS

Mathias Willoughby Haney was born in Bristol, Harrison County, West Virginia on April 27, 1883. He was commissioned a 1st Lieutenant of Infantry in the Officers Reserve Corps on November 27, 1917 and assigned to the 39th Infantry at Camp Greene, North Carolina. He sailed to France as a 1st Lieutenant and Commanding Officer of Company A 39th Infantry 4th Division aboard the Italian troopship *Dante Alegheiri* on May 10, 1918. As his emergency contact, he listed his wife Margaret in Philadelphia, Pennsylvania.

Haney was awarded the French Croix De Guerre with gilt star for his actions on July 18, 1918, the first day of the Aisne-Marne Offensive when he led his Company through the Cresnes Wood (Buisson de Cresnes) near the village of Noroy-sur-Ourcq. Assaulting through heavy machine gun fire from the front and flank, he and his Company using hand grenades destroyed one German artillery piece, captured three others, and took fourteen enemy gunners as prisoners.

He served in the Vesle Sector, the Toulon Sector, and the St. Mihiel Offensive.

Haney was awarded the Distinguished Service Cross for his actions in the Meuse-Argonne Offensive. On September 26, 1918, the first day of the Offensive, Haney led Company A as part of 1st Battalion 39th Infantry, advanced through a dense fog, got lost, and he and his Company found themselves up against the German positions at Montfaucon Hill. When stopped at the hill by formidable defenses, Haney sent two platoons from Company A through a trench on the right, the platoons flanking the hill and capturing more than a hundred prisoners. Haney then personally led the rest of the Company plus two other platoons, one borrowed from Company C and another from Company D and flanked the hill again. In the fighting Haney lost as casualties all of his platoon leaders, all of Headquarters Platoon except one man, and all non-commissioned officers except six Sergeants. However, a large amount of enemy machine guns had been eliminated and Company A's sacrifice had helped to make possible the capture of the village of Montfaucon by another unit.

About midday, the 4th Division advance was halted when

the Corps objective had been reached, in order to wait for the flanking Divisions to catch up. This halt gave the Germans a chance to regroup and inflict heavy casualties upon the 4th Division's forward elements when the attack was resumed. Late in the afternoon, the Battalion Commander Major Roy W. Winton was wounded. Even though he was only a 1st Lieutenant, command of 1st Battalion fell upon Haney. He directed the Battalion to dig in for the night. The next day, September 27, 1918, the attack was led by 3rd Battalion with Haney and 1st Battalion moving close behind in support. The two Battalions were stopped at the Nantillois road by heavy machine gun fire but before they could dig in, German artillery poured a withering fire into their ranks producing heavy casualties, demoralizing the soldiers, and causing a spontaneous general withdrawal.

Haney managed to hold most of Company A, a number of soldiers from other Companies, and a portion of the 11th Machine Gun Battalion in the forward positions, thereby helping to stop the retreat from turning into a complete rout. He attempted to reorganize the command and bring the soldiers back into the front line. The Commander of the 39th Infantry, Colonel Frank C. Bolles, came up and together with Haney was able to reform the right side of the attacking line and bring it back up to the forward elements. They were aided in this reorganization by the Brigade Commander Brigadier General Benjamin Poore, who moved about the left side of the battlefield and personally led troops back to the head of the advance. Bolles received the Distinguished Service Cross for his actions that day. Poore received the Distinguished Service Cross for his actions that day and his actions in a later engagement.

On September 28, 1918, the attack was resumed and while 2nd and 3rd Battalions led the advance, Haney commanded 1st

Battalion as the Brigade reserve force following some distance behind. The going was tough, as the 79th Division on the left was not advancing as fast as the 4th Division, causing the 39th Infantry to be subjected to heavy flanking fire. Haney led his Battalion forward a distance of one kilometer and established a defensive position which he held against enemy counterattacks. After two hours of heavy bombardment from German artillery across the Meuse River, the 7th Brigade including the 39th Infantry was pulled back to await relief by the 8th Brigade.

Haney was promoted to Captain on November 17, 1918. He was awarded the French Knight of the Legion of Honor (Chevalier de la Légion d'Honneur) and a second French Croix De Guerre, this time with palm. He returned to the United States several months ahead of his Regiment as a Captain of the 39th Infantry and part of St. Aignan Casual Company No. 2485 aboard the *U.S.S. New Jersey*, an older battleship being used to transport sick and wounded soldiers back home, leaving Brest, France April 10, 1919 and arriving at Boston, Massachusetts April 23, 1919. He was discharged on April 25, 1919 at Camp Devens, Massachusetts.

The 1920 Census showed Haney and his wife Margaret living in Syracuse, New York where he was employed as the manager of a lumber company. The 1925 New York State Census showed Haney and Margaret living in Hempstead, Nassau County, New York where he was vice-president of a lumber company. Mathias Willoughby Haney died at the age of 48 on July 17, 1931 and is buried in Arlington National Cemetery.

Samuel H. Hanna

Sergeant, United States Army
Army Service Number: 574104
Company B 12th Machine Gun Battalion 4th Division
Date of Action: September 30, 1918
Note: The date of Hanna's award action and death is given in his Citation as September 30, 1918. The Register of Burials of Deceased American Soldiers, 1917 – 1922 from the Office of the Quartermaster General gives his date of death as September 29, 1918. The date of September 29 more closely matches the events described in his Citation.
Location: At Bois de Fays, France
War Department General Orders 5 (1920)
Medal # 5772

CITATION

When Company C 58th Infantry was temporarily halted by heavy machine gun fire Sergeant Hanna exposed himself to enfilading fire in order to place his guns in position to execute a covering fire for the Infantry. With the aid of the fire from the machine guns under his command, the advance was resumed. In the performance of this deed, he was mortally wounded.

DETAILS

Samuel Hancock Hanna was born in Waukegan, Lake County, Illinois on September 21, 1888, the son of David and Mary Hanna. Before entering the Army, he worked as a clerk for the City of Los Angeles, California. He sailed to England as a Private in Company B 12th Machine Gun Battalion 4th Division aboard the British troopship *RMS Aquitania* on May 7, 1918. As his emergency contact, he listed his sister Jennie in Waukegan, Illinois. From England Hanna and the 12th Machine Gun Battalion were sent over to France where they were issued their machine guns. At some time after reaching France, he was promoted to Sergeant. Hanna served in the Aisne-Marne Offensive, the Vesle Sector, the Toulon Sector, and the St. Mihiel Offensive.

He was awarded the Distinguished Service Cross for his actions in the Meuse-Argonne Offensive. The 58th Infantry spearheaded the 4th Division attack against the wooded area known as the Bois de Fays on September 29, 1918. As part of 8th Brigade, the 12th Machine Gun Battalion was supporting the 58th Infantry. The 1st Battalion 58th Infantry which included Company C reached the southern edge of the wooded area known as the Bois des Ogons south of the wooded area known as the Bois de Fays. The soldiers of Company C fought their way into the woods of the Bois des Ogons but were stopped by heavy enemy machine gun fire coming from the Bois de Fays. Under that heavy fire, Hanna moved his machine guns to a forward position ahead of the Infantry in order to cover Company C. He directed the two guns of his section to lay down a heavy suppressive fire on the Germans which then allowed the Company to continue its asssault. He was killed while exposing himself to enemy fire so that he could better position and direct the fire of his guns. His sacrifice had helped Company C 58th Infantry to gain a foothold in the Bois des Ogons. He also received a Citation Star.

Hanna was buried in Plot 1 Section C Grave # 18 in the American battlefield cemetery # 561 at the village of Nantillois.

His Distinguished Service Cross was presented to his sister, Miss Jennie Hanna.

On May 10, 1919 he was reinterred in Plot 4 Section 59 Grave # 163 in the Argonne American Cemetery # 1232. Samuel H. Hanna was reinterred and laid to his final rest on November 8, 1921 in Plot A Row 18 Grave 22 in the Meuse-Argonne American Cemetery, Romagne-sous-Montfaucon, Departement de la Meuse, Lorraine, France.

Roy Harris

Private, United States Army
Army Service Number: 569359
Company F 4th Engineer Regiment 4th Division
Date of Action: August 5, 1918
Location: West of Fismes, France
War Department General Orders 145 (1918)
Medal # 1083

CITATION

Private Harris was a member of a small detachment of engineers which went out in advance of the front line of the infantry through an enemy barrage from 77 millimeters and one-pounder guns to construct a footbridge over the River Vesle. As soon as their operations were discovered, machine gun fire was opened up on them but undaunted the party continued at work, removing the German wire entanglements and completing a bridge which was of great value in subsequent operations.

DETAILS

Roy Harris was born in Quitman, Brooks County, Georgia on August 29, 1891, the son of John S. Harris. (His Georgia World War One service card and Veterans Bureau service card give his year of birth as 1892.) He enlisted as a Private in the Regular Army on June 4, 1917 at Fort Logan, Colorado and was assigned to Company F 4th Engineers. He was considered the shortest man in the Regiment and was known in the 4th Engineers as "Shorty" Harris. He was promoted to Private 1st Class on February 2, 1918 and to Corporal on April 1, 1918. Harris sailed to France as a Corporal in Company F 4th Engineers 4th Division aboard the troopship *U.S.S. Martha Washington* (seized Austrian liner) on April 30, 1918. As his emergency contact, he listed his father John in Jacksonville, Florida. He served in the Aisne-Marne Offensive. On August 1, 1918 he was reduced to the rank of Private.

Harris was awarded the Distinguished Service Cross for his actions in the Vesle Sector. On August 5, 1918, a strong attempt was made by Infantry elements of the 4th Division to cross the Vesle River. That day Harris was a member of a small party of Engineers who built and emplaced one of the footbridges across the river west of the village of Fismes and closer to the village of Ville Savoye which allowed elements of the 58th and 59th Infantry to cross the river. He and his detachment particularly distinguished themselves by remaining at their work under the heaviest of fire and completing that bridge for the Infantry. He also received a Citation Star.

(See the entry for William B. Beach for an overview of the kinds of activities carried out by the soldiers of the 4th Engineers involved in building the bridges across the Vesle River.)

Harris served in the Toulon Sector, the St. Mihiel Offensive, and the Meuse-Argonne Offensive. On November 15, 1918, he was promoted to Private 1st Class. He was promoted to the rank/position of Corporal/Mechanic on March 15, 1919.

Harris was the only soldier of the 4th Division and one of only 469 soldiers of the American Expeditionary Forces to be awarded the Belgian Croix De Guerre in World War One. He also received the French Croix De Guerre.

Harris served on occupation duty in Germany with the 4th Division and returned to the United States with his Company aboard the troopship *U.S.S. Von Steuben* (seized German liner *Kronprinz Wilhelm*) leaving Brest, France July 21, 1919 and arriving at Hoboken, New Jersey July 29, 1919. He was discharged at Camp Gordon, Georgia on August 6, 1919.

Roy Harris died at the age of 44 in Lane County, Oregon on April 4, 1936 and is buried in Raffety Cemetery, North Plains, Washington County, Oregon. His headstone arrangements were handled by Banks Post No. 90 American Legion in Banks, Washington County, Oregon. The Oregon Death Index indicated that at the time of his death he had a wife named Elizabeth. Though he was a highly decorated soldier his simple headstone is inscribed only with his name, home State, rank, date of death and that he was a member of the 4th Engineers 4th Division.

Michael D. Belis

Edward D. Haskew

Wagoner, United States Army
Army Service Number: 571647
33rd Ambulance Company 4th Sanitary Train 4th Division
Date of Action: October 6, 1918
Location: Between Septsarges and Fromereville, France
War Department General Orders 128 (1918)
Medal # 7348

CITATION

Wagoner Haskew was on duty with his ambulance carrying wounded from a battalion aid station. He left with four stretcher cases and went about two kilometers south of Gercourt; while ascending a hill his ambulance was struck by a shell, he received multiple shell wounds of hands, left thigh, and feet. Although seriously wounded, he bravely remained at his post and continued on with his ambulance along a shell-swept road to the crest of the hill near an aid station, when he turned his ambulance off the road and sought assistance for his wounded.

DETAILS

Edward Davy Haskew was born in Wilton, Fairfield County, Connecticut on June 18, 1894, the son of Walter and Polly Haskew. Before entering the Army, he was a carpenter in Gladstone, New Jersey. He entered the Army on July 30, 1917. Haskew sailed to England as a Private 1st Class in Ambulance Company #33 4th Division aboard the Australian/British troopship *Hororata* on May 19, 1918. As his emergency contact, he listed his mother Polly in Gladstone, New Jersey. From England he and his unit were sent over to France. Haskew served in the Aisne-Marne Offensive, the Vesle Sector, the Toulon Sector, and the St. Mihiel Offensive.

He was awarded the Distinguished Service Cross for his actions in the Meuse-Argonne Offensive. On October 6, 1918, the 4th Division's advance had been halted by stiff German resistance and all forward movement had been stopped. The Division was still suffering casualties however as the forward elements that were established in the wooded areas known as the Bois de Fays and the Bois de Brieulles north of the village of Septsarges were under constant artillery and machine gun fire and frequent attacks by enemy aircraft as well.

Haskew picked up a load of wounded at an aid station near the front and headed with them through enemy artillery fire, trying to get them to Mobile Hospital No. 1 at Fromereville in the rear. While moving up the high ground southeast of Septsarges between the villages of Gercourt and Bethincourt his ambulance was struck by a German artillery round. Though seriously wounded by shrapnel in several places across his body, Haskew ignored his wounds, continued on through more fire, and brought the wounded soldiers he was transporting to an aid station before seeking aid for himself.

Edward D. Haskew was the only soldier from the 33rd Ambulance Company to receive the Distinguished Service Cross in World War One.

Because of his wounds, Haskew returned to the United States several months ahead of the Division as a Wagoner in the 4th Sanitary Train and part of Saint Aignan Casual Company No. 411 Regional Section A aboard the troop transport *U.S.S. Finland* leaving St. Nazaire, France December 28, 1918 and arriving at Newport News, Virginia on January 9, 1919. He was discharged on April 15, 1919.

The 1920 Census showed him living with his parents in Peapack-Gladstone, New Jersey where he was employed as a house carpenter. At some time after January 1920, he and his wife Ruth were married. They had a daughter in 1923. The 1930 Census showed Haskew living with his wife, daughter and son in Peapack-Gladstone where he was still a house carpenter and indicated he was a veteran of the World War. Edward D. Haskew died at the age of 61 on April 3, 1956 and is buried in Saint Bernard's Cemetery, Bernardsville, Somerset County, New Jersey.

Morrison Hayes

Corporal, United States Army
Army Service Number: 567487
Company D, 12th Machine Gun Battalion 4th Division
Date of Action: July 19, 1918
Location: Near Hautevesnes, France
War Department General Orders 35 (1920)
Medal # No assigned number discovered

CITATION

Although wounded during an advance, Corporal Hayes refused to be evacuated and led his squad forward with the Infantry, placing the gun in action in the front line. Exposed to intense fire, he maintained his gun in action until he received a second wound which later proved fatal. When ordered to withdraw he assisted in moving the gun back to another position, inspiring his men by his personal heroism.

DETAILS

Morrison Hayes was born in Wellsville, Allegany County, New York on November 12, 1894, the son of Clark and Elizabeth Hayes. On August 5, 1917, he enlisted as a Private in the Regular Army and was assigned to Company B 47th Infantry as the Regiment was forming up at Camp Syracuse, New York. He went with the 47th Infantry as it moved to Camp Greene, North Carolina and became part of the 4th Division. He was promoted to Corporal on November 2, 1917. On December 10, 1917, he was transferred to Company B 10th Machine Gun Battalion which was also in the 4th Division. He sailed to

Michael D. Belis

France as a Corporal in Company B 10th Machine Gun Battalion 4th Division aboard the French troopship *Rochambeau* on May 7, 1918. As his emergency contact, he listed his father Clark in Wellsville, New York. In France on June 7, 1918, Hayes was transferred to Company D 12th Machine Gun Battalion, still in the 4th Division.

He was awarded the Distinguished Service Cross for his actions in the Aisne-Marne Offensive. On July 19, 1918, the second day of the Offensive his Company was supporting 1st Battalion 59th Infantry and was located between the village of Courchamps and the wooded area known as the Bois de l'Orme near the village of Hautevesnes. 1st Battalion 59th Infantry was attached to the French 133rd Infantry Regiment and was in the middle of the attacking line. Rather than stay in a rear supporting position, Hayes moved his squad and their machine gun across the battlefield, keeping up with the leading formations of the Infantry. Despite being wounded, he continued to emplace the gun at the front where it could best be utilized as an attacking element providing direct fire. When the advance was stopped by the intensity of the German resistance, he directed the withdrawal of his squad and their gun even after being wounded a second time. His wounds were too severe, and he died from them on July 19, 1918, his second day of battle in the war.

Hayes was buried in Grave # 15 in the French civilian cemetery (designated cemetery # 1717 by the U.S. Graves Registration Services) at the village of Mary-sur-Marne.

His Distinguished Service Cross was presented to his father, Clark R. Hayes.

On September 20, 1919 Morrison Hayes was reinterred in Plot 4 Section U Grave # 164 in the American cemetery # 1764 at Belleau. He was reinterred and laid to his final rest on October 6, 1922 in Plot A Row 11 Grave 8 in the Aisne-Marne American Cemetery, Belleau, France.

Michael D. Belis

William Herren
First Sergeant, United States Army
Army Service Number: 559453
Machine-Gun Company 58th Infantry Regiment 4th Division
Date of Action: August 7, 1918
Location: Near Ville Savoye, France
War Department General Orders 64 (1919)
Medal # 2788

CITATION
Sergeant Herren carried guns and ammunition to the frontline platoons through an intense barrage after several carrying details had failed to get through. He then volunteered to stay with the right flank platoon which was under heavy fire in an exposed position. During the afternoon, he and one other man pushed forward with a captured machine gun and assisted materially in breaking up several hostile counterattacks during the day.

DETAILS

William Herren was born in the Blue Ridge Mountains of Buncombe County, North Carolina on July 2, 1888. He enlisted in the Regular Army in November 1905. Prior to World War One he had served twelve years in the Army, including a tour of duty in the Philippines with the 8th Infantry. He reenlisted in the 4th Infantry on December 11, 1914 and was promoted to Sergeant on January 7, 1915. When the United States declared war on Germany in April 1917, he was on duty with Machine Gun Company 4th Infantry and when elements of the 4th Infantry were used to create the 58th Infantry Herren thus became one of the original members of the 58th Infantry at the camp at Gettysburg, Pennsylvania. He was promoted to 1st Sergeant on June 5, 1917. He went with the Regiment to Camp Greene, North Carolina where it became part of the 4th Division. He sailed to England as a 1st Sergeant with Machine Gun Company 58th Infantry 4th Division aboard the British troopship *Themistocles* on May 11, 1918. As his emergency contact, he listed his mother Sarah in West Asheville, North Carolina. From England he and his Company were sent over to France. Herren served in the Aisne-Marne Offensive.

He was awarded the Distinguished Service Cross for his actions in the Vesle Sector. The 4th Division had been trying for several days to cross the Vesle River under intense enemy artillery and machine gun fire in the vicinity of the villages of Ville Savoye and St. Thibaut. The Germans launched numerous counterattacks of varying strength and on August 7, 1918, Herren's Machine Gun Company of the 58th Infantry was instrumental in repulsing many of those attacks near Ville Savoye. As 1st Sergeant of the Company, Herren could have remained at the Command Post and monitored operations from there.

Instead, he carried guns and ammunition to the forward platoons on several trips through heavy enemy artillery fire after attempts by others had failed to get through. Later he took another man with him and using a captured German Maxim MG 08/15 machine gun he advanced beyond the forward positions and established a post from which he was able to lay down suppressive fire on enemy counterattacks. He also received a Citation Star.

Herren served in the Toulon Sector and the St. Mihiel Offensive. He was wounded during the Meuse-Argonne Offensive on October 4, 1918. He was also awarded the French Croix De Guerre. When the artist Joseph Cummings Chase travelled the European theater of war in 1918-1919 and painted portraits of American soldiers, he chose Herren as one of his subjects. Herren requested that Chase paint him with his gas mask which Herren affectionately termed as his best friend. In 1920 Chase published his portraits in a book entitled " *SOLDIERS ALL: PORTRAITS AND SKETCHES OF THE MEN OF THE A.E.F.*" In that book the text accompanying Herren's portrait expanded upon the description of the actions Herren performed which earned him the Distinguished Service Cross:

"WILLIAM HERREN, First Sergeant,

Machine Gun Company, 58th Infantry. 'For extraordinary heroism in action near Ville-Savoye, August 7, 1918.'

This soldier showed great bravery and devotion to duty throughout this action. On the morning of August 7, 1918, Herren supplied the Company with spare machine-guns and ammunition through a deadly artillery barrage after several carrying details had failed to get through. The company had lost four machine-guns and was practically without ammunition at this time. After distributing machine-guns and ammunition to the different platoons under terrific machine-gun fire, he showed extraordinary heroism by pushing forward on the right flank with a captured light German Maxim machine-gun and repulsing a counterattack. At this time, the battalion on the right flank had fallen back, leaving that flank unprotected. Sergeant Herren showed complete disregard for personal safety and displayed great resourcefulness throughout the entire action."[13]

Herren was promoted to Color Sergeant on December 2, 1918 and transferred to Headquarters Company 58th Infantry. He served on occupation duty in Germany with the 4th Division until leaving Europe. On March 3, 1919 he was transferred to Headquarters Company 47th Infantry still in the 4th Division. Herren was reduced to the rank of Private on March 13, 1919 and was transferred to Supply Company 58th Infantry on April 9, 1919. He was transferred to Company G 58th Infantry on April 12, 1919, promoted to Sergeant on April 14, 1919 and to 1st Sergeant on April 21, 1919. He returned to the United States as 1st Sergeant of Company G 58th Infantry aboard the troopship *U.S.S. Mount Vernon* (captured German liner *Kronprinzessin Cecilie*) leaving Brest, France July 24, 1919 and arriving at Hoboken, New Jersey August 1, 1919.

Herren married Alphabel Philpott sometime in 1919. He reenlisted on November 18, 1919. The 1920 Census showed him stationed at Camp Dodge, Iowa with his marital status indicated as married. The 1930 Census showed him stationed at Fort Omaha, Nebraska with his marital status indicated as married and his rank indicated as Private. He retired from the Army in 1932. After 1932 he was awarded the Purple Heart for the wounds he received in World War One. The 1940 Census showed Herren, his wife and son living in Santa Cruz, California. During World War Two he was recalled to active duty and served as a 1st Lieutenant from July 2, 1942 to January 10, 1944 when he was released from duty. No other details could be found of his World War Two service. William Herren died at the age of 65 on August 3, 1953 and is buried in Golden Gate National Cemetery, San Bruno, San Mateo County, California.

William B. Hook

Sergeant, United States Army
Army Service Number: 568260
Company B 4th Engineer Regiment 4th Division
Date of Action: August 9, 1918
Location: Near St. Thibaut, France
War Department General Orders 98 (1919)
Medal # 1793

CITATION

While a member of a party engaged in constructing a bridge across the Vesle River in advance of the Infantry, Sergeant Hook voluntarily plunged into the stream under heavy enemy machine gun and grenade fire, swam with a line to the opposite bank which was held by the enemy and securely tied the end of the bridge to the opposite bank.

DETAILS

William Beaumont Hook was born in Zanesville, Muskingum County, Ohio on November 21, 1877, the son of David and Elizabeth Hook. He served (under the name of Beaumont Hook) as a Private in Battery C 1st Ohio Volunteer Artillery April 26-October 23, 1898 in the continental United States during the War with Spain. He entered the Army on August 13, 1917 at Fort McDowell, San Francisco, California. Hook sailed to France as a Sergeant in Company B 4th Engineers 4th Division aboard the troopship *U.S.S. Martha Washington* (seized Austrian liner) on April 30, 1918. As his emergency contact, he listed his father David in Zanesville, Ohio. Hook served in the Aisne-Marne Offensive.

Michael D. Belis

He was awarded the Distinguished Service Cross for his actions in the Vesle Sector. On August 9, 1918 the 47th Infantry of the 4th Division had elements near the village of Bazoches on the other side of the Vesle River across from the village of St. Thibaut. Receiving heavy fire from Bazoches and the high ground north of it, the Infantry fell back to the river and then withdrew across the river. Foot bridges had been needed to allow the Infantry to cross the river earlier, and later to bring the Infantry back to safe positions. Hook braved enemy machine gun fire and grenades to help secure one of those bridges during the initial assault. He swam the river while under heavy fire and tied a line from the footbridge to secure it to the enemy's side of the river. His efforts were instrumental in aiding the Infantry to get across the river.

Hook served in the Toulon Sector, the St. Mihiel Offensive, and the Meuse-Argonne Offensive. While still overseas at some point after August 9, 1918 he was promoted to Sergeant 1st Class. He served on occupation duty in Germany with the 4th Division and returned to the United States with Company B 4th Engineers aboard the troopship *U.S.S. Von Steuben* (seized German liner *Kronprinz Wilhelm*) leaving Brest, France July 21, 1919 and arriving at Hoboken, New Jersey July 29, 1919. He was discharged on August 8, 1919.

The 1920 Census showed Hook living in Zanesville, Ohio with his parents where he was employed as a mechanical engineer. The 1930 and 1940 Census showed him still in Zanesville living with his brother and sister but now unemployed. William B. Hook died at the age of 78 on June 23, 1956 and is buried in Greenwood Cemetery, Zanesville, Muskingum County, Ohio.

Samuel H. Houston

Major, United States Army
Commanding Officer 1st Battalion 58th Infantry Regiment 4th Division
Date of Action: August 4, 1918
Location: Near Ville Savoye, France
War Department General Orders 62 (1919)
Medal # 6904

CITATION

With but 15 minutes in which to prepare his battalion for attack, Major Houston on horseback galloped from flank to flank, fully exposed to deadly artillery fire in order to make the necessary preparations for the advance. After his leading element had started the attack, he was killed by an enemy shell.

DETAILS

Samuel Humes Houston was born in Baltimore, Maryland

Michael D. Belis

on January 22, 1888, the son of Lieutenant Colonel George and Mary Houston. He was commissioned a 2nd Lieutenant in the 28th Infantry (Regular Army) on October 7, 1911 and took part with his Regiment in the Vera Cruz Expedition into Mexico in 1914. Houston was promoted to Captain on July 26, 1917. As were so many experienced personnel at the start of World War One, he was detached from his Regiment to become a cadre to start up or lead other units. He was assigned to the 58th Infantry of the 4th Division at Camp Greene, North Carolina on March 21, 1918. He sailed to England as a Captain in command of Company C 58th Infantry 4th Division aboard the British transport *City of Brisbane* on May 7, 1918. As his emergency contact, he listed his mother Mary in Baltimore, Maryland. From England, Houston and his Company were sent over to France.

While in France he was promoted to the wartime temporary rank of Major and assigned command of 1st Battalion 58th Infantry on June 22, 1918. He led 1st Battalion in its first engagement of the war on July 18, 1918 at Hautevesnes during the Aisne-Marne Offensive when the Battalion was placed under command of the 133rd French Infantry Regiment. He was awarded the French Croix De Guerre with silver star for his courage and leadership while his Battalion worked alongside a French Battalion during the attack of July 19 and 20, 1918.

Houston was awarded the Distinguished Service Cross for his later actions in the Aisne-Marne Offensive. On August 4, 1918 the three Battalions of the 58th Infantry attacked toward the Vesle River with the 2nd Battalion on the left, the 3rd Battalion on the right, and the 1st Battalion directly behind the 3rd. The 3rd Battalion came out of the wooded area known as the Bois de Mont St. Martin southwest of the village of Ville Savoye and moved across the open plateau south of the Montagne de Mont St. Martin with 1st Battalion following about 600 yards behind. The enemy fired an artillery barrage that mostly missed the 3rd Battalion but was adjusted as the 1st Battalion came up. The German artillery then caught 1st Battalion in the open as it was beginning its advance. It was at this point that Houston was killed as he galloped back and forth on his horse along the front line through enemy artillery fire while forming up 1st Battalion and directing it forward.

Houston was the highest-ranking officer of the 58th Infantry to be killed in the war.

He was buried in grave # 240 in the American cemetery # 371 at the village of Saint Gengoulph.

His Distinguished Service Cross was presented to his mother Mrs. Mary Houston.

On June 7, 1919 Samuel H. Houston was reinterred in Plot 4 Section B Grave # 180 in the American cemetery # 1764 at Belleau. On July 1, 1921 his body was disinterred for preparation and shipment. His remains were returned to the United States aboard the United States Army Transport *U.S.A.T. Wheaton,* leaving Antwerp, Belgium August 6, 1921 and arriving at Hoboken, New Jersey August 20, 1921. He was one of 5,759 American dead brought home aboard the *Wheaton* on that journey. Samuel H. Houston was laid to his final rest in Arlington National Cemetery on September 30, 1921.

Michael D. Belis

Henry Howard

Sergeant, United States Army
Army Service Number: 556412
Company A 39th Infantry Regiment 4th Division
Date of Action: September 27, 1918
Location: Near Septsarges, France
War Department General Orders 64 (1919)
Medal # 7667

CITATION

Although seriously wounded during a bombardment which scattered his men and caused his company and battalion to retire behind a ridge in the rear, Sergeant Howard, with about fifteen men, held the advanced position under the continuous fire of machine guns, one-pounders, and artillery until relieved the following day by another battalion. He insisted on remaining with his detachment until the commanding officer of the relieving battalion personally directed his evacuation.

DETAILS

Henry Howard was born in Irvine, Estill County, Kentucky on October 29, 1890, the son of William and Renis Howard. He enlisted as a Private in Company D 28th Infantry (Regular Army) on January 22, 1909. He reenlisted in February 1912 and served in Company C 5th Infantry and then Company A 30th Infantry, being discharged on February 5, 1915. He re-enlisted the same day. When elements of the 30th Infantry were used to create the 39th Infantry in 1917, Howard was then transferred to Company A 39th Infantry and thus became one of the first members of the 39th Infantry. He sailed to France as a Sergeant in Company A 39th Infantry 4th Division aboard the Italian troopship *Dante Alegheiri* on May 10, 1918. As his emergency contact, he listed his wife Kate in Charlotte, North Carolina. Howard served in the Aisne-Marne Offensive, the Velse Sector, the Toulon Sector, and the St. Mihiel Offensive.

He was awarded the Distinguished Service Cross for his actions in the Meuse-Argonne Offensive. On September 26, 1918, the first day of the Offensive, his Company, as part of 1st Battalion 39th Infantry reached its objective near Septsarges but in the fighting at Montfaucon Hill lost all of its Platoon Leaders either killed or wounded. Also lost was all of Headquarters Platoon except for one man. By the end of the day, Howard was one of only six Sergeants in Company A who had not been killed or wounded.

The next day, September 27, the attack was resumed by First and Third Battalions of the 39th Infantry but after the leading elements crossed the Nantillois road the advance was broken up by a tremendous German artillery barrage, causing most of the attacking force to retire to more protected positions. The Regimental Executive Officer was killed by machine gun fire. Howard was severely wounded but did not retreat with his Battalion. He remained with about 15 soldiers at the farthest point reached by the Regiment that day. Under his direction, his small group of men held their position under heavy fire until the next day, September 28, when Second Battalion came up from its position as reserve, led the attack and relieved the little battered command. For his courageous leadership, Howard was awarded the French Croix De Guerre with gilt star. He also received a Citation Star.

Because of his wounds he returned to the United States several months ahead of his Regiment as part of a detachment of sick and wounded in St. Nazaire Convalescent Detachment No. 72 aboard the troopship *U.S.S. Mongolia* leaving St. Nazaire, France February 23, 1919 and arriving at Hoboken, New Jersey March 7, 1919. His entry in the passenger list indicated gunshot wounds to the left leg. He was discharged on November 21, 1919.

The 1920 Census showed him and his wife Kate living with Howard's brother in Valley View, Kentucky where Howard was unemployed. The 1930 Census showed Howard living with his wife, their son and three daughters in Madison County, Kentucky where he was employed as a store merchant and indicated he was a veteran of the World War. After 1932, Howard was awarded the Purple Heart for the wounds he received in World War One. The 1940 Census showed Howard and his family still living in Madison County with Howard unemployed and living on his World War disability pension. Henry Howard died of kidney disease at the age of 49 on September 29, 1940 and is buried in Camp Nelson National Cemetery, Nicholasville, Jessamine County, Kentucky.

William L. Hunter

Private, United States Army
Army Service Number: 2105021
Company D 58th Infantry Regiment 4th Division
Date of Action: September 26 - October 6, 1918
Location: In Bois de Fays, France
War Department General Orders 5 (1920)
Medal # 5666

CITATION

While on duty as a company runner Private Hunter repeatedly carried messages when exposed to artillery and machine gun fire. Although reduced to a state of physical exhaustion he refused to be relieved and continued to perform his duty of maintaining liaison.

DETAILS

William Leslie Hunter was born in Belpre, Ohio on July 1, 1888, the son of Robert and Mary Hunter. On his registration

Michael D. Belis

for the draft in June 1917, he indicated he was single and employed as a farmer in Dunham Township, Ohio. He was married to Alberta Doan on June 27, 1917. Hunter was drafted into the Army as a Private at Camp Sherman, Ohio on October 3, 1917. He was assigned to Co D 58th Infantry at Camp Greene, North Carolina on March 9, 1918. He sailed to England as a Private in Company D 58th Infantry 4th Division aboard the British troopship *City of Brisbane* on May 7, 1918. As his emergency contact, he listed his wife Alberta in Vincent, Ohio. From England, he and his Company were sent over to France. He served in the Aisne-Marne Offensive, the Vesle Sector, the Toulon Sector, and the St. Mihiel Offensive.

Hunter was awarded the Distinguished Service Cross for his actions spanning ten days in the Meuse-Argonne Offensive. During the first ten days of the massive attack, Hunter carried out the dangerous and deadly job of runner, so vital to maintaining communication between his Company and Battalion Headquarters in the era before field radios. In those ten days, the 1st Battalion 58th Infantry which included Hunter's Company D attacked northward from near the village of Esnes toward the wooded area known as the Bois de Fays. Fighting through the villages of Malancourt, Cuisy, near the village of Septsarges and the wooded area known as the Bois de Brieulles 1st Battalion made it into the Bois de Fays and established defensive positions which they held against German counterattacks. Exposed to every kind of deadly fire the enemy could produce, Hunter made his way across the battlefield time after time, only to turn around and do it again, to the point of exhaustion.

Hunter served on occupation duty in Germany with the 4th Division and returned to the United States with his Company aboard the troopship *U.S.S. Mount Vernon* (captured German liner *Kronprinzessin Cecilie*) leaving Brest, France July 24, 1919 and arriving at Hoboken, New Jersey August 1, 1919. He was discharged on August 8, 1919.

The 1920 Census showed Hunter living with his wife at his wife's parents' home in Dunham, Washington County, Ohio where he was employed as a farmer. The 1930 Census showed him living with his wife, daughter and four sons in their own home in Dunham where he was employed as a farmer and indicated he was a veteran of the World War. The 1940 Census showed him living with his wife and six sons in Dunham where he was still employed as a farmer. When he registered for the draft in 1942, he was still in Dunham and still a farmer. William L. Hunter died of arteriosclerosis at the age of 57 on March 28, 1946 and is buried in Rockland Cemetery, Belpre, Washington County, Ohio.

Michael D. Belis

Albert L. J. Ihrke

Private, United States Army
Army Service Number: 2023072
Company B 47th Infantry Regiment 4th Division
Date of Action: August 1, 1918
Location: Near Sergy, France
War Department General Orders 46 (1919)
Medal # 2033

CITATION

Private Ihrke displayed great courage and devotion to duty by remaining in an exposed position under heavy machine gun and shell fire to cover the withdrawal of his company.

DETAILS

Albert Leverne John Ihrke was born in Mayville, Tuscola County, Michigan on October 26, 1895, the son of Frederick and Elizabeth Ihrke. Before entering the Army, he was a farmer in Fremont Township, Michigan. He entered the Army on November 22, 1917. Ihrke sailed to France as a Private in Company B 47th Infantry 4th Division aboard the troopship *U.S.S. Princess Matoika* (seized German liner *Prinzess Alice*) on May 10, 1918. As his emergency contact, he listed his father Fred in Mayville, Michigan.

Ihrke was awarded the Distinguished Service Cross for his actions in the Aisne-Marne Offensive. In late July 1918, during the fighting along the Ourcq River, the village of Sergy was an objective for the 42nd Division. Weakened by losses, the 42nd Division needed help and 1st and 3rd Battalions of the 47th

Infantry from the 4th Division were attached to the 42nd Division for the assault upon Sergy. Ihrke's Company B was part of 1st Battalion 47th Infantry and for the attack was attached to the 167th Infantry Regiment of the 42nd Division.

Company B and Company D 47th Infantry were the frontline elements of the 1st Battalion attack against Sergy and encountered intense hand to hand fighting for the heights outside of the village. The village and the high ground around it were defended by the 93rd Regiment of the crack 4th Prussian Guards Division who counterattacked every gain made by Ihrke's Battalion, turning the battle into a series of maneuvers whereby the Americans would gain ground, fall back, and attack on a different flank, steadily pushing the Germans ever backwards. During one of his Company's withdrawal actions, Ihrke single handedly acted as a rear guard for his Company, providing cover for them to pull back and regroup. He also received a Citation Star.

He served in the Vesle Sector, the Toulon Sector, the St. Mihiel Offensive, and the Meuse-Argonne Offensive. He was one of 20 soldiers of the 4th Division and one of only 399 soldiers of the American Expeditionary Forces to be awarded the Italian War Merit Cross (Croce al Merito di Guerra). The Italian War Merit Cross was presented to him on June 11, 1919 in a 4th Division awards ceremony at Remagen, Germany.

Ihrke served on occupation duty in Germany with the 4th Division and returned to the United States as a Private with Company B 47th Infantry aboard the troopship *U.S.S. Mobile* (seized German liner *Cleveland*) leaving Brest, France July 16, 1919 and arriving at Hoboken, New Jersey July 27, 1919. He was discharged on August 6, 1919. The Veterans Bureau gave his rank at the time of discharge as Private 1st Class.

Michael D. Belis

The 1920 Census showed Ihrke living with his parents in Dayton, Michigan and employed as a farm laborer. He married Gladys Snever on February 25, 1920. The 1930 Census showed Ihrke and Gladys with their two daughters living in Dayton, Michigan where Ihrke was employed as a farmer and indicated he was a veteran of the World War. The 1940 Census showed him working as a farmer in Rich Township, Michigan while Gladys and their daughters lived at Watertown, Michigan. Albert L. J. Ihrke died of a heart attack at the age of 55 on February 15, 1951 and is buried in Fremont Township Cemetery, Mayville, Tuscola County, Michigan.

Homer S. Jarvis

First Lieutenant, United States Army
11th Machine Gun Battalion 4th Division
Date of Action: September 26, 1918
Location: Near Nantillois, France
War Department General Orders 138 (1918)
Medal # 1067

CITATION

Lieutenant Jarvis, with another officer and a soldier, using captured German Maxim guns pushed forward to a heavily shelled area from which the infantry had withdrawn and by their accurate and effective fire kept groups of the enemy from occupying advantageous positions. Maintaining fire superiority all afternoon Lieutenant Jarvis withdrew from his dangerous position only when it became too dark to see.

DETAILS

Homer Smith Jarvis was born in Xenia, Clay County, Illinois on August 19, 1889, the son of Elmer and Mary Jarvis. On February 14, 1917 he married Bessie Kinney. He was commissioned a 1st Lieutenant in Headquarters Company 2nd Infantry Idaho National Guard on March 26, 1917. He became the Executive Officer of Machine Gun Company 2nd Infantry Idaho National Guard. When his National Guard Regiment was called into Federal service it was moved to Camp Greene, North Carolina in September 1917 where the Machine Gun Company became part of the 147th Machine Gun Battalion. Jarvis sailed to France as a 1st Lieutenant in Company A 147th Machine Gun

Michael D. Belis

Battalion 41st Division aboard the troop transport *U.S.S. Covington* (seized German liner *Cincinnati*) on December 13, 1917. As his emergency contact, he listed his wife in Caldwell, Idaho.

Once in France, the 41st Division was used as a replacement Division. Its artillery regiments were attached to other organizations and the rest of the Division's personnel were used as replacements for front line units. At an unknown date, Jarvis was assigned to the 11th Machine Gun Battalion in the 4th Division.

Jarvis was awarded the Distinguished Service Cross for his actions on the first day of the Meuse-Argonne Offensive. On September 26, 1918, the 7th Brigade of the 4th Division, including the 11th Machine Gun Battalion, had reached the Corps objective which was on a line that ran approximately east and west through the village of Nantillois. The Divisions on either side had not reached the Corps objective and the advance of the 4th Division was halted for hours waiting for the flanking units to catch up. When at 5:30 p.m. the other Divisions had still not reached the objective, the order was given for the 4th Division to resume the attack. The halt had given the enemy time to regroup and rebuild his defenses.

After resuming the attack and achieving a minimal gain, without the flanking Divisions there to protect its flanks, the 4th Division pulled back to the line of the Corps objective and dug in for the night. Using captured German Maxim machine guns (most likely MG 08/15 models) Jarvis, along with 1st Lieutenant Earl M. McKinley and Sergeant Glenn M. Grove, also from 11th Machine Gun Battalion moved to an exposed forward position and covered the withdrawal of the 4th Division Infantry in their area and prevented the enemy from launching any successful counterattack. They maintained their

Soldiers Steadfast and Loyal

fire upon the enemy until it became too dark to see. Only then did they pull back and join the withdrawn units. McKinley and Grove were also awarded the Distinguished Service Cross for the action.

At an unknown date after September 26, 1918 Jarvis was promoted to the wartime temporary rank of Captain. Also, at an unknown date he was transferred to the Army Service Corps and assigned to Headquarters Base Section #2 A.E.F. He returned to the United States as a Captain in the Army Service Corps and part of a detachment of Casual Officers aboard the aging protected cruiser *U.S.S. Chicago* leaving Bordeaux, France July 17, 1919, and arriving at New York City July 29, 1919. Jarvis was discharged on August 18, 1919.

The 1920 Census showed Jarvis living with his wife and son in Malheur County, Oregon where he was a rancher. The 1930 Census showed him living in Pocatello, Idaho where he was employed as an automobile salesman and indicated he was a veteran of the World War. He and his wife had three children, two boys and a girl. In May 1930 he was divorced. After 1932 he was awarded the Purple Heart for wounds he received in World War One. When he registered for the draft in 1942, he indicated he lived in Akron, Summit County, Ohio and was employed by a construction company in Akron. He moved about the country for years and returned to Idaho in the 1950's. Homer S. Jarvis died of lung cancer at the age of 72 on January 24, 1962, and is buried in Morris Hill Cemetery, Boise, Ada County, Idaho.

Michael D. Belis

Frank Jaworski

Corporal, United States Army
Army Service Number: 569251
Company F 4th Engineer Regiment 4th Division
Date of Action: August 5, 1918
Location: West of Fismes, France
War Department General Orders 145 (1918)
Medal # 1086

CITATION

Corporal Jaworski was a member of a small detachment of engineers which went out in advance of the front line of the infantry through an enemy barrage from 77-mm. guns and one-pounder guns to construct a footbridge over the River Vesle. As soon as their operations were discovered, machine gun fire was opened up on them but undaunted the party continued at work, removing the German wire entanglements, and completing a bridge which was of great value in subsequent operations.

DETAILS

Frank Jaworski was born in Chicago, Illinois on December 14, 1893, the son of Lawrence and Mary Jaworski. Before entering the Army, he was a steam fitter in Hammond, Indiana. He entered the Army sometime after June 1917. Jaworski sailed to France as a Private 1st Class in Company F 4th Engineers 4th Division aboard the troopship *U.S.S. Martha Washington* (seized Austrian liner) on April 30, 1918. As his emergency contact, he listed his father Lawrence in West Hammond, Indiana.

Jaworski was awarded the Distinguished Service Cross for

his actions in the Aisne-Marne Offensive. After several days of trying to cross the Vesle River under fire, and only managing to put across small numbers of men, on August 5, 1918, the 4th Division made a determined effort to get across the river. The 4th Engineers including Jaworski worked under fire incessantly to build and emplace footbridges which the Infantry could utilize to make a mad dash across the river. On August 5 he was part of a small detachment of Engineers who particularly distinguished themselves by remaining at their work under the heaviest of fire and completing a bridge for the Infantry west of the village of Fismes and near the village of Ville Savoye.

(See the entry for William B. Beach for an overview of the kinds of activities carried out by the soldiers of the 4th Engineers involved in building the bridges across the Vesle River.)

Jaworski was one of three soldiers in the 4th Division to be awarded the Belgian Military Decoration (Décoration Militaire) during World War One. All three soldiers were Corporals from Company F 4th Engineers and all received the Belgian Military Decoration and the Distinguished Service Cross for their actions at the Vesle River on August 5, 1918. Jaworski was one of only 20 soldiers of the American Expeditionary Forces to receive the Belgian Military Decoration.

He was one of 20 soldiers of the 4th Division and one of only 399 soldiers of the American Expeditionary Forces to be awarded the Italian War Merit Cross (Croce al Merito di Guerra).

He also received a Citation Star in 4th Division General Orders 76 (1918).

Jaworski served in the Vesle Sector, the Toulon Sector, the St. Mihiel Offensive, and the Meuse-Argonne Offensive. He served on occupation duty in Germany with the 4th Division and returned to the United States with his Company aboard the troopship *U.S.S. Von Steuben* (seized German liner *Kronprinz Wilhelm*) leaving Brest, France July 21, 1919 and arriving at Hoboken, New Jersey July 29, 1919. The date of his discharge could not be found.

Jaworski went back to live with his parents in West Hammond, Cook County, Illinois where he worked as a police officer. Frank Jaworski died at the age of 25 on November 26, 1919 and is buried in Holy Cross Cemetery and Mausoleums, Calumet City, Cook County, Illinois.

Reuben L. Johnson

Private First Class, United States Army
Army Service Number: 2101807
Company B 47th Infantry Regiment 4th Division
Date of Action: September 28, 1918
Location: Near the Bois de Brieulles, France
War Department General Orders 98 (1919)
Medal # 1832

CITATION

Although he had been painfully wounded in the back by a bursting shell, Private First Class Johnson continued to perform his duties as a runner under heavy artillery and machine gun fire thereby enabling his company commander to maintain control of the company. He remained on duty until late in the night when he was ordered to the dressing station.

DETAILS

Reuben Leonard Johnson was born in Donovan, Iroquois County, Illinois on January 25, 1896, the son of Charles and Anna Johnson. Before entering the Army, he lived at Ashton, South Dakota where he worked as a farm laborer. He entered the Army on September 18, 1917. Johnson sailed to England as a Private and replacement in the Camp Pike June Automatic Replacement Draft Company Number 1 aboard the British transport *S.S. Delta* on June 20, 1918. As his emergency contact, he listed his father Charles in Paxton, Illinois. From England he was sent over to France. After undergoing further training in France, he was assigned to Company B 47th Infantry 4th

Division. At some time after reaching France, he was promoted to Private 1st Class.

Johnson was awarded the Distinguished Service Cross for his actions in the Meuse-Argonne Offensive. On September 27, 1918, the second day of the Offensive, the attack toward the wooded area known as the Bois de Fays was resumed. Johnson's Regiment, the 47th Infantry fought almost all the way through the barbed wire and tangled undergrowth of the wooded area known as the Bois de Brieulles. The enemy fire from the Bois de Fays was too strong however, and the 47th Infantry had to stop at the northern edge of the Bois de Brieulles and dig in. On September 28, the 47th Infantry pulled back to another wooded area known as the Bois de Septsarges where the Regiment remained in reserve yet under almost constant German artillery and machine gun fire. It was at this phase of the battle that Johnson earned his award. Though wounded by shrapnel in his back he continued to perform his duties as a runner for his Company until ordered to the aid station later that night. He also received a Citation Star.

He served on occupation duty in Germany with the 4th Division and returned to the United States with his Company aboard the troopship *U.S.S. Mobile* (seized German liner *Cleveland*) leaving Brest, France July 16, 1919 and arriving at Hoboken, New Jersey July 27, 1919. He was discharged on August 4, 1919.

Johnson went to live with his parents in Paxton City, Patton Township, Ford County, Illinois where he was employed as a concrete laborer. Reuben L. Johnson died at the age of 32 on May 25, 1928 and is buried in Glen Cemetery, Paxton, Ford County, Illinois.

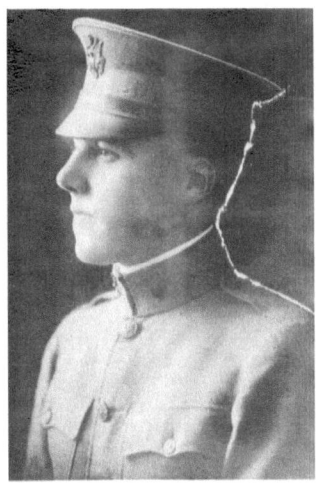

Thomas W. Kearns

First Lieutenant, United States Army
Headquarters Company 47th Infantry Regiment 4th Division
Date of Action: July 29 - 30, 1918
Location: Near Sergy, France
War Department General Orders 71 (1919)
Medal # 1190

CITATION

When a counterattack was impending, First Lieutenant Kearns successively carried 15 wounded men across a shell-swept area in full view of the enemy, taking them to a place of safety and preventing them from being captured by the enemy. Later he reorganized groups of stragglers and led them into combat.

DETAILS

Thomas Wilfred Kearns was born in Boston, Massachusetts on

Michael D. Belis

August 19, 1894, the son of William and Angela Kearns. He enlisted as a Private in the Massachusetts National Guard on June 20, 1916. From June to November 1916, he served with the Massachusetts National Guard on Mexican Border Service in Texas and New Mexico. Kearns enlisted in the Army on April 6, 1917 and attended Officers Training Camp. He was commissioned a 2nd Lieutenant of Infantry in the Officers Reserve Corps on August 15, 1917 and was with the 47th Infantry at Camp Greene, North Carolina when it became a part of the 4th Division. Kearns sailed to France as a 2nd Lieutenant with Headquarters Company 47th Infantry 4th Division aboard the troopship *U.S.S. Princess Matoika* (seized German liner *Prinzess Alice*) on May 10, 1918. As his emergency contact, he listed his father William in Dorchester, Massachusetts.

Kearns was awarded the Distinguished Service Cross for his actions on two days in the Aisne-Marne Offensive during the battle for the village of Sergy, near the Ourcq River. The area in front of Sergy was blanketed with German artillery and small arms fire. In their fight to take and hold the village, from July 29-31, 1918, the 1st and 3rd Battalions of the 47th Infantry lost 25 officers and 462 enlisted men either killed, wounded, or gassed and six men missing in action. On July 29, Kearns ventured out across the battlefield under fire fifteen times, each time bringing back to safety and medical attention a wounded American soldier who otherwise would have been killed or captured by the enemy should a counterattack be mounted. The Germans did counterattack twice during the night of July 29, making Kearns' rescue of those soldiers all the more instrumental to their survival. On July 30 he organized groups of men from shot-up units and led them in coordination with the Regiment in continuing the attack.

Kearns was promoted to 1st Lieutenant on November 1, 1918. The official history of the 47th Infantry in World War One indicated that he was gassed and returned to duty but did not give any dates. He served on occupation duty in Germany with the 4th Division and returned to the United States with Headquarters Company 47th Infantry aboard the troopship *U.S.S. Mobile* (seized German liner *Cleveland*) leaving Brest, France July 16, 1919 and arriving at Hoboken, New Jersey July 27, 1919. He was discharged on August 1, 1919.

The 1920 Census showed Kearns living at his parents' address in Boston, Massachusetts where he was employed as a clerk for a construction company. He returned to the Massachusetts National Guard and served as a Private from March 11 to May 8, 1924. He was commissioned a 2nd Lieutenant, National Guard in the Officers Reserve Corps on May 13, 1924, promoted to 1st Lieutenant May 21, 1925 and to Captain on October 8, 1926. Sometime between 1920 and 1928 he married Cecilia Barter. The 1930 Census showed him living with his wife Cecilia and two daughters in Swampscott, Massachusetts where he was employed as a Manager at Victrola Records and indicated he was a veteran of the World War. The 1940 Census showed him living with his wife, three daughters and mother-in-law in Del Mar, California where he was employed as a Stable Superintendent at a racetrack. When he registered for the draft in 1942, he indicated he was employed by the Consolidated Aircraft Company in San Diego, California. His wife died in 1954 and he later remarried. Thomas W. Kearns died at the age of 77 on November 21, 1972 and is buried in Fort Rosecrans National Cemetery, San Diego, San Diego County, California.

Michael D. Belis

Charles Kelly
Sergeant, United States Army
Army Service Number: 562563
Company C 12th Machine Gun Battalion 4th Division
Date of Action: September 29, 1918
Location: Not given
Note: The location of his award action was in the Bois des Ogons.
War Department General Orders 19 (1920)
Medal # 5787

CITATION
Sergeant Kelly led his platoon in the attack with great bravery against strongly held enemy trenches. Shortly after reaching his objective, he was wounded in the throat. He refused to be evacuated but continued to actively command his men until the night of October 1, by which time, due to his wound he had lost the power of speech.

DETAILS
Charles C. Kelly was born in Hinsdale, Berkshire County, Massachusetts. The date of his entry into the Army could not be found. He sailed to England as a Sergeant in Company C 12th Machine Gun Battalion 4th Division aboard the British troopship *RMS Aquitania* on May 7, 1918. As his emergency contact, he listed his uncle Fred O. Kelly in Dalton, Massachusetts. From England, he and his Battalion were sent over to France where they were issued their machine guns. He served in the Aisne-Marne Offensive, the Vesle Sector, the Toulon Sector, and the St. Mihiel Offensive.

Kelly was awarded the Distinguished Service Cross for his actions in the Meuse-Argonne Offensive. On September 29, 1918, the 8th Infantry Brigade, which included the 12th Machine Gun Battalion spearheaded the attack by the 4th Division toward the wooded area known as the Bois de Fays. One wooded area known as the Bois des Brieulles was cleared of Germans and taken. Company C 58th Infantry supported by Kelly and his machine guns pushed through the woods of the Bois des Brieulles and then advanced through the wooded area known as the Bois des Ogons until within sight of Madeleine Farm which was their planned objective but the fire coming from the Bois de Fays and also from across the Meuse River proved to be far too formidable. At this point in the battle, Kelly was wounded in the throat. The American attack was halted, and the soldiers of Company C and Kelly's platoon pulled back to the southern edge of the Bois des Ogons, dug in and held their positions until relieved two nights later. Despite his wounds, Kelly continued to direct his men as they pulled back to the southern edge of the woods and led them in the defense of their position until they were relieved during the night/early morning of October 1-2, 1918. By that time because of his wounds, he had lost the ability to speak.

Kelly served on occupation duty in Germany with the 4th Division and returned to the United States as a Sergeant in Company C 12th Machine Gun Battalion with his Company aboard the troopship *U.S.S. Von Steuben* (seized German liner *Kronprinz Wilhelm*) leaving Brest, France July 21, 1919 and arriving at Hoboken, New Jersey July 29, 1919. No further details for Charles Kelly could be found.

Michael D. Belis

Orval Kline
Second Lieutenant, United States Army
Company A 11th Machine Gun Battalion 4th Division
Date of Action: October 12, 1918
Location: Northeast of Nantillois, France
War Department General Orders 138 (1918)
Medal # 3543

CITATION

While the infantry was falling back 200 meters to take cover from heavy artillery and machine gun fire, he, with his platoon sergeant, stayed at their one remaining machine gun which they continued to operate for 45 minutes until the infantry position was re-established. They not only successfully covered withdrawal of the infantry, but also captured a German machine gun and three prisoners.

DETAILS

Orval Lee Kline was born in Martinsburg, West Virginia on September 13, 1893, the son of Howard and Cornelia Kline. He enlisted in the Regular Army on March 12, 1915. On an unknown date he was promoted to Corporal. He was promoted to Sergeant on June 12, 1917 and to Mess Sergeant in Machine Gun Company 3rd Infantry on September 10, 1917. On December 1, 1917 Kline was transferred to the Detachment of the 13th Machine Gun Battalion at Camp Logan, Texas. He sailed to England as a Sergeant in 13th Machine Gun Battalion 5th Division aboard the troopship *U.S.S. Leviathan* (seized German liner *Vaterland*) on March 4, 1918. As his emergency contact, he listed his mother in Bloomington, Maryland. From England, he and his Battalion were sent over to France.

Kline was transferred to Company D 14th Machine Gun Battalion (still in the 5th Division) on May 9, 1918. He served in the Anould Sector and in the St. Die Sector. He attended an Army Candidate School (Officer) in France from July 26 to September 30, 1918. On that date he was given a discharge in order to accept a commission. Kline was commissioned a 2nd Lieutenant of Infantry on October 1, 1918 and assigned to the 11th Machine Gun Battalion 4th Division.

He was awarded the Distinguished Service Cross for his actions in the Meuse-Argonne Offensive. On October 12, 1918, the 39th Infantry occupied a line in the northern edge of the wooded area known as the Bois de Forêt, north/northeast of the village of Nantillois and very near the village of Brieulles-sur-Meuse. Machine gun sections from the 11th Machine Gun Battalion including Kline and his guns were operating in support of the 39th Infantry and twice during the day had stopped German movements on nearby Hill 299. The enemy counter-attacked several times that morning and supported those attacks with heavy artillery fire and machine gun fire from nearby high ground positions. An enemy artillery barrage at half past noon caused the American Infantry to withdraw out of the area of the barrages. During the withdrawal, Kline remained at the forward position with his platoon Sergeant Leo D. Roberts, laying down suppressive fire with the one machine gun they had left and maneuvered their gun about the battlefield, resulting in their capture of an enemy machine gun and several prisoners. Their actions allowed the Infantry to withdraw in an orderly fashion and establish a new defensive perimeter further back. Although cited in different General Orders, both Kline

Michael D. Belis

and Roberts were awarded the Distinguished Service Cross for their actions.

Early in the morning of the next day, October 13, 1918, the enemy attempted a counterattack, but it was repulsed by the machine gunners of the 11th Machine Gun Battalion using their two remaining machine guns and one captured gun (presumably the gun Kline had captured the day before.) Kline was wounded during this encounter. He was in hospital from October 13, 1918 to February 17, 1919 when he rejoined his Company. He was awarded the French Knight of the Legion of Honor (Chevalier de la Légion d'Honneur) and the French Croix De Guerre with palm.

Kline served on occupation duty in Germany with the 4th Division and returned to the United States as part of an officers' detachment of the 11th Machine Gun Battalion aboard the troopship *U.S.S. Imperator* (surrendered German liner) leaving Brest, France August 3, 1919, and arriving at Hoboken, New Jersey August 10, 1919. The 1920 Census showed that on January 12, 1920 Kline was still with the 4th Division at Camp Dodge, Iowa. His date of discharge could not be found but most likely occurred during the demobilization of the Division in 1920-1921.

Sometime after 1920 he married his wife, Minnie. Kline served again on active duty in the Infantry section of the Officers Reserve Corps from February 4, 1924 to August 11, 1925. On May 21, 1929, his name was placed on the Emergency Officers' Retired List where it remained until he died. After 1932, he was awarded the Purple Heart for the wounds he received in World War One. When he registered for the draft in 1942, he indicated he lived in Seattle, Washington and was employed by the American Automobile Association. A city directory showed

Kline and Minnie living in Portland, Oregon in 1953 where he was employed as a salesman. Orval Lee Kline died at the age of 63 on October 10, 1956 and is buried in Willamette National Cemetery, Portland, Multnomah County, Oregon.

Max S. Kos
Private, United States Army
Army Service Number: 2004446
Company K 47th Infantry Regiment 4th Division
Date of Action: August 8 - 9, 1918
Location: Near St. Thibaut, France
War Department General Orders 15 (1919)
Medal # 1071

CITATION
Private Kos volunteered to patrol the valley along the railroad tracks north of St. Thibaut for the purpose of locating machine gun nests. He was wounded early in the morning, but he remained in the valley until the next night, securing the information for which he was sent and killing two Germans.

DETAILS

Max Samuel Kos was born in Fort Wayne, Indiana on February 11, 1895, the son of Harry and Josephine Kos. Before entering the Army, he was employed as an insurance agent in Toledo, Ohio and Indianapolis, Indiana. He was drafted into the Army as a Private on March 29, 1918 and took his initial training at Camp Taylor, Kentucky. He was assigned to Company K 47th Infantry at Camp Greene, North Carolina on April 27, 1918. Kos sailed to France as a Private in Company K 47th Infantry 4th Division aboard the Italian troopship *Caserta* on May 10, 1918. As his emergency contact, he listed his mother Josephine in Columbus, Ohio. He served in the Aisne-Marne Offensive.

Kos was awarded the Distinguished Service Cross for his actions on two days in the Vesle Sector. From August 7-12, 1918, the 47th Infantry made a determined attempt to cross the Vesle River and seize the village of Bazoches, directly across the river from the village of St. Thibaut. Progress was impeded not only by German artillery but also by heavy concentrations of enemy machine guns skillfully concealed in the buildings and ruins in Bazoches and in the area between the village and the river. At the time of his award action, Kos' Battalion was dug in along the sunken road east of St. Thibaut, across the river from Bazoches. On August 8, 1918, Kos volunteered to scout for German machine gun positions and was wounded early that morning, yet he continued into the next day and night to search for and record locations of enemy guns in order to aid his Battalion in their advance. During his scouting mission he also killed two German soldiers.

Kos was awarded the French Croix De Guerre with gold star. He was promoted to Private 1st Class on November 12, 1918. He served on occupation duty in Germany with the 4th

Division and returned to the United States as a Private 1st Class in Company K 47th Infantry and a member of the Brest Special Casual Company # 1725 of soldiers destined for early discharge aboard the troopship *U.S.S. Zeppelin* (surrendered German liner) leaving Brest, France, June 17, 1919 and arriving at Hoboken, New Jersey June 27, 1919. He was discharged on July 3, 1919.

Kos married Brenta Higman in either October or November 1920. The 1930 Census showed him divorced and living in Indianapolis, Indiana where he was employed as an insurance salesman and indicated he was a veteran of the World War. In 1935 he was living in Chicago, Illinois. The 1940 Census showed Kos living in the Veterans Administration Facility in Jefferson, Montgomery County, Ohio. When he registered for the draft in 1942, he indicated he was living in Columbus, Franklin County, Ohio where he was unemployed and had a scar on his right thigh (presumably from his wound in World War One). Max S. Kos died at the age of 57 on June 27, 1952 and is buried in South Mound Cemetery, New Castle, Henry County, Indiana.

Jacob Kreis

Private, United States Army
Army Service Number: 2024430
Company I 47th Infantry Regiment 4th Division
Date of Action: August 10, 1918
Location: At St. Thibaut, France
War Department General Orders 147 (1918)
Medal # 6087

CITATION

Accompanied by another soldier Private Kreis penetrated the enemy's lines and patrolled a sector from the north bank of the River Vesle to the town of Bazoches. These two men entered an enemy dugout and killed two Germans, at the same time locating a machine gun emplacement.

DETAILS

Jacob Kreis was born in Russia on February 21, 1894. In his registration for the draft in June 1917 he indicated that he was single, lived in Sheboygan, Wisconsin and worked as a laborer for the Garton Toy Company in Sheboygan. He also indicated he had declared his intention to become a naturalized American citizen. The date of his entry into the Army could not be found. He sailed to France as a Private 1st Class in Company I 47th Infantry 4th Division aboard the Italian troopship *Caserta* on May 10, 1918. As his emergency contact, he listed a friend, Henry Gross in Sheboygan, Wisconsin. Kreis served in the Ainse-Marne Offensive.

Though cited in different orders, Kreis and Corporal George

N. Brigham also of Company I 3rd Battalion 47th Infantry both were awarded the Distinguished Service Cross for actions they performed together on August 10, 1918, in the Vesle Sector. After an unsuccessful attack against the village of Bazoches on August 9, the Battalion pulled back across the Vesle River near the village of St. Thibaut and dug in. During the day of August 10, aerial reconnaissance indicated the enemy may have evacuated Bazoches. Regimental headquarters directed 3rd Battalion to send out five patrols on the night of August 10 to confirm if that was the case. Each patrol consisted of a non-commissioned officer and a private. The men in the patrols were equipped with pistols, gas masks, and canteens only. They were to cross the river at five different points and scout toward the main road (Route Nationale 31) to the north of Bazoches. Four of the patrols could not cross the river due to enemy defenses at their assigned crossing points.

Brigham and Kreis were the only patrol to make it across the river and were on the extreme right of the 47th Infantry's area of responsibility and into the area of the 59th Infantry. They made it to about 300 yards south of Route Nationale 31 and were prevented from going further by heavy enemy traffic on the road and enemy troop movements in and around Bazoches. On their return to the river, while moving through some woods, Brigham and Kreis encountered a German dugout and fought with the inhabitants. Brigham was shot through the neck during the engagement. He and Kreis made their way into the lines of the 59th Infantry where Brigham received medical treatment for his wound and remained at the aid station.

Kreis reported back to their Company with the information they had obtained during their patrol. Kreis had a German Luger pistol he had captured. He also had a shoulder strap he tore off the uniform of an enemy soldier he had shot which showed the German's unit marking. The details he reported to his Battalion Headquarters about the enemy's troop dispositions, machine gun emplacements, and movements were also dispatched to the 59th Infantry on their right and the French Army unit on their left. The next morning the enemy attacked 3rd Battalion, who were ready, thanks to Brigham and Kreis' patrol, and therefore were able to repulse the attack.

Kreis was also awarded a Citation Star and at some time after his award actions was promoted to Corporal. He served in the Toulon Sector and the St. Mihiel Offensive. He is indicated in the official records of the Quartermaster General as dying from wounds received in action on September 28, 1918 in the Meuse-Argonne Offensive. On September 28, the 47th Infantry led the right side of the 4th Division attack for the third day in a row and that day battled its way through the tangled underbrush of the wooded area known as the Bois De Brieulles. Enemy machine gun nests were assaulted and eliminated, and the Regiment stopped on a line close to the northern edge of the woods. Some time during the engagement in the Bois De Brieulles, Kreis received mortal wounds.

Kreis was buried in Plot A Grave # 54 in the American battlefield cemetery # 687 at the village of Souhesme-la-Grande.

His Distinguished Service Cross was presented to Henry Gross, Administrator. (The Register of Burials of Deceased American Soldiers, 1917 – 1922, Office of the Quartermaster General indicated that Henry Gross was the Great Uncle of Jacob Kreis.)

On April 5, 1921, Kreis was reinterred in Plot B Grave 29 in cemetery # 687. On October 25, 1921 his body was disinterred for preparation and shipment. His remains were returned

Michael D. Belis

to the United States aboard the United States Army Transport *U.S.A.T. St. Mihiel* leaving Antwerp, Belgium December 4, 1921 and arriving at New York City December 16, 1921. The Office of the Quartermaster General indicated that his remains were received on January 5, 1922 by John Ballhorn, Undertaker at Sheboygan, Wisconsin. No further details for Jacob Kreis could be found.

Ralph E. Ladue

Second Lieutenant, United States Army
Company A 11th Machine Gun Battalion 4th Division
Date of Action: July 19, 1918
Location: Near Chouy, France
War Department General Orders 1 (1934)
Medal # No assigned number discovered

CITATION

When the advance of the infantry regiment to which he was attached was temporarily halted by direct machine gun frontal fire and by strong enfilading fire from an enemy machine gun nest in a stone tower on a flank, Lieutenant Ladue voluntarily made a reconnaissance in advance of the front-line, crawling from shell hole to shell hole under heavy enemy fire. He then gallantly led two squads of machine guns into a position of great danger about 100 yards in front of the firing line. By his extraordinary coolness and tactical skill, so directed the fire of his guns that the machine gun nest was silenced, thereby permitting the advance of the entire firing line.

DETAILS

Ralph Ellsworth Ladue was born in Stillwater, New York City, New York on August 13, 1890, the son of William and Emma Ladue. He graduated from Colgate University and Albany Law School. Ladue enlisted as a Private in Company D 7th Infantry New York National Guard on October 19, 1914 and served with that organization on Mexican Border Service in 1916. Prior to serving on the Mexican Border, he married Mary Gaffney in 1916. He rose to Corporal in the National Guard and was discharged on June 11, 1917. He attended Officers Training Camp at Plattsburg, New York and on August 15, 1917, was commissioned a 2nd Lieutenant in the Officers Reserve Corps. Ladue was assigned to the 39th Infantry as it was forming up at Syracuse, New York. After he and the Regiment moved to Camp Greene, North Carolina he was transferred to the 11th Machine Gun Battalion. Ladue sailed to France as a 2nd Lieutenant in Company A 11th Machine Gun Battalion 4th Division aboard the troopship *U.S.S. Rijndam* (former Dutch liner) on May 10, 1918. As his emergency contact, he listed his brother Fred in Washington, D.C. In France in June 1918, the Companies of his Battalion were issued their machine guns.

Ladue was awarded the Distinguished Service Cross for his actions in the Aisne-Marne Offensive. On July 19, 1918, the second day of the Offensive, Ladue and his Company were operating in direct support of 3rd Battalion 39th Infantry. In the advance, 3rd Battalion captured the village of Noroy-sur-Ourcq and attacked in the direction of the village of Chouy. Enemy machine gun nests held up the Americans while enemy artillery inflicted casualties. To make matters worse, a French barrage which was intended to precede the American attack was late in being fired. By the time it was fired, the 39th Infantry had already taken the positions the artillery was directed to hit, and the shells fell amidst the ranks of the American soldiers. The Battalion moved forward out of the barrage and as it advanced, Ladue and his machine guns were instrumental in clearing the way for the Infantry.

At one point, the advance of the 39th Infantry was stopped by direct fire from a German machine gun nest and the American soldiers sought cover. Ladue crawled out in front of the halted Infantry, and under heavy enemy fire identified the location of an enemy machine gun which was preventing the Infantry from advancing. He then returned under fire to his section of two squads with their French Model 1914 Hotchkiss machine guns and led them into an exposed forward position from which they were able to fire upon the enemy and eliminate the German gun. His actions allowed the American attack to resume.

Ladue was awarded the French Croix De Guerre with gilt star. On July 31, 1918 he was lightly wounded near the wooded area known as the Forêt de Fère. He served in the St. Mihiel Offensive and was awarded a Citation Star. At an unknown date he was promoted to 1st Lieutenant. Ladue was severely wounded on September 26, 1918, the first day of the Meuse-Argonne Offensive. Because of his wounds he was evacuated back to the United States as a 1st Lieutenant of the 11th Machine Gun Battalion aboard the *U.S.S. Zeelandia* (former Dutch passenger ship) arriving at Newport News, Virginia on November 8,

1918. On the return voyage as his emergency contact, he listed his wife in Brooklyn, New York. Ladue was discharged on March 12, 1919.

He returned to New York City and went to work for the John David Company. He and his wife Mary had a son born in 1924 and another son born in 1929. The 1930 Census showed him, his wife, and two sons living in Yonkers, New York where he was employed as a manager at the John David Company and indicated he was a veteran of the World War. In April 1932, Ladue's Citation Star was converted to the Silver Star Medal. That same year he was awarded a Purple Heart with oak leaf cluster for the wounds he received in the war. On August 18, 1932, he was awarded the Conspicuous Service Cross (#2195) by the State of New York for his World War One service.

Ladue did not receive the Distinguished Service Cross until 1933. The reason his decoration was so long in coming was because the recommendation for it had been destroyed in the fighting at the Vesle River during the war. Through the efforts of the 4th Division Association, new affidavits substantiating the original recommendation were submitted to the Army who then approved the award. The medal was pinned on him in a ceremony at Fort Jay on Governor's Island, New York on December 4, 1933, by retired Major General George H. Cameron, who commanded the 4th Division during World War One. Many veterans of the 4th Division from the State of New York were present at the ceremony. About one month later, the official orders for the decoration were published in the first General Orders of the War Department for 1934. Ladue was also awarded the French Knight of the Legion of Honor (Chevalier de la Légion d'Honneur) by the French government.

The 1940 Census showed Ladue and his family still in

Yonkers, New York where he was employed as a merchant. He eventually became Vice President and then President of the John David Company. He was back in Army uniform during World War Two as a Colonel and was the Director of Personnel Services Division of 2nd Service Command operating in New York. For his service in World War Two he received the Legion of Merit. Ralph E. Ladue died at the age of 86 on March 13, 1977 at Delray Beach, Palm Beach, Florida.

Michael D. Belis

Edward R. Lawless
Sergeant Major, United States Army
Army Service Number: 556118
Headquarters Company 39th Infantry Regiment 4th Division
Date of Action: July 18, 1918
Location: Near Tröesnes, France
War Department General Orders 44 (1919)
Medal # 1060

CITATION

When it had become necessary to send an urgent message to the battalion base company, Sergeant Major Lawless, though under fire for the first time, voluntarily took the message across an open field, a distance of 500 yards. It seemed almost impossible to get through the murderous fire but knowing the importance of the message, Sergeant Major Lawless ventured through, rather than take the longer yet safer route. He completed his mission, returning over the same course.

DETAILS

Edward Russell Lawless was born in Leominster, Worcester County, Massachusetts on September 28, 1892, the son of John and Katherine Lawless. When he registered for the draft in May 1917, he indicated that he was at the time a candidate at the Reserve Officers Training Camp at Plattsburgh Barracks, New York. Lawless did not graduate from the Camp. He entered the Army as an enlisted man on September 22, 1917. Lawless sailed to France as a Battalion Sergeant Major in Headquarters Detachment 39th Infantry 4th Division aboard the Italian troopship *Duca D'Aosta* on May 10, 1918. As his emergency contact, he listed his father John in Leominster, Massachusetts.

Lawless was awarded the Distinguished Service Cross for his actions on the first day of the Aisne-Marne Offensive. The Regiment was attached to the French 33rd Division and was ordered to attack from east of the village of Tröesnes, cross the narrow but deep Saviéres River, and advance toward the wooded area known as the Buisson de Cresnes. The French Infantry led the attack out of the trenches and before the 39th Infantry could follow as the second wave of the assault, the Germans opened up an intensive artillery and mortar barrage on the American positions. The result was the two Battalions of the 39th Infantry which were to lead the advance for the Regiment did not leave the trenches together but instead about an hour apart. This led to considerable difficulty in maintaining communications between the two Battalions and also with the Battalion in reserve, which was to follow later.

Though he could have delegated a junior rank to deliver a vital message, Lawless performed the dangerous job himself, setting a courageous example to the men of the Regiment on their first day of battle. Instead of trying to find a safer but longer route around the deadly enemy fire he charged right through it to carry out his mission and returned through that same fire to get back to his unit, awing and inspiring the soldiers of the 39th Infantry with his bravery.

Lawless served in the Vesle Sector, the Toulon Sector, the St. Mihiel Offensive, and the Meuse-Argonne Offensive. He served on occupation duty in Germany with the 4th Division and returned to the United States as one of two Regimental Sergeants Major of the 39th Infantry with Headquarters Detachment 39th Infantry aboard the troopship *U.S.S. Leviathan* (seized German liner *Vaterland*) leaving Brest, France July 30, 1919 and arriving at Hoboken, New Jersey August 6, 1919. He was discharged on August 12, 1919.

The 1920 Census showed him living in Leominster, Massachusetts with his mother and sister where he and his sister were employed as bookkeepers at a comb factory. The 1930 Census showed him still living at the family home and still employed as a bookkeeper and indicated he was a veteran of the World War. The 1940 Census showed Lawless living at his sister and brother in law's residence at Leominster, Worcester, Massachusetts where he was employed as a salesman in the heating and plumbing industry. When he registered for the draft in 1942, he indicated he was living at his own residence in Leominster and was employed by the American Felt Company. Edward R. Lawless died at the age of 64 on November 5, 1961.

John Legnosky

First Sergeant, United States Army
Army Service Number: 560215
Company L 58th Infantry Regiment 4th Division
Date of Action: October 4, 1918
Location: In the Bois de Malaumont, east of Brieulles
War Department General Orders 16 (1920)
Medal # 5801

CITATION

Although painfully wounded in the foot on 4 October 1918, First Sergeant (then Sergeant) Legnosky remained on duty with his company. After his company commander had been killed, he assumed command of the company and efficiently led it in action until the unit was relieved on October 7, 1918. He repeatedly exposed himself to heavy fire in order to better control his men.

Michael D. Belis

DETAILS

John Legnosky was born in Leisenring, Fayette County, Pennsylvania on January 21, 1900, the son of Alex and Elizabeth Legnosky. He lied about his age to enlist as a Private (at the age of 16) in Company L 4th Infantry (Regular Army) on November 18, 1916. When elements of the 4th Infantry were used to create the 58th Infantry in June 1917, Legnosky thus became one of the first members of the 58th Infantry. He was promoted to Corporal on August 4, 1917 and to Sergeant on January 6, 1918. He sailed to England as a Sergeant in Company L 58th Infantry 4th Division aboard the British troopship *Themistocles* on May 11, 1918. As his emergency contact, he listed his mother in Hadley, Pennsylvania. From England he and his Company were sent over to France. Legnosky served in the Aisne-Marne Offensive, the Vesle Sector, the Toulon Sector and the St. Mihiel Offensive.

He was awarded the Distinguished Service Cross for his actions in the Meuse-Argonne Offensive. On October 4, 1918, Company L 58th Infantry attacked through the wooded area known as the Bois de Malaumont east of the village of Brieulles-sur-Meuse. Legnosky was wounded by shrapnel in his left foot that day but remained with his Company. The Executive Officer of the Company, 2nd Lieutenant Thurman G. Flanagan, was also wounded that day and died the next day October 5. On October 5 the Company Commander, Captain William H. Humphreys, was killed and the last remaining officer of Company L, 1st Lieutenant Floyd E. Fewell, was wounded. Eighteen-year-old Sergeant John Legnosky, though wounded, assumed command of Company L 58th Infantry on October 5 and led the Company in the defense against German counterattacks, consistently moving about in the open under fire to coordinate the Company's defenses until relieved two days later.

On January 31, 1919, Legnosky was promoted to 1st Sergeant. He served on occupation duty in Germany with the 4th Division and returned to the United States as 1st Sergeant of Company L 58th Infantry aboard the troopship *U.S.S. Mount Vernon* (captured German liner *Kronprinzessin Cecilie*), arriving at Hoboken, New Jersey on August 1, 1919.

He was discharged on November 7, 1919 in order to reenlist. The 1920 Census showed Legnosky serving as a 1st Sergeant with the Provisional Regiment of the 4th Division at an unnamed camp at Gary, Indiana. The date of his final discharge could not be found but occurred most likely during the demobilization of the Division in 1920-1921. In the 1920's, Legnosky completed two years of studies at Pennsylvania State College in Agricultural Economics. He married sometime after January 1920. His wife Katherine died in 1934.

Legnosky was back in uniform during World War Two serving in the United States Navy Seabees as a Carpenter's Mate from May 1, 1942 to July 7, 1945. He served in two deployments overseas, first with the 28th U.S. Naval Construction Battalion in Iceland and then with the 141st U.S. Naval Construction Battalion in Hawaii, Samar, and Kwajalein in the Pacific. He left the Navy as a Chief Carpenter Mate in 1945.

John Legnosky died of a heart attack at the age of 56 on July 6, 1956 and is buried in Sheakleyville Cemetery, Sheakleyville, Mercer County, Pennsylvania.

Michael D. Belis

Fred Adcook Lieuallen

Captain, United States Army
Medical Detachment 47th Infantry Regiment 4th Division
Date of Action: July 28 – 31 and August 6 - 12, 1918
Location: Neary Sergy, France and at St. Thibaut, France
War Department General Orders 53 (1920)
Medal # 6250

CITATION

Captain Lieuallen operated a dressing station for two days under unusually heavy enemy fire. Our front line was for a time bent back by the enemy, thus exposing his position to capture by the enemy; he refused to leave the dressing station and continued to attend to the needs of 100 wounded men until the lost ground was retaken by our troops. This officer performed gallant service also at St. Thibaut, France, August 6 - 12, 1918 while maintaining a dressing station with the advanced elements under heavy enemy fire.

DETAILS

Fred Adcook Lieuallen was born in Adams, Umatilla County, Oregon on November 6, 1882, the son of James and Lucy Lieuallen. He married Myrtle Hanks on November 14, 1908. He was a physician before entering the Army and received a commission as a 1st Lieutenant in the Officers Reserve Corps (Medical Department) on June 4, 1917. Lieuallen entered active duty on June 23, 1917. At some time before going overseas he was promoted to Captain. He sailed to France as a Captain in Medical Detachment 47th Infantry 4th Division aboard the

Italian troopship *Caserta* on May 10, 1918. As his emergency contact, he listed his wife Myrtle in Portland, Oregon.

Lieuallen was awarded the Distinguished Service Cross for his actions in two separate engagements. The first occurred in the Aisne-Marne Offensive during the battle for the village of Sergy July 28-31, 1918. Casualties in the 47th Infantry Regiment were extremely heavy in the attack against Sergy and Lieuallen set up his aid station very close to the front lines in order to allow the wounded to be carried the shortest possible distance before receiving surgical aid. This placed him and his assistants in range of enemy artillery and machine gun fire and at risk of enemy counterattacks but saved numerous lives among the wounded. At Sergy, he turned the basement of an old mill in the village into a temporary hospital in which at one point he had over 200 wounded being treated (twice the number as mentioned in his Citation) while Americans and Germans fought each other in the streets outside. During an enemy counterattack, the American Infantry had to pull back, but Lieuallen refused to be evacuated with them and instead remained at his aid station/hospital caring for the wounded, completely exposed to being captured by the Germans until his Regiment was able to counterattack and retake their positions.

The second action mentioned in his Citation was in the Vesle Sector. In the 4th Division attack at the Vesle River and in the attack against the village of Bazoches on the other side of the river August 6-12, 1918 his Regiment again suffered a high rate of casualties. Lieuallen set up his aid station at the village of St. Thibaut which was near the river and under constant fire from the Germans during the entire battle. Of the fourteen men serving under Lieuallen in his section of the Medical Detachment of the 47th Infantry, three were killed during the war, two of them being killed in the above actions. Seven others (which included Lieuallen) were wounded, for a total casualty rate in his detachment of 66.6% killed or wounded, attesting to the fact that Lieuallen kept his detachment operating at the front, right alongside the Infantry.

He was also awarded a Citation Star for his actions at Sergy and at St. Thibaut. At an unknown date after the actions in his Citation, Lieuallen was wounded and pulled out of frontline service. Because of his wounds, he returned to the United States several months ahead of his Regiment as part of a detachment of sick and wounded in St. Aignan Casual Company # 456 aboard the troop transport *U.S.S. Mongolia* leaving St. Nazaire, France January 20, 1919 and arriving at Hoboken, New Jersey January 30, 1919. He was discharged on February 20, 1919.

Lieuallen resumed his medical practice in 1920 in Pendleton, Umatilla County, Oregon, living there with his wife Myrtle, his daughter and his son. The 1930 Census showed them living in Bend City, Deschutes County, Oregon where he was a physician and surgeon and indicated he was a veteran of the World War. After 1932, he was awarded the Purple Heart for wounds he received in World War One. The 1940 Census showed him and Myrtle still in Bend City where his employment was indicated as a medical doctor in private practice. Fred Adcook Lieuallen died at the age of 72 on April 2, 1955 and is buried in Pilot Butte Cemetery, Bend, Deschutes County, Oregon.

Joe Limon

Private, United States Army
Army Service Number: 2268298
Company M 47th Infantry Regiment 4th Division
Date of Action: August 10, 1918
Location: Near St. Thibaut, France
War Department General Orders 32 (1919)
Medal # 7135

CITATION

Wounded in a scouting mission, Private Limon remained on observation until he had gained valuable information. After having his wound dressed, he returned to duty and made several trips to the flank regiments, each time bringing back valuable information for our own troops and of the enemy. He later voluntarily located a minenwerfer and heavy mortar emplacement from which the enemy was firing on neighboring troops.

DETAILS

Joe Limon was born in Bilboa, Spain on May 2, 1892. Prior to entering the Army, he was employed as a waiter for the Alaska Steamship Company aboard the steamship *S.S. Northwestern*, based out of Seattle, Washington and making journeys up and down the western seaboard from Seattle to ports in Alaska. On his registration for the draft in 1917, he indicated he was single, a member of the Cooks & Waiters Union, resided in Seattle, and was not an American citizen. Limon sailed to France as a Private in Company M 47th Infantry 4th Division aboard the Italian troopship *Caserta* on May 10, 1918. As his emergency contact, he listed a friend Antone Uhling in Seattle, Washington. Limon served in the Aisne-Marne Offensive.

He was awarded the Distinguished Service Cross for his actions in the Vesle Sector. On August 10, 1918 the Germans fired an extremely heavy artillery barrage against the 47th Infantry who were located in and around the village of St. Thibaut at the Vesle River. The day was also marked by increased enemy aerial activity. The Americans failed to take the village of Bazoches, across the river from St. Thibaut and the Germans attempted to reorganize and reconsolidate their positions in the area. Limon scouted the enemy and, though wounded, he remained on his mission gathering vital information about German defensive positions and troop strength. After returning and getting his wound bandaged, he carried out more scouting missions and located an enemy mortar emplacement containing both minenwerfer (trench mortar) and heavy mortar weapons. His report allowed accurate counterfire to be dropped upon the German position.

The Citation recorded for the Distinguished Service Cross for Limon in the official history of the 47th Infantry in World War One is more expanded and detailed than the official War Department Citation

"LIMON (2268298), JOE, Private, Co. M. This man was wounded on a scouting mission at St. Thibaut, on August 10, 1918. He remained in observation until he gained the information for which he had been sent. Upon returning from this mission, his wound was dressed, and he was directed by the surgeon to get on the ambulance and go to the hospital. Instead, he reported to his battalion commander stating he wished to remain on duty. After that he was sent on two occasions to gain liaison with the French regiment on our left. He did this in the most effective and praiseworthy manner, each time bringing back most valuable information of the enemy and of our neighboring troops. Upon two voluntary and self-appointed missions, he crawled along the Vesle River west of Bazoches and located a minenwerfer emplacement and a heavy mortar emplacement from which the enemy was firing heavily on the French troops on our left." [14]

Limon was killed in action during the Meuse-Argonne Offensive on September 30, 1918. On that date his Battalion was holding defensive positions in the wooded area known as the Bois de Brieulles and taking heavy fire from enemy artillery situated across the Meuse River to the northeast. His body was not recovered and his official status with the American Battle Monuments Commission is Missing in Action.

His Distinguished Service Cross was presented to his father Erminic Limon.

The name of Joe Limon is inscribed on the Tablets of the Missing at the Meuse-Argonne American Cemetery, Romagne-sous-Montfaucon, Departement de la Meuse, Lorraine, France.

Michael D. Belis

Luther E. Lindahl
Sergeant, United States Army
Army Service Number: 558271
Company I 47th Infantry Regiment 4th Division
Date of Action: September 28, 1918
Location: Near Bois de Brieulles, France
War Department General Orders 46 (1919)
Medal # 2026

CITATION
Sergeant Lindahl charged an enemy machine gun which was inflicting heavy losses upon our troops and delaying the advance. He wounded the gunners and captured the gun, thereby enabling our advance to continue.

DETAILS
Luther E. Lindahl was born in Sheffield, Warren County, Pennsylvania on July 6, 1899, the son of Charles and Hannah Lindahl. He enlisted as a Private in Company I 47th Infantry (Regular Army) on June 6, 1917 at Fort Slocum, New York. He was promoted to Private 1st Class on August 20, 1917, to Corporal on January 24, 1918, and to Sergeant on April 24, 1918. Lindahl sailed to France as a Sergeant in Company I 47th Infantry 4th Division aboard the Italian troopship *Caserta* on May 10, 1918. As his emergency contact, he listed his father Charles in Sheffield, Pennsylvania.

Lindahl served in the Aisne-Marne Offensive and was awarded a Citation Star for his actions in the Vesle Sector. The Citation read:

"LINDAHL (558271), LUTHER E., Sergeant, Co. I. During the battle of August 8, 1918, Sergeant Lindahl handled his men with exceptional coolness and bravery while under machine gun and artillery fire. He was in command of an advance patrol and did very good work in obtaining information of the enemy. On the night of August 9, 1918, Lindahl, with a patrol of four men, did good work in advancing to the enemy's front and obtaining valuable information of the enemy. This was a voluntary act."[15]

He served in the Toulon Sector and the St. Mihiel Offensive.

Lindahl was awarded the Distinguished Service Cross for his actions in the Meuse-Argonne Offensive. On September 28, 1918, the third day of the Offensive, the 47th Infantry fought its way through the tangled brush of the wooded area known as the Bois de Brieulles. In the brutal fighting in the woods, Lindahl single handedly assaulted and eliminated a German machine gun nest which had been holding up his Company's advance.

Lindahl was one of 20 soldiers of the 4th Division and one of only 399 soldiers of the American Expeditionary Forces to be awarded the Italian War Merit Cross (Croce al Merito di Guerra). The Italian War Merit Cross was presented to him on June 11, 1919 in a 4th Division awards ceremony at Remagen, Germany.

He served on occupation duty in Germany with the 4th Division and was promoted to Supply Sergeant on June 15, 1919. Lindahl returned to the United States with his Company aboard the troopship *U.S.S. Mobile* (seized German liner *Cleveland*) leaving Brest, France July 16, 1919 and arriving at Hoboken, New Jersey July 27, 1919. He was discharged at Camp Dix, New Jersey on August 2, 1919.

Michael D. Belis

The 1920 Census showed him living with his parents in Sheffield, Pennsylvania where he was employed as a railroad worker. The 1930 Census showed him living in Clarendon, Warren County, Pennsylvania with his wife Alma and their daughter where he was employed as a laborer at an oil refinery and indicated he was a veteran of the World War. The 1940 Census showed him, his wife and daughter still in Clarendon where he was employed as a timekeeper at an oil refinery. His wife Alma died in 1955 and he later married Ethel Burritt. Luther E. Lindahl died of a coronary occlusion at the age of 60 on May 24, 1960 and is buried in Oakland Cemetery, Warren, Warren County, Pennsylvania.

Clyde H. Lindsey

Private, United States Army
Army Service Number: 3489185
Company A 59th Infantry Regiment 4th Division
Date of Action: September 29, 1918
Location: Near Bois de Brieulles, France
War Department General Orders 46 (1919)
Medal # 3311

CITATION

Disregarding personal safety, Private Lindsey, in the performance of his duties as runner carried repeated messages across greatly exposed area which was subjected to fierce artillery and machine gun fire. He succeeded after another runner had been killed in the attempt.

DETAILS

Clyde Holloway Lindsey was born in Maywood, Marion County, Missouri on September 23, 1894, the son of George and Angeline Lindsey. Before entering the Army, he was a farm laborer living in Cambridge, Illinois. On November 21, 1917 he married Bessie Leonard. He entered the Army on May 28, 1918. Lindsey sailed to England as a Private in Company 30 (Infantry) of the Camp Gordon July Automatic Replacement Draft aboard the British transport *Orca* on July 22, 1918. As his emergency contact, he listed his wife in Maywood, Missouri. From England he was sent over to France where he underwent more training and was eventually assigned as a replacement to Company A 59th Infantry 4th Division.

Lindsey was awarded the Distinguished Service Cross for his actions in the Meuse-Argonne Offensive. On September 29, 1918 the 8th Brigade, including the 59th Infantry, which had been held in reserve for the first two days of the Offensive now spearheaded the advance. Lindsey's Battalion which was on the right flank of the attack met strong resistance in the northern edge of the wooded area known as the Bois de Brieulles. Skirting the edge of the woods, his Battalion reached the enemy trenches on the slope of Hill 281 and were stopped from any further advance by heavy fire from across the Meuse River. Though a previous runner had been killed in an attempt to cross the battlefield, Lindsey assumed the mission and succeeded in repeated trips of delivering vital messages and keeping the lines of communication established.

Lindsey was one of three soldiers of the 4th Division who received the French Military Medal (Médaille Militaire) during World War One and one of only 304 soldiers of the American Expeditionary Forces to be given that award. He was twice awarded the French Croix De Guerre with palm.

At some time after his award action, he was promoted to Private 1st Class. He served on occupation duty in Germany with the 4th Division and returned to the United States as a Private 1st Class in Company A 59th Infantry aboard the troopship *U.S.S. Mount Vernon* (captured German liner *Kronprinzessin Cecilie*) leaving Brest, France July 24, 1919 and arriving at Hoboken, New Jersey August 1, 1919. He was discharged on August 9, 1919.

The 1920 Census showed Lindsey and his wife Bessie living in Osco, Henry County, Illinois where he was employed as a farm laborer. The 1930 Census showed him and Bessie with their two daughters and two sons living in Lacon, Marshall

County, Illinois where he worked as a clerk in a grocery store and indicated he was a veteran of the World War. The 1940 Census showed Lindsey and his family living in Akron, Summit County, Ohio where he worked as a ball punch operator. By 1942 he was working for the Goodyear aircraft factory in Akron, Ohio. Clyde H. Lindsey died at the age of 85 on February 20, 1979 and is buried in Rose Hill Burial Park, Akron, Ohio.

Michael D. Belis

Joseph Longowski
Private, United States Army
Army Service Number: 561582
Company L 59th Infantry Regiment 4th Division
Date of Action: October 4, 1918
Location: Near Bois de Fays, France
War Department General Orders 71 (1919)
Medal # 1328

CITATION
Under heavy fire while performing a mission as Battalion runner he encountered an enemy patrol of four men and forcing them to surrender brought them to the rear.

DETAILS
Joseph Longowski was born in Winona, Winona County, Minnesota on December 25, 1895. A city directory for Winona in 1912 showed Longowski working as a laborer for the Winona Railway and Light Company. He entered the Army on August 6, 1917. Longowski sailed to England as a Private in Company L 59th Infantry 4th Division aboard the armed British troopship *RMS Olympic* on May 5, 1918. As his emergency contact, he listed his mother Rosie Longowski in Winona, Minnesota. From England, he and his Company were sent over to France. Longowski served in the Aisne-Marne Offensive, the Vesle Sector, the Toulon Sector, and the St. Mihiel Offensive.

He was awarded the Distinguished Service Cross for his actions in the Meuse-Argonne Offensive. On October 4, 1918 the 4th Division attack against the wooded area known as the

Bois de Fays was resumed with the 58th Infantry leading the advance. Longowski's Battalion, the 3rd Battalion 59th Infantry which had been held in reserve pushed into the Bois de Fays to assist the 58th Infantry in late morning and established a line of positions from the southeastern part of the woods to the western part. While acting as a runner for 3rd Battalion in the Bois de Fays to Regimental Headquarters back in the southwestern part of the wooded area known as the Bois de Brieulles Longowski captured a German patrol of four men and brought them to his Headquarters as he delivered the messages he was carrying.

At some time after his award action Longowski was promoted to Corporal. He served on occupation duty in Germany with the 4th Division and returned to the United States with Company L 59th Infantry aboard the troop transport *U.S.S. Texan* leaving Brest, France July 24, 1919 and arriving at Norfolk Virginia August 5, 1919. He was discharged on August 12, 1919.

City directories for Winona, Minnesota in 1923 and 1925 showed Longowski and his wife Ella living at 466 High Forest in Winona where he was employed as a laborer for the Chicago and North Western Railway. No further details for Joseph Longowski could be found.

Charles J. Love

Private, United States Army
Army Service Number: 1678783
Company K 59th Infantry Regiment 4th Division
Date of Action: October 5, 1918
Location: Near Bois de Fays, France
War Department General Orders 71 (1919)
Medal # 7109

CITATION

Volunteering for a dangerous liaison mission, Private Love went out alone, crossing an open space for 400 yards which was swept by heavy machine gun fire. Going far in advance of our lines, he obtained the desired information and brought back a German prisoner. He was killed later in the performance of duty by shell fire.

DETAILS

Charles John Love was born in Owasco, Cayuga County, New York on September 27, 1892, the son of John and Etta Love. Before entering the Army, he was employed at a grocery store and then at the N.M. Sargents' Son's chair factory in Boonville, New York. When he registered for the draft in June 1917, he indicated he was employed as a farm laborer in Boonville. He was drafted into the Army as a Private on February 22, 1918 and was assigned to Company K 59th Infantry on May 1, 1918. Love sailed to England as a Private in Company K 59th Infantry 4th Division aboard the armed British troopship *RMS Olympic* on May 5, 1918. As his emergency contact, he listed his mother Etta in Boonville, New York. From England he and his Company were sent over to France. Love served in the Aisne-Marne Offensive, the Vesle Sector, the Toulon Sector, and the St. Mihiel Offensive.

He was awarded the Distinguished Service Cross for his actions in the Meuse-Argonne Offensive. On October 5, 1918, elements of the 80th Division were on the left flank of the 4th Division. Love's 3rd Battalion of the 59th Infantry was holding a position in the wooded area known as the Bois de Fays. The 8th Brigade of the 4th Division, which included Love's Company K was ordered to assist the 80th Division in their advance. However, the 80th Division met too strong a resistance from the Germans and was not able to join up alongside the 4th Division. The day was then largely spent by the 59th Infantry in consolidating their position and scouting the enemy. Love volunteered to conduct a liaison mission to another unit by himself, leaving the woods and crossing an open area through heavy enemy fire. In the performance of his mission, he brought back a prisoner as well. Four days later on October 9, 1918, Love was killed by enemy artillery fire, the same day his Battalion was being withdrawn from the Bois de Fays to move to the rear for a rest.

Love was buried in Grave # 1 in the battlefield cemetery # C-230 at the village of Brieulles-sur-Meuse.

His Distinguished Service Cross was presented to his mother Mrs. Etta M. Love.

On June 7, 1919, Charles J. Love was reinterred in Plot 3 Section 48 Grave # 132 in the Argonne American Cemetery # 1232. On June 22, 1921, his body was disinterred for preparation and shipment. His remains were returned to the United States aboard the United States Army Transport *U.S.A.T.*

Michael D. Belis

Wheaton leaving Antwerp, Belgium August 6, 1921 and arriving at Hoboken, New Jersey August 20, 1921. He was one of 5,759 American dead being brought home aboard the *Wheaton* on that journey. Charles J. Love was laid to his final rest in Boonville Cemetery, Boonville, Oneida County, New York.

Murray R. MacKall

Captain, United States Army
Company C 4th Engineer Regiment 4th Division
Date of Action: August 4 - 5, 1918
Location: West of Fismes, France
War Department General Orders 147 (1918)
Medal # 1085

CITATION

Captain MacKall reconnoitered a section of the River Vesle in advance of the front line of infantry under continuous fire from machine guns and one-pounders. Proceeding alone for about one kilometer along the stream, despite the fact that German machine guns were located near the opposite bank, he continued his reconnaissance and selected several suitable sites, one of which was used the next night. Captain MacKall guided the working party through the enemy's barrage.

DETAILS

Murray Randolph MacKall was born in East Liverpool, Ohio on November 22, 1890, the son of Adam and Rosa MacKall. He attended Stanford University in California and became a lawyer. He was admitted to practice in the Supreme Court of California on May 23, 1916. MacKall was commissioned a 1st Lieutenant in the Officers Reserve Corps on July 19, 1917. He married Helena Pearsall on September 29, 1917. His marriage license indicated he was in military service stationed at Vancouver Barracks, Washington. He is mentioned in the official history of the 4th Engineers in World War One as being in Company

Michael D. Belis

A at least by January 1, 1918. On March 14, 1918 MacKall was promoted to Captain and given command of Company C 4th Engineers. He sailed to France with Company C 4th Engineers 4th Division aboard the troopship *U.S.S. Martha Washington* (seized Austrian liner) on April 30, 1918. As his emergency contact, he listed his wife in Enumclaw, Washington.

MacKall was awarded the Distinguished Service Cross for his actions on two days in the Aisne-Marne Offensive when the 4th Division was attempting to cross the Vesle River. On August 4, 1918, he personally scouted along the Vesle River west of the village of Fismes near the village of Ville Savoye in order to determine possible locations at which to emplace bridges, all the while under fire from German machine guns across the river. Though as a Captain he could have delegated this mission to a lesser rank, he undertook it personally. He carried out the mission alone and one of the locations he recorded was used to emplace a footbridge for the Infantry the next night. On August 5, he guided an engineer detachment through an artillery barrage to the river to work on emplacing that bridge. He was also awarded a Citation Star for his actions of August 4-5 in 4th Division General Orders 76 (1918).

(See the entry for William B. Beach for an overview of the kinds of activities carried out by the soldiers of the 4th Engineers involved in building the bridges across the Vesle River.)

MacKall was also awarded the Belgian Knight of the Order of Leopold II (Chevalier de L'Ordre Leopold II) as one of only two soldiers in the 4th Division to receive this decoration during World War One. (The other soldier was Sergeant Raymond D. Robertson, also from the 4th Engineers and also a Distinguished Service Cross recipient.) MacKall was one of only 47 soldiers of the American Expeditionary Forces to receive the Belgian Order of Leopold II.

MacKall served on occupation duty in Germany with the 4th Division and returned to the United States as a Captain of the 4th Engineers in a detachment of officers from various units aboard the troopship *U.S.S. Zeppelin* (surrendered German liner) leaving Brest, France July 19, 1919 and arriving at Hoboken, New Jersey July 29, 1919. He was discharged on August 14, 1919.

The San Francisco city directory for 1920 showed MacKall and his wife Helena living in San Francisco, California where he was employed as a civil engineer for the California Railroad Commission. The 1930 Census showed him living with his wife and two daughters in Corte Madera, Marin County, California where he was still employed as a civil engineer for the California Railroad Commission and indicated he was a veteran of the World War. The San Francisco city directory for 1932 showed him living in Corte Madera and working as a hydraulic engineer for the California Railroad Commission. When he registered for the draft in 1942, he was still at the same address and still working for the California Railroad Commission. Murray R. Mackall died at the age of 72 on March 23, 1963 and is buried in Golden Gate National Cemetery, San Bruno, San Mateo County, California.

Robert A. Madden

Private, United States Army
Army Service Number: 2004650
Company I 47th Infantry Regiment 4th Division
Date of Action: July 29 - 30, 1918
Location: Near Sergy, France
War Department General Orders 74 (1919)
Medal # 7246

CITATION

Passing through heavy machine gun and artillery fire, Private Madden maintained liaison with adjacent units, displaying marked heroism in his work. He was mortally wounded in the performance of duty.

DETAILS

Robert Arthur Madden was born in Indianapolis, Marion County, Indiana on January 22, 1893, the son of Robert and Gertrude Madden. As a teenager he was a court room page. In 1915, he became a City Fireman in Indianapolis. He was drafted into the

Army as a Private on March 28, 1918 and was assigned to the 47th Infantry on April 27, 1918. Madden sailed to France as a Private in Company I 47th Infantry 4th Division aboard the Italian troopship *Caserta* on May 10, 1918. As his emergency contact, he listed his father Robert in Indianapolis, Indiana.

Madden was awarded the Distinguished Service Cross for his actions on two days in the Aisne-Marne Offensive. From July 29 - August 1, 1918, the 47th Infantry was engaged in heavy fighting at the village of Sergy. Madden was employed by his Company as a runner, doing the dangerous and important job of carrying messages and maintaining communications between units on the battlefield in the days before field radios. He had to move by himself across the battle zone, subjected to mortar, artillery, machine gun, and rifle fire in order to relay commands and information.

Madden's 3rd Battalion 47th Infantry was attached to the 168th Infantry Regiment of the 42nd Division for the battle at Sergy. On July 29, 1918, his Company I along with Company L was in the attacking wave while the other two Companies of the Battalion followed behind in support. On that day, Madden's Battalion had four successive officers commanding the Battalion. The first three were wounded and the fourth was killed but the Battalion occupied Sergy. The next day the Battalion attacked out of the village to the north, but a German counterattack drove them back through the village to the Ourcq River. All of the ammunition for the Battalion's French Chauchat automatic rifles had been expended and some of the soldiers were using captured German rifles and grenades, but the Battalion reorganized, counterattacked, and re-occupied the village. During this phase of the battle, Madden carried messages through deadly enemy fire.

Michael D. Belis

He served in the Vesle Sector. On the morning of August 7, 1918, his Battalion was in a support position on the south side of the Vesle River and dug in east of the village of St. Thibaut while 2nd Battalion 47th Infantry led the attack against the village of Bazoches which was on the north side of the river. Fifteen minutes after the attack began, the Germans blanketed the area with an artillery barrage of high explosive, gas, and shrapnel. The bombardment covered the area in front of Bazoches all the way to the river and across the river into the village of St. Thibaut and lasted intermittently for about two hours. In the afternoon another barrage was fired on the Americans. During the day, between three and four thousand shells were fired by the enemy, one third of which were estimated to be gas. At some time during the day, Madden was wounded. He died of his wounds the next day on August 8, 1918.

Madden was buried in grave # 77 in the American battlefield cemetery # 361 at the village of Chierry.

His Distinguished Service Cross was presented to his father Robert N. Madden.

On June 3, 1919 Robert A. Madden was reinterred in Plot 1 Section P Grave 51 in the American Cemetery # 608 at Seringes-et-Nesles. On May 19, 1921 his body was disinterred for preparation and shipment. His remains were returned to the United States aboard the United States Army Transport *U.S.A.T. Wheaton* leaving Antwerp, Belgium June 19, 1921 and arriving at Hoboken, New Jersey July 2, 1921. He was one of 5,827 American dead brought home by the *Wheaton* on that journey. Robert A. Madden was laid to his final rest in Crown Hill Cemetery, Indianapolis, Marion County, Indiana.

John J. Madore
Private, United States Army
Army Service Number: 558115
Company G 47th Infantry Regiment 4th Division
Date of Action: August 9, 1918
Location: Near Bazoches, France
War Department General Orders 15 (1919)
Medal # 1120

CITATION
Private Madore volunteered to carry a message to an advance squad through heavy machine gun fire. After delivering the message and administering first-aid treatment to wounded men in the squad, he crawled up to the nearest enemy machine gun and put it out of action with a hand grenade.

DETAILS
John James Madore was born in North Sydney, Nova Scotia, Canada on June 1, 1894, the son of Thomas and Anne Madore. He immigrated to the United States through the port of New York on May 26, 1913. He filed a Declaration of Intention to become an American citizen with the U.S. Department of Labor Naturalization Service, District of Massachusetts on June 20, 1916. On his registration for the draft in June 1917 he stated he was living in Malden, Massachusetts, employed as a Steamfitter at the Converse Rubber Shoe Company in Malden, a citizen of Great Britain and that he had declared his intention to become a United States citizen. He entered the Army sometime after June 5, 1917. Madore sailed to France as

a Private in Company G 47th Infantry 4th Division aboard the troopship *U.S.S. Princess Matoika* (seized German liner *Prinzess Alice*) on May 10, 1918. As his emergency contact, he listed a friend Kathryn [sic] Flynn in Malden, Massachusetts. Madore served in the Aisne-Marne Offensive.

He was awarded the Distinguished Service Cross for his actions in the Vesle Sector. On August 9, 1918 elements of Madore's Battalion had crossed the Vesle River and patrols from it had advanced to within one hundred yards of the village of Bazoches. The area in front of the village was a hotbed of German machine gun nests and snipers and heavy fire was placed on the Battalion from that area, from the village itself, and from the heights to the north as well. Eventually the Battalion withdrew to the railroad running along the river and then across the river on the south side to positions near the village of St. Thibaut. Madore's actions took place when the Battalion was still on the north side of the river and near the village of Bazoches. After delivering a message to a squad situated ahead of the attacking line and bandaging their wounded, he single handedly assaulted and destroyed a German machine gun position with a hand grenade.

At unknown dates after his award actions, Madore was wounded twice and returned to duty both times. Also, at an unknown date he was promoted to Sergeant. He served on occupation duty in Germany with the 4th Division and returned to the United States as a Sergeant with Company G 47th Infantry aboard the troopship *U.S.S. Mobile* (seized German liner *Cleveland*) leaving Brest, France July 16, 1919 and arriving at Hoboken, New Jersey July 27, 1919. He was discharged on August 1, 1919.

At some time between 1919 and 1924 he married Catherine

Flynn. John J. Madore became a United States citizen on June 9, 1924. At the time he was issued his Certificate of Naturalization he listed his home address as 718 Salem St., Malden, Massachusetts, and his occupation as Electrician. John J. Madore died at the age of 42 at Boston, Massachusetts on June 24, 1936.

Michael D. Belis

Arthur H.G. Mallet
Second Lieutenant, Army of France
165th Infantry Regiment Army of France
Attached to 2nd Battalion 47th Infantry Regiment 4th Division
Date of Action: August 7, 1918
Location: On the Vesle River near Bazoches
War Department General Orders 4 (1923)
Medal # 1329

CITATION
While serving as liaison officer with the 2nd Battalion 47th U.S. Infantry, which led the attack against the enemy and in the face of stubborn resistance, crossed the Vesle River, seized a critical position north of that stream, and held tenaciously to it throughout the day. Lieutenant Mallet, under a heavy and continuous hostile fire, repeatedly went from one front line combat group to another, assisting materially in the successful conduct of the action by his courageous actions, suggestions, and professional skill. He rendered highly important services at a critical moment when the left of the line was sorely pressed by enemy counterattack and at all times by his soldierly conduct, inspiring courage, and high qualities of leadership was an heroic example to his comrades in arms. He was killed in action late in the afternoon.

DETAILS
Arthur Henri Gérard Mallet was born in Jouy-en-Josas, Yvelines, France on September 20, 1877. He entered the French Army in 1897. Mallet was one of a number of French Army

officers who served as advisors to the U.S. 4th Division during World War One. He was assigned as a Liaison Officer to the 2nd Battalion 47th Infantry.

Mallet was awarded the Distinguished Service Cross for his actions on August 7, 1918 in the Vesle Sector. From August 3-7, 1918, the 39th Infantry had struggled to cross the Vesle River and attack the village of Bazoches without much success and pulled back to the hills outside of the village on St. Thibaut on the south side of the river. The 2nd Battalion 47th Infantry was designated to take over the renewed attack against Bazoches. At about 3:45 a.m. on the morning of August 7, the Battalion moved through St. Thibaut and approached the river. Fifteen minutes later, the Germans opened up a terrible artillery barrage on them, firing gas and high explosive shells which lasted intermittently for about two hours. The Americans suffered considerable losses and fought through that artillery barrage and also machine gun fire, slowly gaining ground.

By noon, one Company of 2nd Battalion had crossed the river and were established in defensive positions facing Bazoches. At 2:15 p.m. the enemy counterattacked but the soldiers of 2nd Battalion held their ground and prevented the Germans from dislodging them. Mallet went across the river with the leading elements and throughout the day braved heavy fire, moved about the battlefield from one group of American soldiers to another and using the knowledge he had gained in four years of frontline service, imparted valuable suggestions and directions which aided immensely in the conduct of a successful operation. The soldiers of the Battalion to which he was attached were inspired by his courage and leadership. Another barrage of artillery was fired against the Americans that afternoon. In the face of heavy artillery, trench mortar and machine gun fire, more men from 2nd Battalion advanced across the river and by five o'clock in the afternoon there were about 350 of them established on the German side of the river. At that point in the fight, Mallet was struck down and killed by enemy fire.

His Distinguished Service Cross was presented to his father Frederic Mallet.

James Manning

Corporal, United States Army
Army Service Number: 568528
Company C 4th Engineer Regiment 4th Division
Date of Action: August 8, 1918
Location: Near St. Thibaut, France
War Department General Orders 46 (1919)
Medal # 1973

CITATION

Corporal Manning was one of four men who volunteered and swam the Vesle River for the purpose of doing work on the opposite bank necessary in the construction of a footbridge. With another soldier he succeeded in felling a large tree in the face of heavy machine gun and one-pounder fire after the remainder of the platoon had withdrawn.

DETAILS

(At least one version of the Citation to James Manning has the notation next to his name of "AKA: FRANK E. BELL". The compendium of Medal of Honor, Distinguished Service Cross, and Distinguished Service Medal awards, AMERICAN DECORATIONS published by the Adjutant General in 1927 lists his Citation under the name of James Manning and also has a one line entry for Frank E. Bell which contains his rank and unit and a statement that his service was rendered under the name James Manning. The official history of the 4th Engineers in World War One has him listed as James Manning and the Army Transportation Service passenger lists have him under the name of James Manning. In The Official Roster of Ohio Soldiers in the World War under the name James Manning is the notation "see Bell, Frank E." with the same serial number for whom the details are given that match all other details found for Manning.)

He was born on October 14, 1894. When he entered the Army, he indicated that he was born James Manning in Youngstown, Mahoning County, Ohio but after leaving the Army he would indicate his name as Frank E. Bell, his place of birth as Webster City, Iowa, and that he was the son of John A. and Rachel Bell.

The following details about his Army service will be given under the name of James Manning.

On his registration for the draft in Missoula, Montana in June 1917 he indicated he was single and his occupation as unemployed Miner. He enlisted as a Private (Regular Army) in Company C 4th Engineers at Fort George Wright, Washington on June 7, 1917. On February 21, 1918 he was promoted to Corporal. Manning sailed to France as a Corporal in Company C 4th Engineers 4th Division aboard the troopship *U.S.S. Martha Washington* (seized Austrian liner) on April 30, 1918. As his emergency contact, he listed his sister Mrs. Arthur Ashurt in Cornucopia, Alberta, Canada. Manning served in the Aisne-Marne Offensive.

He was awarded the Distinguished Service Cross for his actions in the Vesle Sector. After several days of attempting to get across the Vesle River and succeeding in establishing footholds on the north side of the river, the 4th Division on August 8, 1918 consolidated its gains and held in place as the Engineers worked feverishly to construct and maintain several foot bridges for the Infantry. While one Battalion of the 47th Infantry had made it across the river and was searching for possible avenues of approach in which to attack the village of Bazoches, most of the elements of the 4th Division's 7th Brigade were still on the south side of the river near the village of St. Thibaut.

Enemy artillery fire was almost incessant and machine gun fire from Bazoches and the high ground around it made the job of the Engineers a highly dangerous profession. Manning led a detachment of three men who together with him swam the river to the enemy occupied side and continued to work on building a footbridge across the river, even though most of the men in his platoon had pulled back under the heavy fire being brought to bear on their position. He and another soldier managed to chop down a tree large enough to fashion into a rudimentary footbridge and under his direction he and his detachment secured it in place across the river; the whole time under machine gun and one pounder (37mm) fire from the Germans.

He served in the Toulon Sector and the St. Mihiel Offensive. During the Meuse-Argonne Offensive in the first week of October, Manning and several buddies from Company C secured a German artillery gun and a sizeable load of ammunition that had been abandoned by the Germans during their retreat. Even though they were Engineers and not trained artillerymen, they figured out how to fire the gun and knew the general direction of the enemy lines. They put the gun into operation and for several days carried out their own little artillery war against the Germans. Enemy artillery located the gun and began to fire on its position. As the enemy shells began to land too close, Manning and his cohorts abandoned the gun and sought cover. One shell landed squarely on the gun and put an end to the "Engineer artillery." Manning and his fellow Engineers were happy in the knowledge that they had sent many high explosive shells back at the guys who had fired so many at them.

Manning served on occupation duty in Germany with the 4th Division and returned to the United States as a Corporal with his Company aboard the troopship *U.S.S. Von Steuben* (seized German liner *Kronprinz Wilhelm*) leaving Brest, France July 21, 1919 and arriving at Hoboken, New Jersey July 29, 1919. He was discharged on August 4, 1919.

He used the name of Frank Eugene Bell after leaving the Army in 1919. He has two service cards filed with the Veterans Bureau in the 1920's, one under the name of James Manning and another under the name of Frank Eugene Bell. All records found for him after he left the Army are under the name of Frank E. Bell.

He married Waive Drew in 1924. The 1930 Census showed him living with his wife Waive in Altamont, Klamath County, Oregon where he was employed as an electrician and indicated he was a veteran of the World War. In 1933 their son was born. The 1940 Census showed Bell living with his family in Tulelake, Siskiyou County, California where he was a farmer on his own farm. In 1957, he and his family moved to Mesa, Maricopa County, Arizona where he was employed as an electrician. Frank E. Bell died of chronic congestive heart failure at the age of 67 on October 27, 1961 and is buried in Mountain View Funeral Home and Cemetery, Mesa, Maricopa County, Arizona.

Richard Marcella

Bugler, United States Army
Army Service Number: 558037
Machine Gun Company 47th Infantry Regiment 4th Division
Note: The Citation for Richard Marcella gives his unit as Machine Gun Company. He was actually in Company F at the time of his award action and was not transferred to Machine Gun Company until later.
Date of Action: August 9, 1918
Location: Near Bazoches, France
War Department General Orders 46 (1919)
Medal # 6785

CITATION

Responding to a call for volunteers to destroy a hostile machine gun, Bugler Marcella, with two other soldiers, boldly went forward through machine gun fire and accomplished this mission.

DETAILS

Richard Marcella was born in New York City, New York on August 27, 1898, the son of Vincent (Vincenzo) and Gabrael Marcella. He enlisted as a Private in Company F 47th Infantry (Regular Army) at Fort Slocum, New York on June 7, 1917. He was promoted to Private 1st Class on September 13, 1917. He went with his Regiment to Camp Greene, North Carolina as it became a part of the 4th Division. On January 13, 1918 he was promoted to the rank/position of Bugler. Marcella sailed to France as a Bugler in Company F 47th Infantry 4th Division aboard the troopship *U.S.S. Princess Matoika* (seized German liner *Prinzess Alice*) on May 10, 1918. As his emergency contact, he listed his mother Gaby in New York City. He served in the Aisne-Marne Offensive.

Marcella was awarded the Distinguished Service Cross for his actions in the Vesle Sector. On August 9, 1918, Marcella's Battalion had crossed the Vesle River and patrols from it had advanced to within one hundred yards of the village of Bazoches. The area in front of the village was a hotbed of German machine gun nests and snipers and heavy fire was placed on the Battalion from that area, from the village itself and from the heights to the north as well.

Eventually the Battalion withdrew to the railroad running along the river and then pulled back across the river to positions near the village of St. Thibaut. Marcella's actions took place when the Battalion was still across the river and near the village of Bazoches. Marcella, along with Sergeant Louis Scionti of Company F and Cook Henry Garst of Company H, attacked and destroyed a German machine gun nest under heavy fire by the enemy the whole time. Scionti and Marcella also received the Distinguished Service Cross for the action. Marcella also received a Citation Star.

Marcella was awarded the French Croix De Guerre with gilt star. He was transferred to Machine Gun Company on October 9, 1918. He returned to the United States several months ahead of his regiment as a Bugler in Machine Gun Company 47th Infantry and part of Convalescent Detachment No. 46 of sick and wounded aboard the troop transport *U.S.S. Kroonland* leaving St. Nazaire, France February 5, 1919 and arriving at Newport News, Virginia February 18, 1919. He was discharged on March 18, 1919.

The 1920 Census showed Marcella living with his parents, two brothers and four sisters in the Bronx, New York City, New

York where he was unemployed. The 1925 New York State Census showed him still living with his parents in the Bronx where he was employed as a Foreman. When he registered for the draft in 1942, he indicated he was living in the Bronx and was unemployed. At an unknown date he married his wife, Maude. They were living in Daytona Beach, Florida at the time of his death. Richard Marcella died at the age of 57 on July 24, 1958 and is buried in Arlington National Cemetery.

Michael D. Belis

Robert G. Marshall
First Lieutenant, United States Army
Company M 58th Infantry Regiment 4th Division
Date of Action: October 4, 1918
Location: Near the Bois de Fays, France
War Department General Orders 66 (1919)
Medal # 5517

CITATION
When his company's advance was stopped by heavy enfilading machine gun fire, Lieutenant Marshall took seven soldiers and rushed the enemy machine gun nest, killing six of the enemy and capturing 30, including a captain. Lieutenant Marshall accomplished this daring feat without any of his own men becoming casualties.

DETAILS
Robert George Marshall was born in Minneapolis, Minnesota on March 4, 1893, the son of A.G. and Ellen Marshall. Be-

fore the war he was a lumberman employed by Rogers Lumber Company in Minneapolis. Marshall was commissioned a Lieutenant in the Officers Reserve Corps on November 27, 1917. He sailed to England as a 1st Lieutenant in Company M 58th Infantry 4th Division aboard the British troopship *Themistocles* on May 11, 1918. As his emergency contact, he listed his mother in Minneapolis, Minnesota. From England, he and his Company were sent over to France.

Marshall served in the Aisne-Marne Offensive, the Vesle Sector, the Toulon Sector, and the St. Mihiel Offensive.

He was awarded the Distinguished Service Cross for his actions in the Meuse-Argonne Offensive. From September 26 - October 3, 1918, the 4th Division had been prevented from entering the wooded area known as the Bois de Fays by heavy German resistance. On October 4, the attack was resumed and the 58th Infantry left their positions in the northwestern edge of the wooded area known as the Bois de Brieulles and advanced across an open area heading for the southern edge of the Bois de Fays. An enemy machine gun nest held up Company M's attack. Marshall personally took seven of his men and charged the nest. The eight Americans assaulted a German position containing at least thirty-six of the enemy. Under Marshall's direction, the little force killed six of the enemy and captured thirty prisoners, among which was a German Captain, without suffering a single casualty themselves.

At some time after his award action, Marshall was promoted to Captain. The official history of the 58th Infantry in World War One indicates that during his service in France, Marshall was wounded but does not give a date.

He served on occupation duty in Germany with the 4th Division and returned to the United States a month ahead of his

Regiment as Captain of Company M 58th Infantry aboard the troop transport *U.S.S. Pastores* leaving St. Nazaire, France June 17, 1919 and arriving at Hoboken, New Jersey June 26, 1919. On the return voyage, out of the 2,125 soldiers on board the *Pastores*, Marshall was one of 164 soldiers classified as sick and wounded. He was on the passenger list with a description of "PSYCHONEUROSIS OTHER FR ANXIETY STATE." [16]

He was discharged on January 2, 1920. The 1920 Census showed him as a Captain of Infantry and a patient at U.S. Army General Hospital No. 28 at Fort Sheridan, Deerfield, Illinois.

At some time after 1920 he was married. The 1930 Census showed him living in Minneapolis, Minnesota with his wife Marguerite and his mother Ellen with his occupation listed as advertising and indicated he was a veteran of the World War. After 1932 he was awarded the Purple Heart for wounds he received in World War One. The 1936 city directory for Minneapolis showed him living with his wife Meta at 2015 Aldrich Avenue South and his occupation listed as senator. The 1938 city directory for Minneapolis showed him and Meta still at the same address with his occupation listed as salesman for the Bureau of Engraving. Marshall's registration for the draft in 1942 showed him and Meta living in Baltimore, Maryland where he was unemployed. At some time after 1942 he and Meta moved back to the 2015 Aldrich Avenue South address in Minneapolis, Minnesota. Robert G. Marshall died at the age of 64 on March 19, 1957 and is buried in Fort Snelling National Cemetery, Minneapolis, Hennepin County, Minnesota.

Cecil N. Martin

Private, United States Army
Army Service Number: 2100347
Company I 47th Infantry Regiment 4th Division
Date of Action: July 29 - 30, 1918
Location: Near Sergy, France
War Department General Orders 74 (1919)
Medal # 7562

CITATION

Exposing himself to heavy enemy machine gun and artillery fire, Private Martin repeatedly carried messages from his company commander to the battalion post of command. He was killed in performance of this hazardous duty.

DETAILS

Cecil Noah Martin was born in Carmi, White County, Illinois on June 1, 1895, the son of O.M. and Etta Martin. When he registered for the draft in June 1917, he indicated he was a laborer at the Central Refrigerator Company in Lawrenceville, Illinois. He entered the Army on September 18, 1917. He sailed to France as a Private in Company I 47th Infantry 4th Division aboard the Italian troopship *Caserta* on May 10, 1918. As his emergency contact, he listed his mother Ettie in Lawrenceville, Illinois.

Martin was awarded the Distinguished Service Cross for his actions in the Aisne-Marne Offensive while acting as a runner for Company I 47th Infantry during the battle for Sergy on July 29-30, 1918 and (though the date and location is not given in his Citation) also for his actions as a runner in the Vesle Sector during which he was killed.

(See the entry for Robert A. Madden for a description of the experiences of Company I 47th Infantry at Sergy on July 29-30, 1918.)

Martin was killed in action in the Vesle Sector on August 7, 1918 while continuing to serve as a runner. On the morning of August 7, his Battalion was in support while 2nd Battalion 47th Infantry led the attack against the village of Bazoches across the Vesle River from the village of St. Thibaut. Fifteen minutes after the attack began, the Germans blanketed the area with an artillery barrage of high explosive, gas, and shrapnel. The bombardment covered the area in front of Bazoches all the way to the river and across the river into the village of St. Thibaut and lasted intermittently for about two hours. In the afternoon, another barrage was fired on the Americans. During the day between three and four thousand shells were fired by the enemy, one third of which were estimated to be gas. At some time during the day as he made his way across the battlefield carrying messages, Martin was killed by that artillery.

Martin was buried in the American Plot Row A Grave 5 in the American Cemetery # 847 at Bazoches.

His Distinguished Service Cross was presented to his mother Mrs. Ettie Bryant.

On May 19, 1919 Martin was reinterred in Plot 5 Section F Grave # 39 in the American cemetery # 617 at the village of Fismes. On May 24, 1921, his body was disinterred for preparation and shipment. His remains were returned to the United States aboard the United States Army Transport *U.S.A.T. Wheaton* leaving Antwerp, Belgium June 19, 1921 and arriving at Hoboken, New Jersey July 2, 1921. He was one of 5,827

American dead brought home by the *Wheaton* on that journey. Cecil N. Martin was laid to his final rest in Lawrenceville City Cemetery, Lawrenceville, Lawrence County, Illinois. The application for his government headstone dated June 1960 indicates he was awarded the Purple Heart.

Michael D. Belis

Henry F. Martin
Second Lieutenant, United States Army
Note: Although he had been promoted to the temporary rank of 1st Lieutenant two months earlier, his Distinguished Service Cross Citation was awarded at his permanent Regular Army rank which was 2nd Lieutenant at the time of the actions cited.
Company G 47th Infantry Regiment 4th Division
Date of Action: August 7 and 9, 1918
Location: Near Bazoches, France
War Department General Orders 14 (1928)
Medal # No assigned number discovered

CITATION
Early on the morning of August 7, 1918, being ordered to cross the Vesle River, Lieutenant Martin led his platoon under heavy artillery and machine gun fire. Upon reaching the bank of the river, he found no footbridge across the river. He leaped into the river, swam across it, stretched ropes and improvised a pontoon structure. After reaching the north bank of the river, he observed two men in danger of being carried away by the current, going to their aid, and rescuing them while under heavy fire from the enemy. On August 9, 1918, when his company com-

mander was wounded, he fearlessly placed himself at the head of the most exposed unit and led it forward in a determined attack on Bazoches.

DETAILS

Henry Fletcher Martin was born in Green Castle, Pennsylvania on March 6, 1892, the son of John and Mary Martin. He served two years as a Yeoman in the United States Navy. Before entering the Army, he was an attorney at law employed by Cockrell & Cockrell in Jacksonville, Florida. He was commissioned a 2nd Lieutenant of Infantry in the Regular Army on October 26, 1917 and assigned to the 47th Infantry as it moved from Camp Syracuse, New York to Camp Greene, North Carolina. He sailed to France as a 2nd Lieutenant in Company G 47th Infantry 4th Division aboard the troopship *U.S.S. Princess Matoika* (seized German liner *Prinzess Alice*) on May 10, 1918. As his emergency contact, he listed his father John in Palm Beach, Florida. In France he was promoted to the wartime temporary rank of 1st Lieutenant on June 9, 1918. Martin served in the Aisne-Marne Offensive.

He was awarded the Distinguished Service Cross for his actions in the Vesle Sector on two separate dates. The first date in his Citation was August 7, 1918. On that date the 47th Infantry spearheaded the 7th Brigade's advance at the Vesle River. Martin's 2nd Battalion 47th Infantry was ordered to cross the river and lead the attack upon the village of Bazoches. The 4th Engineers had constructed a foot bridge across the river near the village of St. Thibaut, but it had been destroyed by enemy artillery. Martin swam the river and utilizing ropes, logs, and other materials improvised a structure his men could use to help them cross over. He rescued two men who were being swept away by the swift current of the river and got them safely across, all the while under fire from enemy machine guns emplaced in the area in front of Bazoches and in the village itself.

The second date in his Citation was August 9, 1918. Martin took command of Company G when his Company Commander Captain Rupert L. Purdon and the Company Executive Officer 1st Lieutenant Reginald D. Grout were both wounded on August 9, 1918. He inspired his soldiers and led them in a determined attack against Bazoches. During that attack, Company E on his left flank suffered 162 casualties out of 240 men and did not reach the village; yet under Martin's direction Company G did reach Bazoches and were halfway through it before German airplanes bombed both American and German forces in the village, forcing Company G out of Bazoches for the night. For his leadership and daring that day, Martin received a Citation Star. After 1932, that Citation could have been utilized to apply to have it converted to a Silver Star Medal. The Army Registers indicate Martin did not have that done.

Martin was not awarded the Distinguished Service Cross until orders dated 1928, ten years after the end of the war. A number of recommendations for awards in the 4th Division were either lost or destroyed during the fighting in the Vesle River area and his may have been one of those.

Martin served in the Toulon Sector, the St. Mihiel Offensive, and the Meuse-Argonne Offensive. At an unknown date he was wounded and returned to duty. He served on occupation duty in Germany with the 4th Division and returned to the United States as a 1st Lieutenant with Company I 47th Infantry 4th Division aboard the troopship *U.S.S. Mobile* (seized German liner *Cleveland*) leaving Brest, France July 16, 1919 and arriving at Hoboken, New Jersey July 27, 1919. He was promoted to the permanent Regular Army rank of 1st Lieutenant on August 20, 1919.

The 1920 Census showed Martin as an officer with the 4th Division at the Camp at Gary, Indiana. On January 24, 1920 he applied for a passport to travel to Cuba for recreation. He signed the application as 1st Lieutenant 47th Infantry. On that passport application under the category of identifying marks he indicated that he had partial paralysis of his right hand (obviously due to his wounds from the war. The Army Registers show that after 1932 he was awarded the Purple Heart for those wounds.) Martin retired on March 9, 1921. He was recalled to active duty from November 4, 1921 to August 31, 1922 at which time he was finally retired from the Army with a disability in the line of duty at the rank of Captain, date of rank back to March 13, 1922.

Martin returned to his law practice. On September 12, 1922 he married Viola James. The 1930 Census showed him living with his wife, daughter and two sons in Jacksonville, Florida where he was employed as an attorney and indicated he was a veteran of the World War. The 1940 Census showed him living with his wife, two daughters and two sons in Jacksonville, Florida where he was employed as an attorney. Henry F. Martin died at the age of 96 on September 28, 1988 and is buried in Oaklawn Cemetery, Jacksonville, Duval County, Florida.

Michael D. Belis

Forrest L. Martz
Private First Class, United States Army
Army Service Number: 1565239
Company C 12th Machine Gun Battalion 4th Division
Date of Action: October 6, 1918
Location: In the Bois de Fays, near Brieulles, France
War Department General Orders 53 (1920)
Medal # 6247

CITATION

Private Martz and a comrade, under heavy enemy fire went to the rescue of wounded lying in advance of our lines and returned to our lines with two wounded American soldiers. In accomplishing this mission, they advanced to within 75 yards of the enemy lines over an area which the enemy raked with their fire.

DETAILS

Forest Leland Martz was born in Tipton, Indiana on May 27, 1900, the son of Isaac and Laura Martz. At the age of sixteen he lied about his age and enlisted as a Private in Company I 1st Infantry Indiana National Guard on April 24, 1917. On August 15, 1917 he was promoted to Corporal. He was reduced to the rank of Private on February 12, 1918 and promoted to Private 1st Class on March 21, 1918. On May 12, 1918 he was transferred to Machine Gun Company 162nd Infantry in the 41st Division which was already overseas so he shipped out as a replacement. Martz sailed to England as a Private and replacement in the 11th Company June Automatic Replacement

Draft from Camp Shelby along with other units aboard the British troopship *R.M.S. Grampian* on June 11, 1918. As his emergency contact, he listed his father in Tipton, Indiana. From England he was sent over to France. The 41st Division was designated as a replacement Division and on July 24, 1918 Martz was transferred to Company C 12th Machine Gun Battalion 4th Division.

Martz was awarded the Distinguished Service Cross for his actions in the Meuse-Argonne Offensive in the harsh environment that was the wooded area known as the Bois de Fays while supporting Company E 58th Infantry. He and Corporal Arthur M. Hamilton ventured across the battlefield under fire and rescued two wounded comrades, at one point coming as close to the German positions as 75 yards. At the time of that action, Martz was eighteen years old. Both he and Hamilton were awarded the Distinguished Service Cross. Martz also received a Citation Star.

(See the entry for Joseph Bassi for a description of the battlefield conditions in the Bois de Fays upon which Martz and Hamilton ventured to get their injured fellow soldiers.)

He received the Italian War Merit Cross (Croce al Merito di Guerra) as a member of 12th Machine Gun Battalion. He was one of 20 soldiers of the 4th Division and one of only 399 soldiers of the American Expeditionary Forces to be awarded the Italian War Merit Cross.

Martz served on occupation duty in Germany with the 4th Division and sailed home as a Private 1st Class in Company C 12th Machine Gun Battalion aboard the troopship *U.S.S. Von Steuben* (seized German liner *Kronprinz Wilhelm*) leaving Brest, France July 21, 1919 and arriving at Hoboken, New Jersey July 29, 1919. He was discharged on August 6, 1919.

Michael D. Belis

The 1920 Census showed Martz living with his parents and brother in Tipton, Indiana and indicated he was unemployed. He earned a degree from the University of Chicago and then did graduate work at Harvard University. In 1925 he moved to Massachusetts, became an investment banker, and obtained a Reserve Commission as a 2nd Lieutenant in Battery C 241st Coast Artillery of the Massachusetts National Guard. On February 16, 1926 he married Elsa Baalack. In 1927 they had a daughter and in 1929 they had a son. The 1930 Census showed Martz and his family living in Watertown, Middlesex County, Massachusetts where he was an investment salesman and indicated he was a veteran of the World War.

On September 16, 1940 Martz returned to active duty in the Army, was promoted and assigned command of the 26th Training Battalion at Camp Wallace, Texas. He sailed to Europe in 1944 and commanded the 133rd AAA (Anti-Aircraft Artillery) Battalion in France and Germany. After the war he served on occupation duty in Germany (for the second time) and was the military Mayor of Munich. He also served as a member of the court in the Dachau War Crimes Trials. In 1949, he was with the 10th AAA Group at Fort Bliss, Texas.

Martz served 17 months in Korea where he commanded the 76th AAA Battalion during the Korean War. While in Korea he founded the AAA Orphans Home in Seoul, South Korea and worked with the Korean YWCA and the U.S. Army in establishing other homes for lost and abandoned children during the war. He returned to Fort Bliss where he commanded the 4052nd Area Service Unit, Anti-Aircraft Artillery, Guided Missile Section. He served in Saudi Arabia as Chief of Staff of the U.S. Military Mission to Saudi Arabia and retired from the Army as a Colonel on June 30, 1958. Martz served in the combat zone in three wars, participating in eleven campaigns and sectors altogether.

On June 22, 1978, Martz accidentally knocked over an electrical fan in his home which shorted out and caused a fire. He suffered severe burns in the fire and was hospitalized for 35 days before succumbing to his injuries. Forrest L. Martz died from third degree total body surface burns which resulted in respiratory insufficiency at the age of 78 on July 27, 1978 at Brooke Army Medical Center, Fort Sam Houston, Texas. After his death, his body was cremated. The final disposition of his ashes could not be determined.

Roy E. Mathews

Private, United States Army
Army Service Number: 2263417
Company E 58th Infantry Regiment 4th Division
Date of Action: October 5, 1918
Location: In the Bois de Fays, France
War Department General Orders 46 (1919)
Medal # 3310

CITATION

Acting without orders, Private Mathews went through heavy artillery fire to notify his regimental commander that our own barrage was falling short, his bravery and presence of mind thus saving the lives of many American soldiers.

DETAILS

Roy Estes Mathews was born in Fairhaven, Whatcom County, Washington on March 5, 1892, the son of Joseph and Joanna Mathews. He married Valeria Quade on May 9, 1914 and their son was born on June 20, 1915. When he registered for the draft in June 1917 he indicated his occupation as a Laborer at the American Junk Company in Seattle, Washington. He was drafted into the Army on October 3, 1917. He was assigned to Company E 58th Infantry at Camp Greene, North Carolina on February 26, 1918. Mathews sailed to England as a Private in Company E 58th Infantry 4th Division aboard the British transport *S.S. Rhesus* on May 7, 1918. As his emergency contact, he listed his wife in Seattle, Washington. From England he and his Company were sent over to France.

Mathews was wounded on July 18, 1918 the first day of the Aisne-Marne Offensive and then later returned to duty.

He was awarded the Distinguished Service Cross for his actions in the Meuse-Argonne Offensive. On October 5, 1918, the 58th Infantry held much of the wooded area known as the Bois de Fays. It had taken nine days of continued attacks to gain a foothold in the forest and many German counterattacks had been repulsed in order to keep that foothold. During one of those counterattacks, without being ordered to do so, Mathews ran through heavy enemy artillery fire and also through a barrage from American artillery to inform his Regimental commander that the friendly artillery called in to break up the German attack was falling short and into the forward lines of the 58th Infantry. His quick thinking and his courage to dash through both barrages allowed his Regimental Command to get the American artillery fire stopped, thus saving lives. He also received a Citation Star.

On February 1, 1919, Mathews was promoted to Private 1st Class. He was awarded the French Croix De Guerre. He served on occupation duty in Germany with the 4th Division and returned to the United States with his Company aboard the troopship *U.S.S. Mount Vernon* (captured German liner *Kronprinzessin Cecilie*) leaving Brest, France July 24, 1919 and arriving at Hoboken, New Jersey August 1, 1919. On the voyage home he was promoted to Corporal while aboard ship on July 31, 1919. He was discharged on August 15, 1919.

Mathews divorced his wife Valeria sometime before October 1920. The 1920 Census showed him unemployed and living with his son at Valeria's mother's home in Seattle, Washington and Valeria living as a roomer at a different address. Valeria remarried in October 1920. Mathews married his second wife

Jesse sometime before 1930. The 1930 Census showed Mathews and his wife Jesse living in Meadowpoint, King County, Washington where he worked as a lithographer at a print shop and indicated he was a veteran of the World War. After 1932 he was awarded the Purple Heart for the wounds he received in World War One. The 1940 Census showed Mathews living as a lodger in Seattle and employed as a clerk at a mortgage company. Roy E. Mathews died at the age of 65 on February 4, 1958 and is buried in Evergreen-Washelli Memorial Park, Seattle, King County, Washington.

Edward McAndrew

Sergeant, United States Army
Army Service Number: 562392
Company B 12th Machine Gun Battalion 4th Division
Date of Action: September 30, 1918
Note: McAndrew's Citation gives the date of the action and his death as September 30, 1918. The Register of Burials of Deceased American Soldiers, 1917 – 1922, Office of the Quartermaster General gives the date of his death as September 29, 1918. That date more closely corroborates the details and location of the action in his Citation.
Location: Near the Bois des Ogons, France
War Department General Orders 78 (1919)
Medal # 7167

CITATION

Exposing himself fearlessly to enfilading machine gun fire from the enemy, Sergeant McAndrew directed the placing of the guns of his section in such positions as to protect the advance of the infantry and in so doing was fatally wounded. Despite the fact that one-half of his body was paralyzed as a result of his injury, he insisted upon remaining in command of his section until the action was over. He died in a field hospital shortly after being evacuated.

DETAILS

Edward William McAndrew was born in Vincennes, Knox County, Indiana on July 7, 1897, the son of William and Anna McAndrew. On the day the United States declared war on Germany, April 6, 1917 he enlisted as a Private in the Regular Army at Chicago, Illinois. He trained at Brownsville, Texas and was later assigned to Company B 12th Machine Gun Battalion at Camp Greene, North Carolina. McAndrew sailed to England as a Sergeant in Company B 12th Machine Gun Battalion 4th Division aboard the British troopship *RMS Aquitania* on May 7, 1918. As his emergency contact, he listed his mother Mrs. Charles Weyl in Chicago, Illinois. From England he and his Battalion were sent over to France where they were issued their machine guns.

McAndrew served in the Aisne-Marne Offensive, the Vesle Sector, the Toulon Sector and the St. Mihiel Offensive.

He was awarded the Distinguished Service Cross for his actions in the Meuse-Argonne Offensive. On September 29, 1918, the 8th Brigade of the 4th Division, which included the 12th Machine Gun Battalion, moved through the 7th Brigade's positions south of the wooded area known as the Bois des Ogons and continued the Division's attack in the direction of the wooded area known as the Bois de Fays. McAndrew was in charge of a machine gun section consisting of 18 men and two machine guns. He exposed himself to German machine gun fire in order to get his guns set up in positions where they could cover the American Infantry attack and in doing so was severely wounded by enemy fire. Though he couldn't move half of his body due to his wounds, he remained in command of his guns until the attack was completed. He died from his wounds after reaching a field hospital.

McAndrew was buried in Row A Grave # 9 in the American cemetery # 840 at the village of Cuisy.

His Distinguished Service Cross was presented to his mother Mrs. Charles Weyl.

On May 8, 1919 McAndrew was reinterred in Plot 3 Section 42 Grave # 144 in the Argonne American Cemetery # 1232. On June 15, 1921 his body was disinterred for preparation and shipment. His remains were returned to the United States aboard the United States Army Transport *U.S.A.T. Cantigny* leaving Antwerp, Belgium July 21, 1921 and arriving at Hoboken, New Jersey August 1, 1921. He was one of 1,400 deceased American soldiers brought home by the *Cantigny* on that voyage. Edward W. McAndrew was laid to his final rest on August 19, 1921 in Rock Island National Cemetery, Rock Island, Rock Island County, Illinois.

Arnot L. McArthy

Note: In several versions of his Citation, his last name is given as McArty. His actual last name was McArthy.
Private, United States Army
Army Service Number: 2100809
Company D 59th Infantry Regiment 4th Division
Date of Action: October 3 and October 9, 1918
Location: Near Bois de la Côte Lemont, France and near the Bois de Fay, France
Note: The location of "Near Bois de la Côte Lemont" for the date of October 3, 1918 is incorrect. On October 3, 1918 McArthy and the 59th Infantry were in the Bois de Brieulles on the left side of the 4th Division advance. On October 3, 1918 the 47th Infantry was near the Bois de la Côte Lemont which was on the right side of the 4th Division advance.
War Department General Orders 46 (1919)
Medal # 7620

CITATION

On October 3, while acting in the capacity of company runner, Private McArthy carried messages to two platoons of his company through a heavy fire of machine guns and snipers. He successfully delivered the messages after crawling for a distance of 400 yards. On October 9, in company with one other runner he delivered messages to a platoon which was engaged in combat liaison duty in the Bois de Fay, passing through a severe artillery fire while in the execution of this mission.

DETAILS

Arnot Leslie McArthy was born in Enfield, White County, Illinois on May 1, 1896, the son of Josiah Francis "Frank" and Elizabeth "Lizzie" McArthy. Before entering the Army, he was a farmer in Hamilton County, Illinois. He entered the Army on September 19, 1917. McArthy sailed to England as a Private of Infantry and replacement in the Camp Pike June Automatic Replacement Draft Infantry Company #2 aboard the British/Australian troop transport *S.S. Delta* on June 20, 1918. As his emergency contact, he listed his father Frank in Enfield, Illinois. From England he was sent over to France, where he was eventually assigned to Company D 59th Infantry Regiment 4th Division.

McArthy was awarded the Distinguished Service Cross for his actions on two separate dates in the Meuse-Argonne Offensive. The first date was October 3, 1918. On that date his Regiment was positioned on the left flank of the 4th Division in the wooded area known as the Bois de Brieulles. Heavy German resistance had halted the 4th Division advance and the day was spent consolidating positions and preparing to resume the attack the next day. While performing his duty as a runner, enemy machine gun and sniper fire forced McArthy to resort to crawling 400 yards to deliver his messages.

The second date in his Citation was October 9, 1918. On that date the American advance resumed after a halt of four days and the Germans concentrated artillery, both high explosive and gas and machine gun fire on the Bois de Fay making that area incredibly dangerous for anyone moving about and not protected by cover. McArthy and another runner braved that heavy fire to deliver their messages.

McArthy was also awarded the French Croix De Guerre. At

an unknown date after his award actions, he attended Officers School in France and upon graduation was commissioned a 2nd Lieutenant of Infantry and transferred to the 324th Infantry in the 81st Division. He returned to the United States as a 2nd Lieutenant in Company I 324th Infantry aboard the troopship *U.S.S. Martha Washington* (seized Austrian liner) leaving St. Nazaire, France June 7, 1919 and arriving at Charleston, South Carolina June 18, 1919. He was discharged on July 10, 1919.

The 1920 Census showed McArthy living with his parents in Crook, Hamilton County, Illinois where he was a farmer and partner in the farming business with his father. On August 27, 1922 he married Pearl Draper. The 1930 Census showed him living with his wife and two daughters in Indian Creek, White County, Illinois where he was employed as a schoolteacher and indicated he was a veteran of the World War. The 1940 Census showed him still living in Indian Creek with his wife, two daughters and three sons where he was employed as a farmer. When he registered for the draft in 1942, he gave his residence as Enfield, White County, Illinois. Arnot L. McArthy died at the age of 91 on April 23, 1988 and is buried in Bethel Memorial Cemetery, Mount Vernon, Jefferson County, Illinois.

Howard C. McCall

Captain, United States Army
Company G 59th Infantry Regiment 4th Division
Date of Action: July 19, 1918
Location: Near Chezy, France
War Department General Orders 37 (1919)
Medal # 6730

CITATION

After his company had suffered heavy losses in taking its immediate objective, Captain McCall placed himself at the head of his command and led his men forward in the face of violent shell and machine gun fire until he fell mortally wounded, cheering his men on with his last words.

DETAILS

Howard Clifton McCall was born in Philadelphia, Pennsylvania on April 9, 1891, the son of Joseph and Lenore McCall.

Michael D. Belis

He graduated from the Law School of the University of Pennsylvania and was admitted to the bar but had not yet opened a law practice when war was declared. Before the war he did legal work for the Philadelphia Electric Company. He entered Officers Training Camp at Fort Oglethorpe, Georgia on August 15, 1917 and upon completion of the program was commissioned a Captain of Infantry on November 15, 1917. He was on that date assigned to Company G 59th Infantry at Camp Greene, North Carolina.

McCall sailed to England as a Captain and Commanding Officer of Company G 59th Infantry 4th Division aboard the armed British troopship *RMS Olympic* on May 5, 1918. As his emergency contact, he listed his father Joseph in Philadelphia, Pennsylvania. On the voyage, the *Olympic* rammed, fired on, and sank the German submarine U-103 on May 12, 1918. McCall was onboard the *Olympic* when it rammed the enemy sub, witnessed the action, and wrote a short but detailed account of the event in a letter home that was published in a Philadelphia newspaper. From England he and his Company were sent over to France.

McCall was awarded the Distinguished Service Cross for his actions on the second day of the Aisne-Marne Offensive. The 2nd Battalion of the 59th Infantry, which included McCall's Company G, was attached to the French 164th Infantry Division during the opening phase of the Aisne-Marne Offensive. The Battalion suffered heavy casualties on July 19, 1918 during its morning attack against the German positions in the wooded area known as the Bois de Cobourg near the village of Chezy. The advance was halted, but in the afternoon the Battalion resumed the attack and seized the wooded area. In the advance, McCall personally dispatched three German soldiers with his .45 caliber auto pistol then armed himself with an M1903 Springfield rifle and used the rifle to fire at the enemy as he moved forward. As he led Company G toward successive objectives with French tanks in support, McCall was eventually struck down by machine gun fire.

An account recorded in the Pennsylvania War History Commission echoes the Citation wording that says he cheered his men with his dying words. That account read: "His men narrate his last utterance, so characteristic of the man who fell mortally wounded, shouting: Cheer O, my brave laddies!"[17]

Another account written in a letter from one of his platoon leaders in Company G, 1st Lieutenant John J. Piorkoski talked at length about McCall:

"Captain McCall's company went over the top at 4 a.m. Kept on going for three days. It was hard on the men, but we never did mind, and stuck together to see it through. The hell that was there can never be described...He fought with coolness and deliberation, not looking after his own safety, but that of his men...He led his men forward and kept on all day, keeping the men cheery all the time and fighting every foot of the terrain covered in his advance...In the face of a terrific hail of shell and machine gun fire he kept on going until late in the evening, when he fell. It is believed that death was instantaneous."[18]

An additional account from an unnamed platoon leader from his Company which was published in the Pennsylvania Gazette stated:

"I was with Captain Howard McCall the afternoon of the 19th of July 1918. He took charge of the first wave that went over, advanced behind the tanks. He was armed with a service rifle and used it continually during the advance. The tanks were destroyed, and he advanced ahead of the rest of the battalion. He fell (was shot) within 40 yards of the enemy's line while advancing, and with a few men left, at double time—shot through the neck and face." [19]

Whether he urged his men on with his last breath or died instantly without uttering a word, all accounts agree that McCall was a courageous and outstanding officer who led from the front and was revered by the officers and men in his Company.

McCall was buried in Grave # 16 in the American battlefield cemetery # 356 at the village of Priez.

In a letter dated March 26, 1919 the Adjutant General's Office notified McCall's mother Mrs. Joseph McCall that her son was being awarded the Distinguished Service Cross posthumously and that the Quartermaster General had been directed to send the medal to her.

On June 9, 1919 McCall was reinterred in Plot 2 Section A Grave # 103 in the American Cemetery # 1764 at Belleau. On July 7, 1921 his body was disinterred for preparation and shipment. McCall's remains were returned to the United States aboard the United States Army Transport *U.S.A.T. Wheaton* leaving Antwerp, Belgium August 6, 1921 and arriving at Hoboken, New Jersey August 20, 1921. He was one of 5,759 American dead brought home by the *Wheaton* on that journey. Howard C. McCall was laid to his final rest in Woodlands Cemetery, Philadelphia, Philadelphia County, Pennsylvania.

Howard C. McCall Post No. 20 American Legion of Philadelphia was named for him. McCall was the son of Joseph B. McCall who was the president of Philadelphia Electric Company (PECO.) In 1912 Philadelphia Electric Company had purchased land right outside of Philadelphia and built a baseball field/sports complex on it, naming the ballpark Kelly's Field. On May 15, 1919, Kelly's Field was rededicated to the fifteen PECO employees who died in France during World War One and was renamed Howard McCall Field in honor of Captain Howard C. McCall. The site evolved into a conference center and recreation complex that today is known as the McCall Golf & Country Club.

Michael D. Belis

George C. McCelvey
Captain, United States Army
Note: Most versions of his Citation give his rank as Captain. The official history of the 47th Infantry in the war gives his rank at the time of his award actions as First Lieutenant.
Company H 47th Infantry Regiment 4th Division
Date of Action: August 7 - 9, 1918
Location: Near St. Thibaut, France
War Department General Orders 37 (1919)
Medal # 6616

CITATION
He stood in the swift current of the Vesle River and helped the men of three platoons across. He was pulled into the river twice by drowning men but each time succeeding in bringing them ashore. On succeeding days, he was conspicuously present in places of danger, setting a splendid example to his command.

DETAILS

George Calhoun McCelvey was born in Mount Carmel, McCormick County, South Carolina on June 21, 1888, the son of Patrick and Eunice McCelvey. He graduated from The Citadel Military Academy, Class of 1911. Before the war he was a schoolteacher and Principal. He was commissioned a Lieutenant in the Officers Reserve Corps on November 27, 1917. McCelvey sailed to France as a 1st Lieutenant in Company H 47th Infantry 4th Division aboard the troopship *U.S.S. Princess Matoika* (seized German liner *Prinzess Alice*) on May 10, 1918. As his emergency contact, he listed his father in Mount Carmel, South Carolina. At some time during his service in France he was promoted to Captain, though it is unclear if he was promoted before or after his award actions. He served in the Aisne-Marne Offensive.

McCelvey was awarded the Distinguished Service Cross for his actions in the Vesle Sector. On August 7, 1918, McCelvey's 2nd Battalion spearheaded the 47th Infantry's advance in the attempt to cross the Vesle River near the village of St. Thibaut. Almost as soon as the Battalion moved through the village the Germans opened up with an artillery bombardment of high explosive and gas which lasted on and off for two hours, inflicting heavy casualties. McCelvey and his Battalion continued to the river and began to cross while under fire. McCelvey stood in the chest deep water of the river and assisted more than 150 soldiers of the 47th Infantry (from two platoons of Company H and one platoon of Company G) one by one as they struggled to cross the swift moving current, all the while exposing himself to machine gun and sniper fire. Twice he was dragged into the river by drowning men but managed to break free from them, keep them from drowning, and bring them to shore. Once all these men were across the river, he steadied them in the defense against German counterattacks until August 9 when he was wounded.

McCelvey was also awarded a Citation Star and the French Croix De Guerre. He recovered from his wounds, returned to duty, and was wounded again. Because of his wounds, he was evacuated to the United States ahead of his Regiment as a Captain of Infantry and part of a Casual Detachment of Sick and Wounded consisting of 92 officers and over 1,500 enlisted men from various units aboard the transport *U.S.S. Wilhelmina* leaving Bordeaux, France February 11, 1919 and arriving at Hoboken, New Jersey February 27, 1919. He was discharged on July 18, 1919.

McCelvey left active duty but remained a Captain in the Officer Reserve Corps and went back to teaching. He eventually became an Elementary School and a High School Principal and was never married. After a long life in the education profession, George C. McCelvey died at the age of 84 on March 14, 1973. He is buried in Mount Carmel Cemetery, Mount Carmel, McCormick County, South Carolina. The McCelvey Center in York, South Carolina is named for him. It is home to the Historical Center of York County and the Southern Revolutionary War Institute. The central building of the Institute is the old school which in 1973 was named McCelvey Elementary School in honor of George C. McCelvey who had been a Principal there.

Arno S. McClellan

Second Lieutenant, United States Army
Company B 47th Infantry Regiment 4th Division
Date of Action: August 1, 1918
Location: At Sergy, France
War Department General Orders 70 (1919)
Note: General Orders 70 (1919) rescinded a previous Citation for the Distinguished Service Cross issued to Arno S. McCleallan [sic] in General Orders 138 (1918)
Medal # No assigned number discovered

CITATION

Second Lieutenant McClellan fearlessly led his platoon in locating and successfully attacking German machine guns, thereby facilitating the advance of his company. He also led a combat patrol in front of his position for the purpose of driving out hostile snipers. Later, when his company was forced to retire to a more sheltered zone Lieutenant McClellan with one soldier remained in an exposed position and rendered valuable service by covering the withdrawal with accurate fire from an automatic rifle.

DETAILS

Arno Seals McClellan was born in Scottsburg, Scott County, Indiana on December 30, 1890, the son of Enos and Flora McClellan. He enlisted as a Private in the Marine Corps on April 24, 1914. On July 20, 1914 he was promoted to Corporal and on March 1, 1915 he was promoted to Sergeant. On July 13, 1915 he was given a medical discharge from the Marines with a character rating of excellent. He was married to Cora Schultz on December 4, 1916. At the time he registered for the draft in June 1917, he was married with a child and was the Postmaster in Harveys, Pennsylvania.

On November 27, 1917 McClellan was commissioned a 2nd Lieutenant in the Regular Army at Fort Oglethorpe, Georgia. He was assigned to Company B 47th Infantry at Camp Greene, North Carolina as the 4th Division was being formed up there. He sailed to France as a 2nd Lieutenant in Company B 47th Infantry 4th Division aboard the troopship U.S.S. *Princess Matoika* (seized German liner *Prinzess Alice*) on May 10, 1918. As his emergency contact, he listed his wife in Harveys, Pennsylvania.

McClellan was awarded the Distinguished Service Cross for his actions in the Aisne-Marne Offensive during the bitter fighting at the village of Sergy. For the assault on Sergy, McClellan's Company as part of 1st Battalion 47th Infantry was attached to the 167th Infantry Regiment of the 42nd Division. In four days of battle for the village, McClellan's 1st Battalion along with the 3rd Battalion 47th Infantry suffered horrendous casualties as they fought with elements of the German 4th Prussian Guards Division. Large numbers of enemy machine guns and snipers were encountered all across the area before Sergy. The advance of the 47th Infantry was one of dealing with those threats on a one-by-one basis. McClellan led his platoon forward of his Company's advance and succeeded in destroying or driving out German machine guns. He then personally led a combat patrol ahead of the American lines to hunt for and eliminate enemy snipers. Later he remained in a forward position with another soldier and covered the withdrawal of his Company with a French Model 1915 Chauchat automatic rifle. He was also awarded a Citation Star.

McClellan served in the Vesle Sector. He was promoted to 1st Lieutenant on August 13, 1918 and served in the Toulon Sector and the St. Mihiel Offensive. He suffered wounds from gas on October 10, 1918 in the Meuse-Argonne Offensive and was sent to a hospital to recover. He was promoted to Captain on November 3, 1918. Because of his wounds he returned to the United States early as a Captain of Company C 47th Infantry and part of an Officers' Company of sick and wounded from U.S. Base Hospital No. 6 aboard the troop transport *U.S.S. Aeolus* (seized German liner *Grosser Kurfürst*) leaving Bordeaux, France December 17, 1918 and arriving at Newport News, Virginia December 31, 1918. McClellan was discharged on July 17, 1919 at the U.S. General Hospital #24 at Pittsburgh, Pennsylvania with a rating of 50% disabled.

The 1920 Census showed McClellan living in Graysville, Pennsylvania with his wife and two sons (the Census form is incorrect as it should read one daughter and one son) where he was employed as a Postmaster. From 1926-1929 McClellan was the Sheriff of Greene County, Pennsylvania. The 1930 Census showed him living with his wife, daughter and three sons in Waynesburg, Pennsylvania where he was employed as a Manager of Farm Lands and indicated he was a veteran of the World War. The 1940 Census showed him living with his wife and two sons in Graysville, Pennsylvania where he was employed as a manager with the Social Security Administration.

During World War Two, McClellan was recalled to active duty as a Major with the 2571st Service Unit, Army Service Forces and served in the Continental United States from March 16, 1942 to November 2, 1944, leaving active duty with the rank of Lieutenant Colonel. He divorced in 1947 and on January 31, 1949 married his new wife, Brooks. Arno S. McClellan died

Michael D. Belis

of a cerebral hemorrhage at the age of 59 on October 5, 1950 and is buried in Greene County Memorial Park, Waynesburg, Greene County, Pennsylvania.

Joseph McCollum

Wagoner, United States Army
Army Service Number: 567527
Company B 10th Machine Gun Battalion 4th Division
Date of Action: July 23, 1918
Location: Near Bois du Roi, France
War Department General Orders 71 (1919)
Medal # 7681

CITATION

On duty as a runner, Wagoner McCollum volunteered to re-establish liaison with the French unit to which his company was attached, after several officers and men had failed in the attempt. He performed the mission successfully although exposed to heavy fire. Though knocked down and temporarily stunned by the explosion of a shell, he accomplished a second dangerous mission, remaining on duty until ordered to the rear.

DETAILS

Joseph Augustus McCollum was born in Ludlow, Hampden County, Massachusetts on September 22, 1896, the son of Joseph and Mary McCollum. He entered the Army on July 28, 1917. He sailed to France as a Wagoner in Company B 10th Machine Gun Battalion 4th Division aboard the French troopship *Rochambeau* on May 7, 1918. As his emergency contact, he listed his mother Mary in Springfield, Massachusetts.

The 10th Machine Gun Battalion was the only one of the three Machine Gun Battalions in the 4th Division that was not assigned to a Brigade. It was utilized on the battlefield where it was deemed most needed.

McCollum was awarded the Distinguished Service Cross for his actions in the Aisne-Marne Offensive. For the Offensive, the 10th Machine Gun Battalion was one of the units in the 4th Division attached to the French 164th Division, jumping off in the attack with the French on July 18, 1918. When all other Infantry and Machine Gun units of the 4th Division had been pulled out of the front lines for rest and regroup July 20-21, the 10th Machine Gun Battalion remained at the front supporting the French until July 23, 1918. On July 23, the Battalion had advanced as far as the wooded area known as the Bois du Roi, west of the Château-Thierry-Soissons Road. When several other officers and enlisted men had failed to get across the battlefield to carry messages and re-establish communications between his Company and the French Command, McCollum volunteered and succeeded in doing so, on one occasion even after being knocked down and rendered temporarily senseless by a near miss from German artillery.

He was also awarded the French Croix De Guerre. McCollum returned to the United States as a Wagoner of Infantry and part of the Saint Aignan Casual Company No. 5941 destined for discharge at Camp Devens, Massachusetts aboard the troop transport *U.S.S. Finland* leaving Brest, France June 21, 1919 and arriving at Boston, Massachusetts June 30, 1919. He was discharged on July 2, 1919.

The 1920 Census showed McCollum living with his mother and family in Springfield, Massachusetts where he was employed as a chauffeur. In May 1923, he married Florence Haggerty. The 1930 Census showed McCollum living with his wife and daughter in Springfield where he was employed as a chauffeur and indicated he was a veteran of the World War. Joseph McCollum died at the age of 83 on September 12, 1980 and is buried in Saint Michael's Cemetery, Springfield, Hampden County, Massachusetts.

Michael D. Belis

William H. McGinnis

Corporal, United States Army
Army Service Number: 567996
Company D 12th Machine Gun Battalion 4th Division
Date Of Action: August 10, 1918
Location: Near Chéry-Chartreuve, France
Note: Though His Citation Indicates The Location As Being Near The Village Of Chéry-Chartreuve, The Actual Area Was 10 Miles Or More North Of Chéry-Chartreuve Across The Vesle River From And Nearer To The Village Of Ville Savoye.
War Department General Orders 50 (1919)
Medal # 6997

CITATION

An incendiary shell exploded near a large ammunition dump near which his company was resting, wounding several of his comrades and setting fire to a portion of the dump. While a second explosion was imminent, Corporal McGinnis rushed into the flames and dragged a wounded man to safety.

DETAILS

William Holroyd "Piggy" McGinnis was born in Beckley, Raleigh County, West Virginia on December 25, 1894, the son of William and Sarah McGinnis. He got his nickname by winning a greased pig contest when he was ten years old, at a 4th of July celebration in his hometown. At the age of 16 he lied about his age and enlisted in the Army as a Private in Company 80 of the Coast Artillery Corps on July 18, 1911. He served on duty in Florida until his discharge on July 17, 1914 at Key West

Barracks. He again entered the Army on February 28, 1918. McGinnis sailed to France as a Private in Company A 10th Machine Gun Battalion 4th Division aboard the French troopship *Rochambeau* on May 7, 1918. As his emergency contact, he listed his father in Beckley, West Virginia. At some time in France, he was promoted to Corporal and transferred to Company D 12th Machine Gun Battalion also in the 4th Division. His brother James was a Lieutenant in Headquarters Company 12th Machine Gun Battalion. McGinnis served in the Aisne-Marne Offensive.

He was awarded the Distinguished Service Cross for his actions in the Vesle Sector. On August 7-10, 1918, his Company was working in support of 1st Battalion 59th Infantry as they advanced along the Rouen-Rheims Road and attacked enemy positions in the woods around the Château du Diable. They were in an area on the north side of the Vesle River, across the river from the village of Ville Savoye.

As the enemy retreated, they left behind several ammunition dumps. The Germans fired artillery in the area of the dumps, hoping to blow them up and prevent the Americans from using them. Several of the soldiers from his Company were walking down the road next to one of the dumps when it was struck and exploded, wounding the men, and setting fire to the brush in the area all around. With incoming enemy artillery landing all around him, McGinnis charged into the fire, picked up a wounded soldier who was unable to move by himself, and carried the man to safety at the same time that another dump nearby exploded. Though cited in different orders, this appears to be the same action for which Wallis H. Sturtevant was also awarded the Distinguished Service Cross. Both soldiers were from Company D 12th Machine Gun Battalion, and each rescued a fellow soldier from that burning ammunition dump that day.

McGinnis served in the Toulon Sector, the St. Mihiel Offensive, and the Meuse-Argonne Offensive. At some time after his award action, McGinnis was promoted to Sergeant. He returned to the United States several months ahead of his Battalion as a Sergeant and part of Le Mans Casual Company #1205 aboard the *U.S.S. Michigan*, an older battleship being used to ferry soldiers back from Europe, leaving Brest, France February 18, 1919, and arriving at Newport News, Virginia March 3, 1919. He was discharged on March 17, 1919.

In a letter dated May 29, 1919, McGinnis was officially informed by the Adjutant General's Office that he was awarded the Distinguished Service Cross and that the Army Quartermaster's Department had been directed to mail the medal to him.

The 1920 Census showed him living with his parents in Beckley, Raleigh County, West Virginia where he was employed as a bank clerk. McGinnis received a law degree from West Virginia University in 1925 and became a prosecuting attorney for Raleigh County in 1928. The 1930 Census showed him living in Beckley with his occupation listed as lawyer and indicated he was a veteran of the World War. McGinnis was Assistant Prosecutor of Raleigh County from 1932-1936. On June 3, 1933, he married Dana Snow. The 1940 Census showed McGinnis living with his wife and two daughters in Beckley, Raleigh County, West Virginia where he was employed as Assistant Prosecuting Attorney. From 1940 through 1954, he was a Federal Marshal for southern West Virginia.

In October 1921, two and a half years after leaving the Army, McGinnis enlisted as a Private in the West Virginia National

Guard. On June 30, 1922, he was commissioned a 2nd Lieutenant and ten days later promoted to 1st Lieutenant on July 10, 1922. He was promoted to Captain on April 11, 1934 and discharged from the National Guard in 1940. During his National Guard service, he was Aide-de-Camp to Brigadier General Ellerbe Carter of the Kentucky National Guard. In 1949, McGinnis had to re-enlist as a Private in the National Guard and serve four months in order to complete the necessary total years to qualify for a pension from that organization.

William H. McGinnis died at the age of 73 on May 20, 1968 and is buried in Sunset Memorial Park, Beckley, Raleigh County, West Virginia.

Michael D. Belis

Earl M. McKinley

First Lieutenant, United States Army
Company B 11th Machine Gun Battalion 4th Division
Date of Action: September 26, 1918
Location: Near Nantillois, France
War Department General Orders 142 (1918)
Medal # 1066

CITATION

Lieutenant McKinley, with another officer and a soldier, using captured German Maxim guns pushed forward to a heavily shelled area from which other troops had withdrawn and by their accurate and effective fire kept groups of the enemy from occupying advantageous positions, maintaining fire superiority all afternoon. Lieutenant McKinley withdrew from his dangerous position when it became too dark to see.

DETAILS

Earl Martin McKinley was born in East Liverpool, Ohio on April 10, 1890. He enlisted as a Private in Company I 11th Infantry (Regular Army) on July 8, 1912 at Fort Drum, New York. He listed his civilian occupation as Machinist. He served with his Company on Mexican Border Service in Texas, New Mexico, and Arizona. On July 8, 1916, he was promoted to Corporal. On November 25, 1916, he was promoted to Sergeant and the next day to Supply Sergeant. McKinley was transferred to Company I 51st Infantry on June 16, 1917. He served with that unit until he received a discharge on July 8, 1917 in order to accept a commission.

McKinley was given a temporary wartime commission as a 2nd Lieutenant of Infantry on July 9, 1917 and assigned to the 58th Infantry which was then forming up at Gettysburg, Pennsylvania. He went with the Regiment to Camp Greene, North Carolina where it became part of the 4th Division. He was promoted to 1st Lieutenant on February 12, 1918. At some time before going overseas, he was transferred to the 11th Machine Gun Battalion still in the 4th Division. McKinley sailed to France as a 1st Lieutenant in the 11th Machine Gun Battalion and part of the advance detachment of the 4th Division aboard the troop transport *U.S.S. Finland* on April 30, 1918. As his emergency contact, he listed his sister Mrs. Mabel Walker in Buffalo, West Virginia.

McKinley served in the Aisne-Marne Offensive, the Vesle Sector, the Toulon Sector, and the St. Mihiel Offensive.

He was awarded the Distinguished Service Cross for his actions in the Meuse-Argonne Offensive. On September 26, 1918, the first day of the offensive, the 7th Brigade of the 4th Division including the 11th Machine Gun Battalion had advanced to the Corps objective which was on a line that ran approximately east and west through the village of Nantillois. The Divisions on either side had not reached the Corps objective and the advance of the 4th Division was halted for hours waiting for the flanking units to catch up. When at 5:30 p.m. the other Divisions had still not reached the objective, the order was given for the 4th Division to resume the attack. The halt had given the enemy time to regroup and rebuild his defenses.

After resuming the attack and achieving a minimal gain, without the flanking Divisions there to protect its flanks, the 4th Division pulled back to the line of the Corps objective and dug in for the night. Using captured German Maxim machine guns (most likely MG 08/15 models), McKinley, along with 1st Lieutenant Homer S. Jarvis and Sergeant Glenn M. Grove, also from 11th Machine Gun Battalion, moved to an exposed forward position and covered the withdrawal of the 4th Division Infantry in their area and prevented the enemy from launching any successful counterattack. They maintained their fire upon the enemy until it became too dark to see. Only then did they pull back and join the withdrawn units. Jarvis and Grove were also awarded the Distinguished Service Cross for the action.

McKinley was promoted to Captain on October 30, 1918. He returned to the United States as a Captain of Infantry and part of a Casual Detachment of officers aboard the troopship *U.S.S. Prinz Friedrich Wilhelm* (surrendered German liner) leaving Brest, France August 14, 1919 and arriving at Hoboken, New Jersey August 23, 1919. He was discharged on August 25, 1919.

No further details for Earl M. McKinley could be found.

Jean L. Meurisse

Captain, Army of France
27th Infantry Regiment Army of France
Attached to 58th Infantry Regiment 4th Division
Date of Action: July 18, 1918
Location: Near Chevillon, France
War Department General Orders 81 (1919)
Medal # No assigned number discovered

CITATION

Acting as liaison officer with the 58th American Infantry, he showed marked personal courage under intense fire, setting an example of fearlessness to the officers and men with him. His knowledge of German artillery enabled him to advise methods of approach for our troops which were instrumental in preventing many casualties.

DETAILS

Captain Jean L. Meurisse was one of a number of French Army officers who served as advisors to the U.S. 4th Division during World War One. He was assigned as a Liaison Officer to the 58th Infantry in July 1918.

He was awarded the Distinguished Service Cross for his actions in the Aisne-Marne Offensive. On July 18, 1918, the first day of the Offensive, the 58th Infantry was attached to the 13th Chasseurs of the French 164th Infantry Division. Early in the morning of July 18 the 2nd Battalion 58th Infantry attacked toward the village of Chevillon. Advancing over the crest of a ridge west of the village, the Battalion took two lines of German trenches and then was met by a heavy concentration of enemy machine gun fire and an intense barrage of enemy artillery.

Though it suffered many casualties, the Battalion continued on and in a courageous dash captured the village. It then pushed on under fire to the slopes of Hill 172 where it fought its way up the hill, capturing German trenches in vicious bayonet fighting. By the day's end, the Battalion had endured 2 officers and 93 enlisted men killed, 11 officers and 436 enlisted men wounded, and 1 officer and 66 enlisted men missing for a total casualty count of 609 out of 1,281 in the Battalion. Rather than remain at the rear Command Post monitoring operations, Meurisse was right alongside the American soldiers as they attacked, giving seasoned advice in how to deploy during the advance and providing an example of courage under fire which inspired all around him.

Meurisse listed his emergency contact, as: Madam J. Meurisse, wife, 6 du Laminoir Essonnes (Seine and Oise), France.

Arthur M. Miller

Private First Class, United States Army
Army Service Number: 557794
Company B 47th Infantry Regiment 4th Division
Date of Action: August 1, 1918
Location: Near Sergy, France
War Department General Orders 37 (1919)
Medal # 6704

CITATION

Private Miller was killed while returning with an answer to a very important message which he had voluntarily delivered at a very critical state of the attack. His mission was one of extreme danger, taking him to the most advanced position through a sweeping fire of artillery and machine guns.

DETAILS

Arthur Merrill Miller was born in Thetford Mines, Quebec, Canada on October 13, 1895, the son of Arthur and Mary Miller. He came with his parents to live in the United States when he was only a few months old and when he registered for the draft in 1917, he declared himself to be an alien and not a citizen of the United States. Before entering the Army, he was employed as a Ship Fitter Helper at a shipyard in Philadelphia, Pennsylvania. He enlisted as a Private in Company B 47th Infantry (Regular Army) at Camp Greene, North Carolina on February 14, 1918. Miller sailed to France as a Private in Company B 47th Infantry 4th Division aboard the troopship *U.S.S. Princess Matoika* (seized German liner *Prinzess Alice*) on May

Michael D. Belis

10, 1918. As his emergency contact, he listed his mother Mary in Websterville, Vermont. Miller was promoted to Private 1st Class in France on June 17, 1918.

He was awarded the Distinguished Service Cross for his actions in the Aisne-Marne Offensive. Miller's 1st Battalion 47th Infantry along with 3rd Battalion 47th Infantry fought for four days in the battle at the village of Sergy. Elements of the 4th German Prussian Guards Division made the 47th Infantry suffer heavy casualties to capture and hold the village. Miller and Company B 47th Infantry were at the front of the attacking line during the entire fight as 1st Battalion secured and held the high ground to the north and northwest of Sergy while 3rd Battalion took the village itself. During the performance of his dangerous job as runner, Miller was killed on August 1, 1918 traversing the battlefield under heavy fire. He also received a Citation Star for his bravery.

Miller was buried in Plot # 1 Section N Grave # 34 in the American battlefield cemetery # 608 at the village of Seringes-et-Nesles.

His Distinguished Service Cross was presented to his mother Mrs. Mary Miller.

On May 2, 1921 Miller's body was disinterred for preparation and shipment. His remains were returned to the United States aboard the United States Army Transport *U.S.A.T. Wheaton* leaving Antwerp, Belgium June 19, 1921 and arriving at Hoboken, New Jersey July 2, 1921. He was one of 5,827 American dead brought home by the *Wheaton* on that journey. Arthur M. Miller was laid to his final rest in Wilson Cemetery, Barre, Washington County, Vermont.

Manton C. Mitchell

Major, United States Army
Commanding Officer 2nd Battalion 39th Infantry Regiment 4th Division
Date of Action: August 5, 1918
Location: Near St. Thibault, France
War Department General Orders 60 (1920)
Medal # 6284

CITATION

The attack battalion having been held up by heavy machine gun fire while attempting to cross the Vesle River, Major Mitchell, who was in command of the support battalion, went forward through heavy machine gun fire and encouraged and assisted the advanced troops to cross the river. He was severely wounded in the leg while directing these movements, but he refused to be evacuated and continued in the attack, remaining with the attack battalion until the evening of August 5.

Michael D. Belis

DETAILS

Manton Campbell Mitchell was born in Providence, Rhode Island on November 30, 1887, the son of John and Jessie Mitchell. He entered the U.S. Military Academy at West Point on June 15, 1905 and graduated 100 out of a class of 103 on June 11, 1909. He was commissioned a 2nd Lieutenant in the 1st Infantry. He married Kathleen Carroll on May 15, 1912. Their son was born the next year. Mitchell was transferred to the 24th Infantry on September 1, 1915 and was promoted to 1st Lieutenant on July 1, 1916. He served in the 24th Infantry on the Punitive Expedition into Mexico in 1916. He was promoted to Captain on May 15, 1917. Mitchell was assigned to the 39th Infantry, 4th Division at Camp Greene, North Carolina on February 8, 1918. He sailed to France as a Captain in Headquarters 39th Infantry aboard the Italian troopship *Duca D'Aosta* on May 10, 1918. As his emergency contact, he listed his wife Kathleen in Providence, Rhode Island.

Mitchell was promoted to the wartime temporary rank of Major on June 7, 1918 and given command of 2nd Battalion 39th Infantry. He led 2nd Battalion in the attack near the Ourcq River on July 18, 1918, the first day of the Aisne-Marne Offensive.

He was awarded the Distinguished Service Cross for his actions eighteen days later, still in the Aisne-Marne Offensive. On August 5, 1918, a determined effort to cross the Vesle River near the village of St. Thibaut was made by the 39th Infantry. Mitchell personally brought up Company G from his Battalion to the line occupied by Company I on the bank of the river and directed elements from Company G, Company H, and Company I in an attempt to move across the river and dig in on the opposite side. About thirty-eight men made it across. American artillery had been keeping most of the enemy across the river under cover and when the barrage lifted just before noon, the Germans came out and concentrated machine gun and rifle fire at the crossing point, forcing the rest of 2nd Battalion to remain on the St. Thibaut side of the river.

Enemy snipers were effective in preventing supporting soldiers from getting to the river with timber needed to use as footbridges, and also prevented reinforcements from getting to the riverbank. Mitchell moved about in the open in an effort to direct activities at the river and was struck in the leg by a sniper's bullet. Rather than be evacuated, he remained for the rest of the day at the river directing his Companies in a continuing attempt to get more men across the river until finally his wounds forced him to turn over his command and retire from the field.

Mitchell was awarded a Citation Star and the French Croix De Guerre with palm. He served on occupation duty in Germany with the 4th Division and was promoted to the wartime temporary rank of Lieutenant Colonel on May 6, 1919. On July 5, 1919, he was transferred to the 47th Infantry still in the 4th Division and the next day he was assigned as Commanding Officer of that Regiment. Mitchell returned to the United States in command of the 47th Infantry aboard the troopship *U.S.S. Mobile* (seized German liner *Cleveland*) leaving Brest, France July 16, 1919 and arriving at Hoboken, New Jersey July 27, 1919.

He went with the 4th Division to Camp Dodge, Iowa and on September 10, 1919 was relieved from his detail to the 47th Infantry and given command of the 39th Infantry. He continued in this assignment until the demobilization of the 4th Division began in 1920. He was returned to his Regular Army rank of Captain on June 30, 1920 and was promoted to the permanent rank of Major on July 1, 1920.

Mitchell was an instructor at the Infantry School at Camp Benning, Georgia from June 1920 to July 1921. He was Secretary of the Infantry School from September 1922 to July 1923. He graduated as a Distinguished Graduate from the Command & General Staff School in 1924. He graduated from the Army War College in 1927. On September 30, 1928 he was assigned to the General Staff in Washington, D.C.

Manton C. Mitchell died at the age of 41 on October 26, 1929 and is buried in Arlington National Cemetery. He was a member of the Rhode Island Society of the Sons of the American Revolution.

Hans E. Morgan

Private, United States Army
Army Service Number: 2023257
Company B 47th Infantry Regiment 4th Division
Date of Action: August 1, 1918
Location: Near Sergy, France
War Department General Orders 37 (1919)
Medal # 7480

CITATION

After all the other members of his automatic rifle squad had been wounded and evacuated and he himself wounded three times, Private Morgan remained at his post operating his automatic rifle against a machine gun nest until his supply of ammunition was exhausted. He then turned his rifle over to another squad before being evacuated.

DETAILS

Hans Edwin Morgan was born in Cherry Grove/Cadillac, Wexford County, Michigan on October 9, 1891, the son of Hans Christian and Martha Morgan. Before entering the Army, he worked on his father's farm in Axin, Michigan. He entered the Army on November 21, 1917. Morgan sailed to France as a Private in Company B 47th Infantry 4th Division aboard the troopship *U.S.S. Princess Matoika* (seized German liner *Prinzess Alice*) on May 10, 1918. As his emergency contact, he listed his father Hans in Axin, Michigan.

Morgan was awarded the Distinguished Service Cross for his actions in the Aisne-Marne Offensive during the bitter fighting at the village of Sergy. In four days of battle for the village, Morgan's 1st Battalion 47th Infantry, along with the 3rd Battalion 47th Infantry, suffered horrendous casualties as they fought with elements of the German 4th Prussian Guards Division. Large numbers of enemy machine guns and snipers were encountered all across the area in front of Sergy and the advance of the 47th Infantry was one of dealing with those threats on a one-by-one basis. On August 1, 1918, after everyone in his squad was wounded and removed from the battlefield, even though wounded himself three times, he remained alone at his forward position firing his French Model 1915 Chauchat automatic rifle until he ran out of ammunition. Only then did he allow himself to be evacuated. Before leaving, he made sure his rifle went to another automatic rifle squad who could put it to good use.

For his gallantry and dedication, Morgan also received a Citation Star. He was one of 20 soldiers of the 4th Division and one of only 399 soldiers of the American Expeditionary Forces to be awarded the Italian War Merit Cross (Croce al Merito di Guerra).

Because of his wounds, he returned to the United States several months ahead of his Regiment. He was transferred to Company D 338th Infantry 85th Division which was scheduled to be demobilized at Camp Custer, Michigan (his home State) soon after its arrival in the United States and sailed with that unit aboard the troopship *U.S.S. Leviathan* (seized German liner *Vaterland*) leaving Brest, France March 26, 1919 and arriving at Hoboken, New Jersey April 2, 1919. He was discharged on April 11, 1919.

Morgan was married to Helen Hector on September 20, 1919. In 1920 he was living in Flint, Michigan with his wife and a son who was born that year. In 1922, his daughter was born and in 1924 his second son was born. The 1930 Census showed him living with his family in Flint, Genesee County, Michigan where he was employed in Maintenance at an auto factory and indicated he was a veteran of the World War. The 1940 Census showed him and his family living in Flint where he was employed as a Foreman at an automobile factory. When he registered for the draft in 1942, he indicated he was still living in Flint and was employed by the Buick Motor Division Factory #25. Hans E. Morgan died at the age of 86 on November 27, 1977 and is buried in Flushing City Cemetery, Flushing, Genesee County, Michigan.

Michael D. Belis

Robert H. Murdoch

First Lieutenant, United States Army
Medical Detachment 47th Infantry Regiment 4th Division
Date of Action: July 29 – 31, 1918 and August 6 - 12, 1918
Location: At Sergy, France and at St. Thibaut, France
War Department General Orders 133 (1918)
Medal # No assigned number discovered

CITATION

Accompanying his battalion in the attack on Sergy Lieutenant Murdoch advanced for more than a mile under heavy shell fire and as soon as the southern half of the town had been taken, he established his dressing station, maintaining it during the three days of fighting under constant and severe bombardment. When his battalion went into action at St. Thibaut, this faithful officer again displayed heroic devotion to duty by working in his dressing station under the most trying conditions for six days while the town was bombarded with gas and high-explosive shells.

DETAILS

Robert Harrison Murdoch Jr. was born in Wilkes Barre, Pennsylvania on July 18, 1890, the son of Robert and Nancy Murdoch. He graduated from the Hahnemann Medical College of Philadelphia: Homeopathic Medical College of Pennsylvania in 1913 and became a physician that year. He was commissioned a 1st Lieutenant in the Officers Reserve Corps (Medical Department) on August 10, 1917 and assigned to the 47th Infantry at Syracuse, New York on August 28, 1917. He went with the Regiment to Camp Greene, North Carolina on October 26, 1917. He sailed to France as a 1st Lieutenant in Medical Detachment 47th Infantry 4th Division aboard the troopship *U.S.S. Princess Matoika* (seized German liner *Prinzess Alice*) on May 10, 1918. As his emergency contact, he listed his mother Nancy in Wilkes Barre, Pennsylvania.

He was awarded the Distinguished Service Cross for his actions on two separate occasions. The first action mentioned in Murdoch's Citation was in the Aisne-Marne Offensive while he was attached to 3rd Battalion 47th Infantry. On July 29, 1918, in the attack against the village of Sergy, he advanced with the Infantry and while the contest for the village was still going on he set up an improvised aid station at the southern edge of the village to treat the wounded as close to the front as possible. That first day Murdoch had some eighty wounded gathered in his aid station and he worked on them through enemy artillery and mortar fire. He continued to treat the wounded at his dangerously exposed aid station in Sergy until July 31, 1918.

The second action mentioned in his Citation was in the Vesle Sector. From August 6-12, 1918, the 47th Infantry was engaged in crossing the Vesle River and attempting to seize the village of Bazoches which lay across the river from the village of St. Thibaut. Elements of the 47th Infantry crossed the river but were pushed back by heavy German resistance. Murdoch maintained his aid station near the river in the village of St. Thibaut which was under almost constant bombardment by the enemy the entire time.

Murdoch also was awarded a Citation Star. He served in the Toulon Sector and the St. Mihiel Offensive. He was killed in action near the wooded area known as the Bois de Brieulles on September 26, 1918, the first day of the Meuse-Argonne Offensive and received a posthumous promotion to Captain.

Murdoch was buried in Plot 1 Grave # 11 in the American cemetery # 1342 at the village of Dannevoux.

His Distinguished Service Cross was presented to his mother Mrs. N. Ophelia Murdoch.

On June 3, 1919 Murdoch was reinterred in Plot 3 Section 46 Grave # 132 in the Argonne American Cemetery # 1232. Robert H. Murdoch was reinterred and laid to his final rest on November 22, 1921 in Plot E Row 25 Grave 22 in the Meuse-Argonne American Cemetery and Memorial, Romagne-sous Montfaucon, Departement de la Meuse, Lorraine, France.

Martin Nelson

Corporal, United States Army
Note: Though his Citation gives his rank as Corporal, Nelson was actually a Private at the time of his award actions and was not officially promoted to Corporal until later.
Army Service Number: 2106070
Company H 58th Infantry Regiment 4th Division
Date of Action: July 18 and 19, 1918
Location: Near Chezy, France
War Department General Orders 16 (1920)
Medal # 6122

CITATION

On the morning of the 18th of July, Corporal Nelson was wounded in the hip by a piece of shrapnel. A few hours later he was wounded in the arm by a bullet. He refused to be evacuated but continued forward in the attack. On the 19th, he was wounded in the left knee. In spite of his wounds, this noncommissioned officer continued with his organization throughout the campaign.

Michael D. Belis

DETAILS

Martin Leonard Nelson was born in Roseau, Minnesota on July 29, 1894. Before the war he was a farmer in North Dakota. He was drafted into the Army on September 21, 1917. He took his initial training with Company G 347th Infantry at Camp Dodge, Iowa. On March 9, 1918, he was transferred to Company H 58th Infantry at Camp Greene, North Carolina. Nelson sailed to England as a Private with Company H 58th Infantry 4th Division aboard the British troopship *S.S. Rhesus* on May 7, 1918. As his emergency contact, he listed his father Nicolia in Roseau, Minnesota. From England he and his Company were sent over to France.

Nelson was awarded the Distinguished Service Cross for his actions on two days in the Aisne-Marne Offensive. On July 18, 1918, the first day of the Offensive, Nelson's 3rd Battalion was working in support of the French 152nd Infantry Regiment near the village of Chezy. That first day, 3rd Battalion reached the southern slope of Hill 172 east of the village of Chevillon and dug in. Even though Nelson had been wounded twice, he remained with his Company. The next day, July 19, when the advance along the line continued, Nelson and his unit gave supporting fire from their position on the hill. Nelson was wounded that day for the third time in two days. He still refused to be evacuated and stayed with his Company until they were relieved two days later on July 21. He also received a Citation Star.

Nelson served in the Vesle Sector, the Toulon Sector, and was promoted to Corporal on September 8, 1918. He served in the St. Mihiel and Meuse-Argonne Offensives. Nelson served on occupation duty in Germany with the 4th Division and returned to the United States as a Corporal with Company H 58th Infantry aboard the troopship *U.S.S. Mount Vernon* (captured German liner *Kronprinzessin Cecilie*) leaving Brest, France July 24, 1919 and arriving at Hoboken, New Jersey August 1, 1919. He was discharged on August 8, 1919.

At an unknown date after being discharged he married his wife, Mattie. After 1932, he was awarded the Purple Heart with two oak leaf clusters for the wounds he received in World War One. The 1940 Census showed him living in Grand Forks, North Dakota and indicated he was a World War Veteran involved in government work. When he registered for the draft in 1942, he indicated he was still living in Grand Forks and was employed at the Club Room of the Court House there. Martin L. Nelson died at the age of 64 on February 19, 1959 and is buried in Greenwood Cemetery, Bemidji, Beltrami County, Minnesota.

Michael D. Belis

Francis K. Newcomer

Lieutenant Colonel, United States Army
Executive Officer 4th Engineer Regiment 4th Division
Date of Action: August 5, 1918
Location: Near Fismes, France
War Department General Orders 143 (1918)
Medal # 1092

CITATION

Lieutenant Colonel Newcomer made a reconnaissance along the south bank of the Vesle River in advance of the front lines for the purpose of selecting a bridge site. He then led a small party of engineers, assisted in the work of removing the German entanglements and constructing a foot bridge across the Vesle River, completing this work in the face of fire of great intensity. His coolness and personal bravery afforded an inspiring example to the men of his command.

DETAILS

Francis Kosier Newcomer was born in Byron, Ogle County, Illinois on September 14, 1889, the son of Brigadier General Henry C. and Rebecca Newcomer. He entered the U.S. Military Academy at West Point on March 1, 1909 and graduated first in a class of 93 cadets on June 12, 1913. At the Academy he was a Lieutenant in Company C of the Corps of Cadets and was the top Distinguished Cadet for 1913. Upon graduation, Newcomer was commissioned a 2nd Lieutenant in the Corps of Engineers. He was promoted to 1st Lieutenant on February 27, 1914 and to Captain on August 12, 1916. He was assigned to the 4th Engineers at Vancouver Barracks on June 3, 1917 as one of the first officers in the Regiment and was promoted to the war time temporary rank of Major on August 5, 1917. He was assigned command of 2nd Battalion of the 4th Engineers on January 1, 1918. Newcomer sailed to France as a Major with Regimental Headquarters 4th Engineers aboard the troopship *U.S.S. Martha Washington* (seized Austrian liner) on April 30, 1918. As his emergency contact, he listed his wife Mary in Pittsburgh, Pennsylvania. He was promoted to Lieutenant Colonel (temporary) on August 1, 1918.

Newcomer was awarded the Distinguished Service Cross for his actions in the Aisne-Marne Offensive. His Citation indicates the action occurred near Fismes, but it was actually closer to the village of Ville Savoye to the west of Fismes. All during the day of August 4, 1918, the Engineers had brought their bridging material close to the Vesle River at the village of Chéry-Chartreuve, southwest of Ville Savoye. The next day, August 5, 1918, a determined effort was made by elements of the 4th Division to cross the Vesle River. Newcomer advanced under fire across the battlefield in front of the American lines in order to scout for locations to emplace footbridges to aid the Infantry in getting across, although as Regimental Executive Officer he could have delegated the responsibility to a junior officer. He then personally led a small group of Engineers to the river where under fire they worked to remove barbed wire entanglements the Germans had placed along the riverbanks and in the chest deep water as well. Newcomer and his party also succeeded in installing a footbridge across the river. The presence of such a senior officer working right alongside the men under machine gun, mortar, and artillery fire was an inspiration to his Regiment.

From August 12, 1918 to February 1, 1919 Newcomer was Assistant Commandant of the U.S. Army Engineer School at Langres, France. He then rejoined the 4th Engineers on occupation duty in Germany and returned to the United States with 4th Engineer Headquarters aboard the troopship *U.S.S. Von Steuben* (seized German liner *Kronprinz Wilhelm*) leaving Brest, France July 21, 1919 and arriving at Hoboken, New Jersey July 29, 1919.

He reverted back to his permanent rank of Captain in 1919. He was promoted to Major on July 1, 1920. Newcomer was a professor at the U.S. Military Academy, an instructor at the Engineering School and served in various positions of high importance in the Corps of Engineers. On August 1, 1935 he was promoted to Lieutenant Colonel and on July 1, 1942 he was promoted to Colonel. Newcomer served in the China-Burma-India theatre of operations during World War Two and received the Legion of Merit.

In 1944, he was assigned to the Panama Canal command organization and in November 1944 he was promoted to Brigadier General. He was appointed Governor of the Panama Canal Zone in May 1948. He retired from the Army in September 1949 and continued as Governor of the Panama Canal Zone until 1952. Francis K. Newcomer died at the age of 78 on August 16, 1967 and is buried in Arlington National Cemetery.

Michael D. Belis

John H. Norton
Captain, United States Army
Note: Though his rank is given as Captain, John H. Norton was actually a 1st Lieutenant at the time of his award actions.
Commanding Officer Company K 47th Infantry Regiment 4th Division
Date of Action: July 29 - 30, 1918
Location: At Sergy, France
War Department General Orders 70 (1919)
Medal # 6968

CITATION
When the company on the left of his own had fallen back, leaving a gap through which the enemy was approaching for a counterattack, Captain Norton, with the remnants of two squads formed an automatic rifle post and successfully covered the withdrawal of the remainder of his command to a stronger line of resistance. Though his small group was almost annihilated by hostile fire, he held this position until the arrival of reinforcements, inflicting heavy losses on the enemy.

DETAILS

John Henry "Jack" Norton was born in West Springfield, Massachusetts on April 15, 1897, the son of Captain Paul and Mabel Norton. He entered the U.S. Military Academy on June 15, 1914 and due to the United States entry into the war was graduated early with his class on August 30, 1917, finishing 90 out of a class of 152. He was commissioned a 2nd Lieutenant in the 47th Infantry, immediately promoted to 1st Lieutenant and joined his Regiment at Camp Syracuse, New York on October 9, 1917. He moved with the 47th Infantry to Camp Greene, North Carolina on October 27, 1917 where he was assigned command of Company I. From November 27, 1917, to January 15, 1918, he was at Fort Sill, Oklahoma attending a course in machine gun and automatic rifle firing. Norton sailed to France as a 1st Lieutenant in the 47th Infantry and part of the advance detachment of the 4th Division aboard the troop transport *U.S.S. Finland* on April 30, 1918. As his emergency contact, he listed his mother in Springfield, Massachusetts. At some time in France before his Regiment saw its first action, Norton was assigned to command of Company K 47th Infantry.

He was awarded the Distinguished Service Cross for his actions on two days in the Aisne-Marne Offensive. During the battle of Sergy, Norton's Battalion, 3rd Battalion 47th Infantry was attached to the 168th Infantry Regiment of the 42nd Division. On July 28, 1918, the 168th Infantry had fought for the village of Sergy and had taken and lost it several times during the day. On the morning of July 29, it was driven out of Sergy again. Norton and his Battalion were then directed to attack Sergy. Norton led his Company in the advance through point blank fire from the German artillery positions.

With Company I on the right and Company L on the left leading the advance, the 47th Infantry pushed about halfway through the village taking heavy casualties among their ranks. At that point the Germans fired a terrific artillery barrage upon the village and Company L proceeded to withdraw. Norton's Company K had been acting in support of the two attacking Companies and when Company L retreated, it exposed Norton's left flank. At that location, the Germans began forming for a counterattack. As Norton directed his Company to join the withdrawal, he assembled a handful of men and using French M1915 Chauchat automatic rifles set up a position from which he and his small party covered the retreat, all the while under heavy enemy fire until a machine gun Company from the 42nd Division arrived to assist. The next morning, July 30, at 9:00 in the morning the advance resumed. Norton's Battalion was down to a total of about 300 men with which to mount their attack. They were met by severe machine gun and artillery fire followed by a German counterattack. All of the ammunition for the Battalion's French Chauchat automatic rifles had been expended the day before and some of the soldiers were using captured German rifles and grenades, but the Battalion held their ground.

Norton was wounded on August 6, 1918 near the Vesle River. He was sent to Base Hospital No. 20 at Guyon where he remained until rejoining his Company in the middle of October and served in the Meuse-Argonne Offensive. He was promoted to the wartime temporary rank of Captain on November 5, 1918 and assigned as Regimental Adjutant. Norton became ill and was admitted to Base Hospital No. 87 at Toul, France on November 19, 1918. He died in the hospital two days later on November 21, 1918 of lobar pneumonia.

Norton was buried in Grave # 867 in the American military cemetery # 91 at Toul on November 23, 1918.

His Distinguished Service Cross was presented to his mother Mrs. Mabel C. Norton.

On February 13, 1922 Norton's body was disinterred for preparation and shipment. His remains were returned to the United States aboard the United States Army Transport *U.S.A.T. Cambrai* leaving Antwerp, Belgium March 19, 1922 and arriving at Brooklyn, New York March 29, 1922. He was one of 1,065 American dead being returned on the *Cambrai* on that voyage. John H. "Jack" Norton was laid to his final rest in the United States Military Academy Post Cemetery, West Point, Orange County, New York.

Michael D. Belis

John W. Norton
Sergeant, United States Army
Army Service Number: 557080
Company I 39th Infantry Regiment 4th Division
Date of Action: August 6, 1918
Note: Though his Citation gives the date of August 6, 1918, the actual date of John W. Norton's award action is August 5, 1918.
Location: Near St. Thibault, France
War Department General Orders 50 (1919)
Medal # 7058

CITATION
While leading his platoon toward the Vesle River, Sergeant Norton encountered extreme machine gun fire. Exposing himself to determine the exact location from which this fire was being made, he was seriously wounded but he continued to direct the fire of his men even after he was no longer able to move with them. His action greatly aided his platoon to advance and join the remainder of the company.

DETAILS
(In half of the records found for him his name is listed as John W. Norton. In the other half he is recorded simply as John Norton.)

John W. Norton was born in Central Falls, Rhode Island on December 24, 1875. He enlisted in the 26th U.S. Volunteer Infantry in 1899 and served in the Philippine Insurrection, being discharged in May 1901. In August 1901 he enlisted in the Regular Army and during the next sixteen years he served in the 27th, 13th, 20th, and 8th Infantry before being assigned to the 39th Infantry when it was formed at Syracuse, New York in 1917. Norton sailed to France as a Sergeant in Company I 39th Infantry 4th Division aboard the French troopship *Espagne* on May 8, 1918. As his emergency contact, he listed his sister Mrs. Mary McGhee in Central Falls, Rhode Island.

Norton was awarded the Distinguished Service Cross for his actions in the Aisne-Marne Offensive. On August 5, a determined effort to cross the Vesle River near the village of St. Thibaut was made by the 39th Infantry Regiment. German artillery, minenwerfer (trench mortar) and machine gun fire caused tremendous casualties among the attacking Americans. Captain Ralph Slate led Company I 39th Infantry through dense barbed wire entanglements with Norton in charge of a platoon and the Company reached the riverbank and dug in, all the while under intense enemy fire. The official history of the 39th Infantry in World War One picks up the narrative:

> "While advancing through the wire entanglements, Captain Slate was wounded, but refused to be evacuated, and led his men forward to the river. Sergeant John W. Norton, commanding the fourth platoon, was also wounded during this advance, having his right leg shot off. Despite the seriousness of his wounds, Sergeant Norton refused to be carried to the rear and directed the movement of his platoon until it reached the riverbank. For the heroism displayed in this action he was awarded the Distinguished Service Cross." [20]

For his valor and stirring example of leadership Norton was also awarded the French Croix De Guerre with gold star. Because of the severity of his wounds, once he was medically stable, Norton was evacuated to the United States aboard the

Michael D. Belis

troopship *U.S.S. Madawaska* (seized German liner *König Wilhelm II*) leaving France in late September 1918 and arriving at Newport News, Virginia on October 5, 1918. He spent considerable time in Army hospitals before being medically retired on November 6, 1919.

The 1930 Census showed Norton living with his wife Elizabeth in Central Falls, Providence County, Rhode Island where he was unemployed and indicated he was a veteran of the Spanish War and the World War. (At the time the Philippine Insurrection was considered an extension of the War with Spain.)

John W. Norton died at the age of 61 on November 1, 1937 and is buried in Saint Patricks Cemetery, Cumberland, Providence County, Rhode Island. His plain and unadorned grave marker gives no indication that he was a veteran of two wars or that he was awarded the Distinguished Service Cross, the Philippine Campaign Medal, the Victory Medal with battle clasp, the Croix De Guerre and entitled to the Purple Heart. It simply has a Christian cross, gives his name, home state, date of death, and states he was a Sergeant in the 39th Infantry of the 4th Division.

Robert William Norton

Captain, United States Army
Commanding Officer 2nd Battalion 39th Infantry Regiment 4th Division
Date of Action: October 11, 1918
Location: Near Cunel, France
War Department General Orders 46 (1919)
Medal # 3335

CITATION

During the action in the Bois De Forêt, Captain Norton, with another officer, braved the hazardous fire by going out into "No Man's Land" and capturing 20 Germans at the point of his pistol. Although he lost two of the enemy during the encounter, he personally conducted the remaining back to our lines.

DETAILS

Robert William Norton was born in Newark, New Jersey on November 12, 1895. At the age of eighteen he enlisted as a Pri-

Michael D. Belis

vate in the Medical Department of the Regular Army on May 15, 1914. During his enlistment, he was promoted to Private First Class, Sergeant, Sergeant 1st Class, and Mess Sergeant. On July 30, 1917, he was discharged in order to accept a commission as a 2nd Lieutenant of Infantry with date of rank back to June 14, 1917. He was immediately promoted to 1st Lieutenant and assigned to the 39th Infantry at Camp Syracuse, New York.

Norton sailed to France as a 1st Lieutenant in command of Company H 39th Infantry 4th Division aboard the Italian troopship *Dante Alegheiri* on May 10, 1918. As his emergency contact, he listed a friend Mrs. Mary B. Neenan in East Bloomfield, New York. (On his return voyage home, he would list her as his foster mother.) In France, on June 28, 1918, Norton was promoted to the wartime temporary rank of Captain with date of rank back to January 2, 1918.

He led Company H 39th Infantry in the attack on July 18, 1918 near the Ourcq and Savieres Rivers on the first day of the Aisne-Marne Offensive and was slightly wounded in action at some time between July 18 and July 21, 1918. He then led the Company in the Vesle Sector, the Toulon Sector, and the St. Mihiel Offensive. By the time of the Meuse-Argonne Offensive, Norton had been assigned as Commander of 2nd Battalion 39th Infantry.

Norton was awarded the Distinguished Service Cross for his actions in the Meuse-Argonne Offensive. On October 10, 1918, Norton led his Battalion in the attack toward Hill 299 east of the village of Cunel and north of the wooded area known as the Bois de Malaumont. As soon as the advance began, Norton narrowly averted death or serious injury when he was targeted by a German sniper whose shot missed Norton but knocked the pipe out of his mouth. The next day, October 11, the attack was resumed with Norton and 2nd Battalion leading the way. Moving into the wooded area of the Bois de Forêt to the east of Cunel, 2nd Battalion took the German defenses with a bayonet charge. Norton was with his men at the front of the attack and using his .45 caliber auto pistol, along with another officer, charged an enemy position, captured twenty soldiers and personally conducted them back to the American lines, losing two on the way.

Norton served on occupation duty in Germany with the 4th Division. At Brest, France as his Regiment gathered for an awards ceremony before leaving for home, Norton was presented with the French Knight of the Legion of Honor (Chevalier de la Légion d'Honneur) and was twice awarded the French Croix De Guerre with palm. He returned to the United States in command of Company H 39th Infantry 4th Division aboard the troopship *U.S.S. Leviathan* (seized German liner *Vaterland*) leaving Brest, France July 30, 1919 and arriving at Hoboken, New Jersey August 6, 1919.

The 1920 Census showed Norton with the 4th Division at the camp at Gary, Indiana in January 1920. He was promoted to the permanent rank of Captain on July 1, 1920. He left the 4th Division and served in Hawaii. Norton was retired on November 11, 1922 but was recalled to active duty on October 3, 1924. He was finally retired from the Regular Army at the rank of Captain with a disability in the line of duty on August 15, 1932. He served as a Captain in the 3rd Regiment of the New York National Guard from November 25, 1940 until he was transferred to the retired list on June 23, 1943. Robert W. Norton died at the age of 82 on October 13, 1978 and is buried in Beaufort National Cemetery, Beaufort, Beaufort County, South Carolina.

Michael D. Belis

Cornelius J. O'Brien
Private, United States Army
Army Service Number: 568759
Company D 4th Engineer Regiment 4th Division
Date of Action: August 11, 1918
Location: Near Ville Savoye, France
War Department General Orders 16 (1923)
Note: General Orders 16 (1923) rescinded a previous Citation for the Distinguished Service Cross issued to Cornelius J. O'Brien Company E 2d [sic] Engineers in General Orders 142 (1918)
Medal # No assigned number discovered

CITATION
While engaged on the construction of a bridge over the Vesle River Private O'Brien voluntarily left shelter during intense fire and carried one of his wounded officers through a heavy machine gun and artillery barrage to a dressing station.

DETAILS
Cornelius J. O'Brien was born in Ennistymon, County Clare, Ireland on July 28, 1895. He immigrated to the United States at the age of 16 sailing from Queenstown, Ireland aboard the British liner *RMS Franconia* and arriving at Boston, Massachusetts on October 4, 1911. He established residency in the State of Montana on October 10, 1911. On February 24, 1915 he filed a Declaration of Intention to become a naturalized American citizen with the U.S. Department of Labor Naturalization Service. Before entering the Army, he was living in Butte, Montana where he worked as a miner for the Badger State Mining Company.

O'Brien enlisted as a Private in the 4th Engineers (Regular Army) on June 9, 1917. He served with Headquarters Detachment until July 11, 1917 and was then transferred to Company B. He was transferred back to Headquarters Detachment on August 22, 1917 and served with them until December 11, 1917. He then served with Company E until February 14, 1918 when he was transferred to Company D. On February 21, 1918, O'Brien was promoted to Corporal.

O'Brien sailed to France as a Corporal in Company D 4th Engineers 4th Division aboard the troopship *U.S.S. Martha Washington* (seized Austrian liner) on April 30, 1918. As his emergency contact, he listed his mother Kate in Ennistymon, County Clare, Ireland. He was reduced to the rank of Private on June 25, 1918. O'Brien served in the Aisne-Marne Offensive.

He was awarded the Distinguished Service Cross for his actions in the Vesle Sector. On August 11, 1918, elements of the 4th Division were holding positions on both sides of the Vesle River. The 4th Engineers were constantly busy constructing footbridges and artillery bridges across the river. Their efforts were hampered by enemy artillery and by German machine guns and mortars situated on the north side of the river. O'Brien was part of a detachment building an artillery bridge across the Vesle River on August 11. During a period of intense enemy fire O'Brien left a position of safety and carried a wounded officer to a first aid station near the village of Ville Savoye at the risk of his own life. He was one of four enlisted men from Company D 4th Engineers who along with their officer were awarded the Distinguished Service Cross for their actions during the building of that artillery bridge across the Vesle River under fire on August 11, 1918.

O'Brien was promoted to Corporal on August 15, 1918.

He served in the Toulon Sector and the St. Mihiel Offensive. He was promoted to Sergeant on September 15, 1918. He was awarded a Citation Star for his actions in the Meuse-Argonne Offensive on September 30, 1918 in 4th Division General Orders Number 81 (1918).

O'Brien died on October 19, 1918 the day when all units of the 4th Division except for the Artillery and the Ammunition Train were pulled out of the front lines. He is recorded in the official history of the 4th Engineers in World War One as dying of causes other than wounds. However, the Register of Burials of Deceased American Soldiers, 1917 – 1922, Office of the Quartermaster General has him recorded as dying from wounds received in action.

O'Brien was buried in Plot F Row 1 Grave # 3 in the American cemetery # 624 at Rampont.

His Distinguished Service Cross was presented to his mother Mrs. Kate Mullins O'Brien.

On March 28, 1922 Cornelius J. O'Brien was reinterred and laid to his final rest in Plot H Row 4 Grave 15 in the Meuse-Argonne American Cemetery and Memorial, Romagne-sous Montfaucon, Departement de la Meuse, Lorraine, France.

Morton Osborn

Sergeant, United States Army
Army Service Number: 558182
Company H 47th Infantry Regiment 4th Division
Date of Action: August 7 - 9, 1918
Location: Southeast of Bazoches, France
War Department General Orders 137 (1918)
Medal # 7159

CITATION

Wounded in the head and shoulder, Sergeant Osborn rejoined his platoon as soon as his wounds had been dressed and remained with it until the command was relieved, displaying rare qualities of leadership and judgment under heavy machine gun and rifle fire.

DETAILS

Morton Osborn was born in Load, Greenup County, Kentucky on February 7, 1891, the son of Nathan and Mollie Osborn. Before entering the Army, he was a farmer. He enlisted as a Private in Company M 4th Infantry on March 21, 1912. He served with the Regiment on Mexican Border Service in 1913-1914 and on the Vera Cruz Expedition in 1914. He was discharged on March 20, 1915. Osborn once again entered the Army on April 6, 1917. He sailed to France as a Sergeant in Company H 47th Infantry 4th Division aboard the troopship *U.S.S. Princess Matoika* (seized German liner *Prinzess Alice*) on May 10, 1918. As his emergency contact, he listed his father Nathan in Load, Kentucky. Osborn served in the Aisne-Marne Offensive.

He was awarded the Distinguished Service Cross for his actions spanning three days in the Vesle Sector. From August 7-9, 1918, Osborn's 2nd Battalion 47th Infantry was designated as the spearhead of the attack against the village of Bazoches across the Vesle River from the village of St. Thibaut. Casualties were heavy from enemy artillery fire but by the end of August 7, about 350 soldiers of 2nd Battalion had made it across the river, some on a footbridge that was built by the 4th Engineers, some using trees felled at the river and laid across it, and some by wading or swimming across the chest deep water. Machine guns in the buildings and on the rooftops in Bazoches, in positions in front of the village and artillery on the high ground behind the village took a murderous toll on the 47th Infantry. Gas masks which got wet in crossing the river were useless and it was estimated that one third of the German artillery shells fired at the Americans were gas.

On August 8, the enemy fired about seven thousand artillery rounds in the area around St. Thibaut and at the Americans who were across the river in front of Bazoches. Most of the day was spent digging in and consolidating positions. The footbridge at the river was hit by German mortars and rebuilt again and more of 2nd Battalion got across. On August 9, elements of 2nd Battalion got as close as one hundred yards from Bazoches but were withdrawn since the units on their flanks could not make it that far. During those three days, Osborn stayed with his platoon even though wounded in the head and shoulder and presented his men with a stirring example of bravery and leadership. He was also awarded a Citation Star.

Osborn recovered from his wounds, returned to duty, and was wounded again. His injuries caused him to be evacuated from the theatre several months ahead of his Regiment. He returned to the United States as a Sergeant of the Transportation Corps and part of the St. Aignan Casual Company No. 431 (destined for early discharge at Camp Devens, Massachusetts) aboard the British/Canadian troop transport *RMS Canada* leaving Brest, France January 10, 1919 and arriving at Boston, Massachusetts January 21, 1919. He was discharged on January 22, 1919.

The 1920 Census showed Osborn living with his parents in Greenup County, Kentucky where he was unemployed. At some time between 1920 and 1925 he married his wife, Kansas. The 1930 Census showed him living with his wife, two sons and daughter next door to his parents in Greenup County where he was a farmer on a truck farm and indicated he was a veteran of the World War. His registration for the draft in 1942 indicated he was living in Load, Greenup County, Kentucky and was self employed. Morton Osborn died at the age of 77 in Greenup County, Kentucky on October 13, 1968.

Michael D. Belis

Paul J. Pappas
Private, United States Army
Army Service Number: 2659513
Company M 39th Infantry Regiment 4th Division
Date of Action: October 12, 1918
Location: Near Argonne Forest, France
Note: The location mentioned in his Citation of Argonne Forest is inaccurate. At the time of his award actions, Pappas and his Company were operating in the wooded area known as the Bois de Forêt some sixteen miles to the northeast of the area known as the Argonne Forest.
War Department General Orders 44 (1919)
Medal # 3178

CITATION
When his company withdrew from their position, Private Pappas, with one other soldier saw the enemy forming for a counterattack and without thought of their danger, refused to withdraw but held this part of the line for several hours by the efficient use of an automatic rifle, subject to withering machine gun fire during the entire time.

DETAILS

Paul John Pappas was born in Turkey on May 22, 1895. He immigrated to the United States in 1913. When he registered for the draft in 1917, he indicated he had declared his intention to become a naturalized American citizen. He was drafted into the Army as a Private on May 26, 1918 and listed his home of record as Niles, Ohio. He was promoted to Corporal on June 15, 1918. Pappas sailed to England as a Corporal of Infantry and a replacement in the Camp Gordon July Automatic Replacement Draft Company #7 of Infantry aboard the British troopship *Canopic* on July 21, 1918. As his emergency contact, he listed a friend Peter Harlampy in Niles, Ohio. From England, Pappas was sent over to France where on August 14, 1918 he was assigned to Company M 39th Infantry 4th Division. On August 18, 1918 he was reduced to the rank of Private. Pappas served in the Toulon Sector and the St. Mihiel Offensive.

He was awarded the Distinguished Service Cross for his actions in the Meuse-Argonne Offensive. On October 12, 1918, the 39th Infantry was holding positions south of Hill 299 in the wooded area known as the Bois de Forêt. This would be the farthest north the 4th Division would reach during the Offensive. At half past noon, the Germans laid a heavy artillery barrage on the front lines causing about 300 casualties in the 39th Infantry and forcing a withdrawal of the Americans back to the southern edge of the woods. As his Company pulled back, Pappas and another soldier observed the enemy gathering for an attack and remained at their position, covering the withdrawal of their Company by using an automatic rifle. Pappas held his position under heavy fire by the Germans for more than two hours until the barrage ended, and his Company moved forward to resume their previously abandoned positions.

Pappas was also awarded a Citation Star and the French Croix De Guerre with gold star. He was one of 20 soldiers of the 4th Division and one of only 399 soldiers of the American Expeditionary Forces to be awarded the Italian War Merit Cross (Croce al Merito di Guerra).

In the official history of the 39th Infantry in World War One published in 1919, Pappas was one of 16 men in the 39th Infantry chosen by their fellow soldiers as the bravest men in the Regiment. He was described in that history with the following:

> "PRIVATE PAUL J. PAPPAS entered the army from Ohio. In the Argonne fighting displayed great coolness and bravery. As the line was withdrawing, Private Pappas saw the enemy forming for a counterattack and refused to withdraw. Although subjected to a withering machine gun fire, by efficient use of his automatic rifle, he held that section of the line for several hours."[21]

He served on occupation duty in Germany with the 4th Division and returned to the United States with Company M 39th Infantry aboard the troopship *U.S.S. Leviathan* (seized German liner *Vaterland*) leaving Brest, France July 30, 1919 and arriving at Hoboken, New Jersey August 6, 1919. He was discharged on August 20, 1919.

Pappas attended Asbury University in Kentucky and in October 1926 became an ordained Methodist Deacon. He married Margaret Broadfoot in 1928. They became missionaires to Greece. The 1930 Census showed Pappas, his wife and son living in Tarpon Springs, Pinellas County, Florida with his employment indicated as a Preacher for the Greek Mission and indicated he was a veteran of the World War. The 1940 Census showed him living with his wife, two sons, two daughters, his wife's parents, his wife's sister and brother in Tarpon Springs where he was employed as a Clergyman in Evangelistic work. His wife Margaret died in January 1978. Paul J. Pappas died at the age of 83 on July 26, 1978 and is buried in Osceola Memory Gardens, Kissimmee, Osceola County, Florida.

Michael D. Belis

James K. Parsons
Colonel, United States Army
Commanding Officer (Temporary) 39th Infantry Regiment 4th Division
Date of Action: September 27 - October 11, 1918
Location: Near Cuisy, France
War Department General Orders 98 (1919)
Medal # No assigned number discovered

CITATION
Having volunteered to take command of a battalion whose commander had been wounded, Colonel Parsons was knocked down by hostile shell fire, but he succeeded in rallying his men and kept them well organized so as to withstand the heavy fire of the enemy. On the following day, he assumed command of the regiment and commanded it in successful attacks, refusing to be evacuated after being so severely gassed that he was unable to see.

DETAILS

James Kelly Parsons was born in Rockford, Coosa County, Alabama on February 11, 1877, the son of Lewis and Catherine Parsons. During the War with Spain, he was commissioned a 1st Lieutenant in the 3rd Alabama Volunteer Infantry on August 5, 1898. The 3rd Alabama consisted of black soldiers led by white officers and served in the continental United States during the war. On April 10, 1899, Parsons received a Regular Army commission as a 2nd Lieutenant in the 20th Infantry and served with Company M 20th Infantry in the Philippine Insurrection. He was promoted to 1st Lieutenant on February 2, 1901. On July 23, 1904 he married Volinda Henderson. He was promoted to Captain on January 27, 1908. In 1916 he was assigned as Inspector and Advisor for the New York National Guard. Parsons was promoted to Major on May 15, 1917 and on August 5, 1917 he was promoted to the wartime temporary rank of Lieutenant Colonel.

Parsons sailed to England as a Lieutenant Colonel of the 338th Infantry aboard the British troopship *R.M.S. Andania* on December 27, 1917. As his emergency contact, he listed his wife in Columbus, Ohio. The 338th Infantry was then sent over to France where its personnel were used as replacements for other units. Parsons was assigned to the General Staff American Expeditionary Forces Service of Supply where he served for several months. He was promoted to the wartime temporary rank of Colonel on July 30, 1918. In September 1918, he was attached to the 39th Infantry 4th Division as an observer.

Parsons was awarded the Distinguished Service Cross for his actions spanning two weeks in the Meuse-Argonne Offensive. On September 26, 1918, the opening day of the Offensive, 3rd Battalion 39th Infantry led the advance toward the village of Cuisy with 1st Battalion 39th Infantry following behind in support. Major Roy Winton, Commander of 1st Battalion was wounded. Parsons assumed command of 1st Battalion on September 27 and the attack was resumed in the direction of the village of Nantillois. A fierce German artillery barrage was fired upon the 39th Infantry that day and led to a withdrawal of the ground gained by the Americans. Parsons moved across the battlefield under fire and despite being knocked off his feet by an artillery round he directed 1st Battalion to halt the retreat, dig in, and consolidate their positions. On September 28, the 39th Infantry Regimental Commander Colonel Frank C. Bolles was wounded and had to relinquish his command. The Executive Officer of the Regiment, Lieutenant Colonel William Holliday, had been killed the day before on September 27 and thus on September 28 Parsons assumed command of the 39th Infantry Regiment.

For the next two weeks, Parsons led the 39th Infantry through heavy fighting, always at the front of the Regiment and alongside the men as they advanced across no man's land, inspiring his soldiers with his courage and leadership. He took every objective assigned to him. On October 10, 1918, the 39th Infantry moved into the wooded area known as the Bois de Fays and prepared to attack northward the next day. Early on the morning of October 11, before the advance could begin, Parsons and his entire Regimental staff were all wounded in a gas attack. He tried to remain in command, but he was temporarily blinded and had to be removed from the field.

Parsons remained in hospital from October 11 to November 8, 1918. At the end of November, he was assigned as Commander of the Embarkation Center at Saint-Nazaire, France which processed American service members for their post-war return trips to the United States. Parsons was awarded the Distinguished Service Medal for his command of the Embarkation Center and returned to the United States as a Colonel of Infantry and part of a detachment of unassigned officers aboard the transport *U.S.S. Manchuria* leaving St. Nazaire, France July 8, 1919 and arriving at Hoboken, New Jersey on July 18, 1919. Upon the demobilization of the Army after the war, he reverted back to his permanent rank of Major and was promoted to Lieutenant Colonel on July 1, 1920. He was promoted to Colonel on November 27, 1923.

Parsons had a long career in the Army and was an early advocate of the integration of black soldiers in the Army. He also believed in the mechanization of the Army and in the 1920's drew up plans for a mechanized force which was not implemented at the time but was similar in design to the armored infantry forces which later operated during World War Two. Parsons was promoted to Brigadier General on December 1, 1930 and to Major General on June 1, 1936. He commanded the 2nd Infantry Division 1936-1938 and Third Corps Area of the United States 1938-1940.

Parsons was instrumental in bringing the training of Army soldiers up to a wartime footing in the days leading up to World War Two. He is credited as being highly influential in the adoption of the Army's first modern lightweight wind and water-resistant outer garment, the Olive Drab Cotton Field Jacket which was standardized in June 1940, and which was commonly called the Parsons jacket. Upon reaching the mandatory retirement age of 64 he retired from the Army on February 28, 1941. James K. Parsons died at the age of 83 on November 8, 1960 and is buried in Arlington National Cemetery.

Michael D. Belis

Arthur Paulson
Sergeant, United States Army
Army Service Number: 560703
Company A 59th Infantry Regiment 4th Division
Date of Action: September 29, 1918
Location: Near Brieulles, France
War Department General Orders 89 (1919)
Medal # 7363

CITATION
While fearlessly exposing himself by walking along the front line, in order to convey orders to his platoon, Sergeant Paulson was shot three times through the stomach. He nevertheless refused to go to the rear until he had conducted the platoon to its new position, and then declined assistance, walking 500 yards under fire to the dressing station. Upon arriving there, he insisted on sitting up, saying that the stretchers were needed for others. He died shortly afterward, having exhibited exceptional qualities of leadership, courage, and devotion to duty.

DETAILS
Arthur Paulson was born in Cadillac, Wexford County, Michigan on August 7, 1886, the son of Joseph and Anna Paulson. He entered the Army on July 20, 1917 at Columbus, Ohio. Paulson sailed to England as a Corporal in Company A 59th Infantry 4th Division aboard the British troopship *Megantic* on May 3, 1918. As his emergency contact, he listed his mother Anna in Cadillac, Michigan. From England he and his Company were sent over to France. At some time after reaching France, he

was promoted to Sergeant. He served in the Aisne-Marne Offensive, the Vesle Sector, the Toulon Sector, and the St. Mihiel Offensive.

Paulson was awarded the Distinguished Service Cross for his actions in the Meuse-Argonne Offensive. On September 29, 1918, the 8th Brigade of the 4th Division spearheaded the right side of the 4th Division attack in the direction of the wooded area known as the Bois de Fays, to the southwest of the village of Brieulles-sur-Meuse. In the 59th Infantry, Paulson's 1st Battalion along with 2nd Battalion were the assault units while 3rd Battalion followed behind in support. The 1st Battalion, including Paulson's Company A, reached the German trenches on the slope of Hill 281 but were driven back by heavy enemy machine gun fire from the Bois de Fays and enemy artillery mostly from the east side of the Meuse River. Paulson exposed himself to enemy fire in order to steady his men and was wounded by three bullets to his stomach area during this phase of the battle. He refused to go to the rear to seek help for his wounds, remained with his platoon, directed the platoon through the withdrawal and into defensive positions they could hold. Only then did he proceed to the aid station in the rear, insisting he go alone and thereby sparing his men the danger of exposing themselves to enemy fire to assist him. He died of his wounds at the aid station the next day on October 1, 1918.

Paulson was buried in Plot 1 Section A Grave # 41 in the American cemetery # 607 at the village of Senoncourt.

His Distinguished Service Cross was presented to his mother Mrs. Anna Paulson.

On April 15, 1921 Paulson's body was disinterred for preparation and shipment. His remains were returned to the United States aboard the United States Army Transport *U.S.A.T.*

Michael D. Belis

Wheaton leaving Antwerp, Belgium June 19, 1921 and arriving at Hoboken, New Jersey July 2, 1921. He was one of 5,827 American dead brought home by the *Wheaton* on that journey. Arthur Paulson was laid to his final rest in Maple Hill Cemetery, Cadillac, Wexford County, Michigan.

Donald A. Pegg

Private, United States Army
Army Service Number: 11076
Medical Detachment 12th Machine Gun Battalion 4th Division
Date of Action: September 30, 1918
Note: His Citation has the date of his award action and therefore his date of death as September 30, 1918 while the American Battle Monuments Commission and the Register of Burials of Deceased American Soldiers, 1917 – 1922, Office of the Quartermaster General have his date of death as October 1, 1918.
Location: Near the Bois des Ogons, France
War Department General Orders 44 (1919)
Medal # 6701

CITATION

While engaged in administering first aid under terrific machine gun fire, Private Pegg voluntarily went to an especially dangerous position to care for a wounded soldier and in so doing was himself killed.

Michael D. Belis

DETAILS

Donald Arthur Frederick Pegg was born in Indianapolis, Marion County, Indiana on October 14, 1897, the son of George and Georgina Pegg. At the time he entered the Army he gave his home of record as Arlington, New Jersey and indicated he was an Art Student. He enlisted in the medical department of the Enlisted Reserve Corps of the Army on May 22, 1917 and took his training at Governor's Island, New York City, New York. Pegg sailed to France as a Private assigned to U.S. Army Base Hospital #8 aboard the troop transport *U.S.S. Finland* on August 7, 1917. As his emergency contact, he listed his father George at the First National Bank in Arlington, New Jersey. He served for eleven months at Base Hospital #8 at Savenay, France.

On July 17, 1918, he was transferred to the Medical Detachment of the 12th Machine Gun Battalion 4th Division. He served in the Aisne-Marne Offensive, the Vesle Sector, the Toulon Sector, and the St. Mihiel Offensive.

Pegg was awarded the Distinguished Service Cross for his actions in the Meuse-Argonne Offensive. On September 29, 1918, the 8th Brigade including the 12th Machine Gun Battalion spearheaded the right side of the 4th Division attack toward the Meuse River. By the end of the day the advance had stalled, and the Division dug in along a line from the wooded area known as the Bois de Brieulles to the wooded area known as the Bois des Ogons. From September 30 to October 2, progress was slow. The German lines were expertly emplaced, and their defenses were well coordinated. The Americans moved forward through the woods each of those days in a painstaking process of eliminating one enemy machine gun after another, one machine gun nest at a time. The air seemed to be alive with a constant flow of German machine gun bullets and enemy artillery fire was intense as well. As an aid man, Pegg moved about the battlefield under that heavy fire attending to the wounded and saving numerous lives. He went to help a soldier in a particularly exposed location and was struck down and killed.

Pegg was buried in plot 2 Grave # 64 in the American cemetery # 702 at the village of Septsarges.

His Distinguished Service Cross was presented to his father George A. Pegg.

On May 21, 1919 Donald A.F. Pegg was reinterred in Plot 3 Section 63 Grave # 154 in the Argonne American Cemetery # 1232. He was reinterred and laid to his final rest on October 11, 1921 in Plot H Row 11 Grave 31 in the Meuse-Argonne American Cemetery, Romagne-sous-Montfaucon, France.

Michael D. Belis

John C. Persons
Captain, United States Army
Adjutant, 47th Infantry Regiment 4th Division
Date of Action: August 8, 1918
Location: At St. Thibaut, France
War Department General Orders 9 (1923)
Medal # 1918

CITATION
While serving as adjutant, 47th Infantry, Captain Persons was instructed by his regimental commander to deliver a message to the brigade commander. The telephone lines to the rear having been destroyed, he proceeded under intense enemy fire through a narrow pass, accompanied by a corporal and private of his regiment. Exposed to constant enemy fire, he had reached a place of safety when he learned that the corporal had been hit by enemy fire. Immediately returning, he carried the corporal to a dressing station in a storm of machine-gun and rifle fire from the enemy lines, thus saving the soldier's life and in utter disregard for his own safety.

DETAILS
John Cecil Persons was born in Atlanta, Fulton County, Georgia on May 9, 1888, the son of William and Alice Persons. He graduated from the University of Alabama with a law degree in 1913. Upon graduation he married his wife Elonia in June 1913. He and his wife had a son in 1914 who died eighteen months later from pneumonia. Their first daughter was born in 1917. In May of 1917 Persons entered the Officers Training

Camp at Plattsburg, New York. On his registration for the draft in June 1917, Persons asked for exemption due to his dependent family, his occupation as a farmer, and his lumber business which provided lumber for the government. The exemption was not granted and on November 17, 1917, he was commissioned a Captain in the Officers Reserve Corps and assigned to the 47th Infantry at Camp Greene, North Carolina. He sailed to France as a Captain in Regimental Headquarters 47th Infantry 4th Division aboard the troopship *U.S.S. Princess Matoika* (seized German liner *Prinzess Alice*) on May 10, 1918. As his emergency contact, he listed his wife in Tuscaloosa, Alabama. Persons served in the Aisne-Marne Offensive.

He was awarded the Distinguished Service Cross for his actions in the Vesle Sector. Beginning on August 7, 1918, the 47th Infantry led the 7th Brigade attack from the village of St. Thibaut at the Vesle River. Elements of the Regiment made it across the river and established positions near the village of Bazoches. On August 8, enemy artillery attempted to dislodge the Americans from their positions on both sides of the river. During the day, upwards of seven thousand shells fell on St. Thibaut, in the area around the village and on the north side of the river opposite the village. German machine gun and rifle fire from positions across the river from St. Thibaut kept the 47th Infantry under cover and restricted most movement.

The Commander of the 47th Infantry, Colonel Robert H. Peck, needed to relay the details of his situation to the Brigade Commander, Brigadier General Benjamin A. Poore, who was at his headquarters not far from St. Thibaut. Peck instructed Persons to see that the message got to Poore. The artillery barrage had destroyed the phone lines, so the message had to be delivered by hand. Persons understood the importance of getting the information to Poore and undertook the mission, accompanied by two enlisted men. The three men made their way through heavy enemy fire and Persons and one soldier made it to a place of relative safety while the other enlisted man fell wounded. Persons charged back out into the deadly gunfire, picked up the wounded man, threw him over his shoulder, and ran through a fusillade of bullets to bring the soldier to safety and then carried him to a forward aid station.

Persons was promoted to Major on September 23, 1918. The remainder of his service overseas could not be determined. He returned to the United States as a Major of Infantry and one of over 2,000 American soldiers sailing aboard the Dutch liner *S.S. Rotterdam* leaving Brest, France February 8, 1919 and arriving at Hoboken, New Jersey February 17, 1919. He was discharged at Camp Dix, New Jersey on February 19, 1919.

He was vice-president and president of several banks in Alabama from 1922 to about 1943. On June 7, 1924 he was appointed Lieutenant Colonel in the Alabama National Guard. The 1930 Census showed Persons living in Birmingham, Alabama with his wife and two daughters where he was employed as the president of a bank and indicated he was a veteran of the World War.

He was promoted to Brigadier General in the National Guard on August 27, 1930. On November 10, 1940, Persons was promoted to Major General and given command of the 31st Infantry "Dixie" Division (National Guard). The Division was federalized on November 25, 1940. During World War Two, Persons commanded the Division in training in the United States and in March 1944 brought it to the Pacific theatre where he received the Distinguished Service Medal for his leadership of the Division in the New Guinea campaign. In

September 1944, he relinquished command of the Division and returned to the United States where he resumed duties with the Alabama National Guard. He retired as a Lieutenant General in 1948. John C. Persons died at the age of 86 on December 22, 1974 and is buried in Elmwood Cemetery, Birmingham, Jefferson County, Alabama.

Michael D. Belis

Oscar W. Peterson

Sergeant, United States Army
Army Service Number: 560685
Company A 59th Infantry Regiment 4th Division
Date of Action: July 19, 1918
Location: Near Courchamps, France
War Department General Orders 95 (1919)
Medal # 6177

CITATION

Discovering the enemy making a counterattack to the left flank of his platoon, Sergeant Peterson immediately organized a combat group of 25 men and though greatly outnumbered by the Germans, he succeeded in routing them, inspiring his men by his disregard of personal danger. He was severely wounded later in the day, but he refused to go to the rear until he had reorganized his platoon an hour and a half later.

DETAILS

Oscar William Peterson was born in Jamestown, North Dakota on July 18, 1898, the son of Erick and Ellen Peterson. Before entering the Army, he worked as a boilermaker. He enlisted as a Private in Company A 4th Infantry (Regular Army) at Fort Thomas, Kentucky on March 24, 1917. When elements of the 4th Infantry were used to form the 59th Infantry, he was transferred to Company A 59th Infantry on June 12, 1917, and he thus became an original member of the 59th Infantry. He was promoted to Private 1st Class on July 8, 1917, and to Corporal on August 3, 1917. Peterson went with the 59th Infantry to

Camp Greene, North Carolina as it became a part of the 4th Division. He was promoted to Sergeant on March 21, 1918. He sailed to England as a Sergeant in Company A 59th Infantry 4th Division aboard the British troopship *Megantic* on May 3, 1918. As his emergency contact, he listed his mother in Jamestown, North Dakota. From England he and his Company were sent over to France.

Peterson was awarded the Distinguished Service Cross for his actions in the Aisne-Marne Offensive. On July 19, 1918 the second day of the Offensive, 1st Battalion 59th Infantry which included Peterson's Company A was attached to the French 164th Division and attacked from a point eastward of the village of Courchamps. The first day of the Offensive had caught the Germans by surprise but on the second day they had reorganized and met the American advance with determined resistance. Peterson's Battalion got as far as a low crest about 300 meters beyond the Courchamps-Priez road and was stopped by heavy machine gun and artillery fire. The Battalion Commander was wounded, and seven other officers were either killed or wounded.

Peterson gathered a group of 25 men and directed them in repulsing an enemy counterattack, inspiring his men with his courage and leadership. Along the attacking line, the German firepower was overwhelming however, and 1st Battalion had to pull back across the road. Here Peterson was seriously wounded but remained in command of his platoon until he got them dug in and prepared to defend their position. Only then did he allow himself to be helped to the aid station.

Because of his wounds, Peterson was evacuated back to the United States. He sailed as a Sergeant of Company A 59th Infantry and part of a detachment of sick and wounded from various units aboard the troopship *U.S.S. Leviathan* (seized German liner *Vaterland*) leaving Brest, France October 9, 1918 and arriving at Hoboken, New Jersey October 16, 1918. He was discharged as a Sergeant at Camp Dodge, Iowa with a Surgeon's certificate of 20% disability on January 11, 1919.

The 1920 Census showed him living with his parents in Jamestown, North Dakota where he was employed as a retail merchant in the groceries industry. Sometime after 1920 he was married. The 1930 Census showed him living with his wife Clara and their two sons in Jamestown where he was employed as a spices salesman and indicated he was a veteran of the World War. The 1940 Census showed him living with his wife, son, and his mother in Valleyford, Washington where he was a farmer. Oscar W. Peterson died at the age of 75 in 1974 and is buried in Greenwood Memorial Terrace, Spokane, Spokane County, Washington.

James J. Pirtle
First Lieutenant, United States Army
Company I 59th Infantry Regiment 4th Division
Date of Action: October 4 - 5, 1918
Location: In the Bois de Fays, France
War Department General Orders 126 (1919)
Medal # 5559

CITATION
Throughout the engagement in the Bois de Fays, Lieutenant Pirtle led his men with absolute disregard for his personal safety. He walked up and down the lines under intense enemy machine gun and artillery fire, encouraging his men and consolidating his position. His courageous example contributed greatly to the success of the operation in which his organization was engaged. He continued in action until severely wounded in the knee and was carried from the field.

DETAILS

James Julian Pirtle was born in Carlisle, Sullivan County, Indiana on July 12, 1897, the son of George and Sarah Pirtle. (The Army Registers give his year of birth as 1896.) He attended Wabash College in Indiana and dropped out to attend Officers Training Camp at Fort Benjamin Harrison, Indiana in 1917. He was commissioned a 2nd Lieutenant of Infantry in the Officers Reserve Corps on August 15, 1917. On May 1, 1918, he was given a commission in the Regular Army as a 2nd Lieutenant of Infantry and assigned on that date to the 59th Infantry at Camp Greene, North Carolina. Pirtle sailed to England as a 2nd Lieutenant in Company I 59th Infantry 4th Division aboard the armed British troopship *RMS Olympic* on May 5, 1918. As his emergency contact, he listed his father Dr. George W. Pirtle in Carlisle, Indiana. From England he and his Company were sent over to France.

Pirtle served in the Aisne-Marne Offensive. He served in the Vesle Sector where on August 11, 1918 he was wounded in action for the first time. He was promoted to the wartime temporary rank of 1st Lieutenant on September 12, 1918, the first day of the St. Mihiel Offensive.

He was awarded the Distinguished Service Cross for his actions on two days in the Meuse-Argonne Offensive. On October 4, 1918, Pirtle's 3rd Battalion 59th Infantry entered the wooded area known as the Bois de Fays to assist the 58th Infantry which had gained a foothold in the woods. The whole forest was a mess of shattered trees and tangled underbrush through which the Germans had strung barbed wire. Enemy machine gun nests were everywhere and advancement through the Bois de Fays was painstakingly slow and produced a high rate of casualties. Enemy artillery rained down on the Americans and also exploded in air bursts overhead.

Territory gained had to be held against German counterattacks and then once those were repulsed the Infantry would move forward again and slowly gain more ground. During October 4-5, Pirtle was at the front of his men, leading his platoon forward in the attack then moving along the front line exposed to enemy fire, holding them together in the defense against counterattacks until he was severely wounded and had to be brought to an aid station. He spent the next four and a half months in a hospital. He also received a Citation Star.

Because of his wounds, Pirtle returned to the United States several months ahead of the rest of his Regiment as a 1st Lieutenant in Company I 59th Infantry and part of Convalescent Detachment 56 aboard the U.S. Naval Transport *U.S.S. Sixaola* leaving Bordeaux, France February 2, 1919 and arriving at Gravesend Bay, New York on February 17, 1919. The passenger list indicated his condition as gunshot wound to the right leg.

He continued his Army career and was promoted to the permanent rank of Captain on July 1, 1920. He received a Bachelor of Arts Degree from Wabash College in 1925 and married Margaret Clay that same year. They had a daughter in 1927 and a son in 1933. (His son graduated from the U.S. Military Academy in 1955.)

Pirtle was the U.S. Army Drill Master for the University of Indiana marching band in 1929-1930 and then served as an Infantry officer on garrison duty in the Philippines for several years before World War Two. After 1932, he was awarded the Purple Heart with oak leaf cluster for the wounds he received in World War One. He was promoted to Major on August 1, 1935, and Lieutenant Colonel on August 18, 1940. During World War Two, he was the commander of the 131st Infantry Regiment and then Chief of Staff of the 6th Infantry Division, both during training in the United States. In 1944 he sailed to England, then was sent over to France and was the Commander of the 3rd Replacement Depot as it moved across Europe during the war.

Pirtle was promoted to Colonel on March 11, 1948. Also, in 1948 his Citation Star from World War One was converted to a Silver Star Medal. On occupation duty in Germany after the war, he became the Commanding Officer of the Augsburg Military Post in 1950. His final duty in the Army was Inspector General of 2nd Army at Fort Meade, Maryland. He retired from the Army as a Colonel on August 31, 1954. For his service during World War Two, he was awarded the Legion of Merit and the Army Commendation Medal. James J. Pirtle died on his 70th birthday on July 12, 1967 and is buried in Bonaventure Cemetery, Savannah, Chatham County, Georgia.

Richard G. Plumley

Captain, United States Army
Regimental Adjutant 39th Infantry Regiment 4th Division
Date of Action: September 27 - October 11, 1918
Location: Near the Bois de Septsarges, France
War Department General Orders 98 (1919)
Medal # No assigned number discovered

CITATION

On duty as regimental adjutant, Captain Plumley left a place of safety and going forward under heavy fire assisted in reforming the assault battalion which had lost most of its officers and was becoming disorganized. During the following days, he repeatedly crossed areas which had been subjected to heavy gas bombardments and as a result became almost blind and greatly weakened by gas poisoning. He refused to be evacuated however, and remained on duty throughout the night, rendering valuable assistance to the regimental commander who had just taken command.

Michael D. Belis

DETAILS

Richard Gardiner Plumley was born in Hammonton, New Jersey on February 3, 1894, the son of Alexander and Mary Plumley. He attended Connecticut Agricultural College in Storrs, Connecticut where for three years in addition to his studies he received military instruction in the days before the existence of R.O.T.C. After college he went to work as a salesman of office supplies in Hartford, Connecticut. Plumley enlisted as a Private in Company K 1st Infantry Connecticut National Guard on November 1, 1915. He was with that unit at Nogales, Arizona on Mexican Border Service from June 20 - October 12, 1916. While with the National Guard he was promoted to Corporal on August 2, 1916 and then to Sergeant on February 23, 1917. He attended Officer's Training Camp at Plattsburg Barracks, New York from May 15 - August 14, 1917 and on August 15, 1917 was commissioned a 2nd Lieutenant in the Officers Reserve Corps and assigned to Company I 30th Infantry at the camp at Syracuse, New York.

On October 26, 1917, Plumley was given a commission as a 2nd Lieutenant in the Regular Army and immediately promoted to 1st Lieutenant (temporary). On October 30, 1917, he moved with the 30th Infantry to Camp Greene, North Carolina. While at Camp Greene he was transferred to Headquarters Company 39th Infantry 4th Division on December 12, 1917. Plumley sailed to France as a 1st Lieutenant in the 39th Infantry and part of the advance detachment of the 4th Division aboard the troop transport *U.S.S. Finland* on April 30, 1918. As his emergency contact, he listed his uncle Dr. A.M. Phillips in Glenbrook, Connecticut.

Plumley was assigned to Company I 39th Infantry after reaching France. He served in the Aisne-Marne Offensive.

In the fighting in the Vesle Sector, he commanded the Stokes Mortar section of Company I. (The Stokes Mortar was a British 3-inch portable mortar and the forerunner of today's U.S. 81mm mortar.) Plumley served in the Toulon Sector and the St. Mihiel Offensive. On September 12, 1918, he was promoted to the wartime temporary rank of Captain. After his promotion, he was assigned as Regimental Adjutant of the 39th Infantry.

Plumley was awarded the Distinguished Service Cross for his actions spanning a two-week period in the Meuse-Argonne Offensive. On September 27, the 3rd Battalion 39th Infantry led the left side of the 4th Division attack and had crossed the Nantillois road, taking heavy casualties but was stopped at Hill 266. The Battalion Commander was wounded, and most all other officers became casualties as well. The Battalion began withdrawing in a piecemeal fashion. Plumley rushed out across the battleground under fire, took charge of the Battalion, and directed it to dig in at the base of the hill, thereby preventing any further retreat.

Colonel J. K. Parsons assumed command of the 39th Infantry on September 28 after its commander Colonel Frank C. Bolles had been wounded. For the next ten days, the Regiment was positioned in the wooded area known as the Bois de Septsarges and on October 9 the 39th Infantry led another advance into the wooded area known as the Bois de Fays. During those ten days, Plumley constantly moved across the battle zone, under fire and subjected to the effects of gas, in order to insure coordination among the Battalions of the Regiment. On October 11, Parsons and the entire Regimental staff of the 39th Infantry including Plumley were wounded by a gas attack directly upon their position. Parsons remained on the battlefield and attempted to continue his command but had to be evacuated as he was unable to see. He would be awarded a Distinguished Service Cross for his actions during the two weeks he commanded the Regiment.

Lieutenant Colonel Troy H. Middleton from the 47th Infantry was directed to join the 39th Infantry and take command. Plumley, though severely injured by the gas and nearly blind would not leave the battlefield until he had thoroughly acquainted Colonel Middleton with all the information needed to take effective command of the Regiment.

Plumley spent the next two months in three different hospitals in France recovering from the effects of chlorine and mustard gas before returning to his Regiment on December 5, 1918. He served on occupation duty in Germany with the 4th Division and returned to the United States with Regimental Headquarters 39th Infantry aboard the troopship *U.S.S. Leviathan* (seized German liner *Vaterland*) leaving Brest, France July 30, 1919 and arriving at Hoboken, New Jersey August 6, 1919. He resigned from the Army at Camp Dodge, Iowa on September 18, 1919.

The 1920 Census showed him living with his parents in Hartford, Connecticut where he was employed as a clerk at an insurance company. At some time between 1920 and 1924 he married Mabel Sponsel. The 1930 Census showed him living with his wife and their daughter in Darien, Fairfield County, Connecticut where he was employed as assistant comptroller of production at a lock factory and indicated he was a veteran of the World War. The 1940 Census showed him, his wife and daughter, still in Darien where he was employed as manager of the lock factory. Richard G. Plumley died at the age of 82 on September 27, 1976 and is buried in Fairfield Memorial Park at Stamford, Connecticut.

Benjamin A. Poore

Brigadier General, United States Army
Commanding Officer 7th Infantry Brigade 4th Division
Date of Action: September 27 and October 11, 1918
Location: At Bois de Septsarges, France and at Bois de Fays, France
War Department General Orders 44 (1919)
Medal # 1669

CITATION

At Bois de Septsarges on September 27, General Poore personally reformed his disorganized troops who were falling back through lack of command and because of severe casualties. Under heavy fire he led them to the lines and presented an unbroken front to the enemy. Again, on October 11, in the region of Bois de Fays he gathered together troops who were taking refuge from hostile fire and turned them over to the support commander.

Michael D. Belis

DETAILS

Benjamin Andrew Poore was born in Center, Cherokee County, Alabama on June 22, 1863, the son of Marion and Susan Poore. He entered the U.S. Military Academy on September 1, 1882 and graduated 33 out of a class of 77 on July 1, 1886, being commissioned a 2nd Lieutenant of Infantry. One of his classmates at West Point was John J. Pershing who would later be his Supreme Commander in the American Expeditionary Forces in France. On June 20, 1888, he married Adelaid Carlton. Poore was promoted to 1st Lieutenant on September 16, 1892, and was assigned to the 6th Infantry. He received a Citation for gallantry in action in Puerto Rico in 1898 during the War with Spain. Poore was promoted to Captain on March 2, 1899. He received another Citation for gallantry in action in the Philippine Islands in 1899 during the Philippine Insurrection. (After 1932, both Citations would be converted to Silver Star Medals.) He was promoted to Major on October 8, 1908 and assigned to the 22nd Infantry. Poore served as a Battalion Commander in the 22nd Infantry on Mexican Border Service 1910-1912. On August 4, 1914 he was promoted to Lieutenant Colonel unassigned.

Poore was promoted to Colonel on July 1, 1916. On August 5, 1917 he was promoted to the wartime temporary rank of Brigadier General. He joined the 4th Division at Camp Greene, North Carolina on April 4, 1918 and was assigned as Commander of 7th Infantry Brigade of the Division. Poore sailed to France with Headquarters 7th Brigade 4th Division aboard the troopship *U.S.S. Princess Matoika* (seized German liner *Prinzess Alice*) on May 10, 1918. As his emergency contact, he listed his wife in Charlotte, North Carolina.

Poore led his Brigade in the Aisne-Marne Offensive. In the

Vesle Sector he temporarily commanded the 4th Division from August 16-27, 1918. He served with his Brigade in the Toulon Sector and the St. Mihiel Offensive.

He was awarded the Distinguished Service Cross for his actions on two separate dates in the Meuse-Argonne Offensive. On September 26, 1918, the first day of the Offensive, Poore's Brigade spearheaded the left side of the 4th Division attack. The next day, September 27, was the first date of his award actions. On that day, the 39th Infantry Regiment was leading the advance and taking heavy casualties, including its Executive Officer Lieutenant Colonel William E. Holliday who was killed. The Regiment was stopped at Hill 266 by intense German artillery fire and the front line began to retire toward the wooded area known as the Bois de Septsarges. Poore charged out onto the battlefield, halted the retreating troops on the left side of the attacking line, and personally led them through enemy fire and established them into defensive positions from which he directed them in repulsing further enemy counterattacks. At the time, he was 55 years old.

On October 11, 1918, the second date of his award actions, Poore's Brigade was again spearheading the advance, this time out of the wooded area known as the Bois de Fays and toward the wooded area known as the Bois de Forêt. As the troops crossed a large open area, German machine guns and trench mortars opened up a heavy fire, causing the attacking elements to halt, scatter, and seek cover. Once again Poore moved across the battleground under fire and organized the soldiers into a cohesive force which he then turned over to the support Battalion as it moved up into the lines.

Poore was the only General Officer to be awarded the Distinguished Service Cross for actions performed while serving with the 4th Division in World War One. He again temporarily commanded the 4th Division from October 22-31, 1918.

Poore commanded 7th Brigade on occupation duty in Germany. When the Commanding General of the American Expeditionary Forces John J. Pershing visited the 4th Division at Büchel, Germany on March 18, 1919 and awarded decorations to the Division's soldiers, the first awards of the day were bestowed upon Poore, when he received from Pershing the Distinguished Service Cross and the Distinguished Service Medal. Poore also was awarded the Officer of the French Legion of Honor (Officier de la Légion D'Honneur) and the French Croix De Guerre. He was one of 20 soldiers of the 4th Division and one of only 399 soldiers of the American Expeditionary Forces to be awarded the Italian War Merit Cross (Croce al Merito di Guerra).

When the Fourth Division Association was formed in Germany not long before the Division came home, Poore was elected as its first President by the officers and enlisted men of the Division.

Poore returned to the United States as a Brigadier General unassigned aboard the troopship *U.S.S. Mount Vernon* (captured German liner *Kronprinzessin Cecilie*) leaving Brest, France July 24, 1919 and arriving at Hoboken, New Jersey August 1, 1919. He was returned to his permanent rank of Colonel on March 15, 1920 and commanded the 1st Infantry Regiment. He was promoted to Brigadier General on December 21, 1921 and commanded 4th Brigade 2nd Division.

On October 11, 1925 he was promoted to Major General and given command of 7th Corps Area of the United States. He was a member of the court martial board at the trial of Colonel Billy Mitchell in 1925. He retired from the Army as a

Major General on June 22, 1927 after 45 years of service. Benjamin A. Poore died at the age of 77 on August 27, 1940 and is buried in Arlington National Cemetery.

Michael D. Belis

Ernest R. Potter

First Sergeant, United States Army
Army Service Number: 556150
Company D 39th Infantry Regiment 4th Division
Date of Action: August 7, 1918
Location: Near St. Thibaut, France
War Department General Orders 98 (1919)
Medal # 1796

CITATION

When all the officers of his company had become casualties and the morale of the men was sinking, Sergeant Potter assumed command and after reorganizing the company successfully led it in repelling several vicious hostile counterattacks. During the action he was wounded in the shoulder, but he refused to go to the rear until he was ordered to do so by the officer sent to relieve him.

DETAILS

Ernest R. Potter was born in Clarion, Pennsylvania on August 5, 1886, the son of John and Sarah Potter. He enlisted in the United States Marine Corps on June 8, 1908. As a Marine he served in the continental United States, Panama, the Philippines, Cuba, and on sea duty aboard various ships. He served as a Corporal on the Vera Cruz Expedition in Mexico in 1914. He was discharged from the Marines as a Private on September 12, 1916. The next day on September 13, 1916 he enlisted as a Private in the United States Army. He was promoted to Private First Class on September 18, 1916 and to Corporal on December 1, 1916. On October 18, 1917 he was promoted to Sergeant. Potter sailed to France as a Sergeant in Company D 39th Infantry 4th Division aboard the Italian troopship *Dante Alegheiri* on May 10, 1918. As his emergency contact, he listed his sister, Mrs. Anna Wyant in Tarentum, Pennsylvania. He was promoted to First Sergeant in France on June 5, 1918. Potter served in the Aisne-Marne Offensive.

He was awarded the Distinguished Service Cross for his actions in the Vesle Sector. By August 4, 1918, the village of St. Thibaut had been occupied by the 39th Infantry. On August 5, the Regiment attempted to get across the Velse River in order to attack the village of Bazoches. German artillery and machine gun fire was intense and small groups of the 39th Infantry who managed to cross the river had to be withdrawn because flanking support could not be established. A four-hour artillery barrage was laid upon the Germans on August 6, followed by an unsuccessful attempt by the 4th Engineers to emplace a bridge across the river.

During the night, Engineers cut telephone poles, lashed them together, and laid them across the river as a footbridge.

Potter and his Company crossed over this bridge and fought with the enemy on the other side of the river. On August 7, all of the officers of his Company became casualties and Potter took command of the Company. Though wounded himself, he directed the Company in repulsing several German counterattacks and remained in command until a replacement officer could be sent to relieve him.

In the official history of the 39th Infantry in World War One published in 1919, Potter was one of 16 men in the 39th Infantry chosen by their fellow soldiers as the bravest men in the Regiment. He was described in that history with the following:

> "FIRST SERGEANT ERNEST R. POTTER after leaving Tarentum, Pennsylvania, served eight years with the Marines. Snap sparkles from his brown eyes; if their asking does not get results, twelve years of service-hardened muscles will. Wounded in the Vesle fighting but refused to go to the rear until three German counter attacks had been repulsed. His comrades are as proud of their "top kicker's" D. S. C. as he is."[22]

Potter served on occupation duty in Germany with the 4th Division and returned to the United States as First Sergeant of Company D 39th Infantry aboard the troopship *U.S.S. Leviathan* (seized German liner *Vaterland*) leaving Brest, France July 30, 1919, and arriving at Hoboken, New Jersey August 6, 1919. In January 1920 he was at the camp at Gary, Indiana with part of the skeleton organization that was the 4th Division as it began the process of demobilization. He was discharged on September 12, 1920.

Eleven months later he enlisted as a Private in the U.S. Marines on August 6, 1921. He was promoted to Sergeant in 1922. He served in the continental United States and overseas in the Pacific area. In 1926, he was sick in the hospital at the Naval Station at Cavite in the Philippines. He was shipped back to the United States on September 9, 1926, and upon reaching the States immediately entered the Naval Hospital in San Diego, California where he died at the age of 42 on January 2, 1927. Ernest R. Potter is buried in Mount Airy Cemetery, Natrona Heights, Allegheny County, Pennsylvania.

Though he spent 13 years in the Marines his grave marker is inscribed with his Army service. The inscription reads: "1st SERGT. Co. D. 39th INF. ENLISTED SEPT. 13, 1916 DISCHARGED SEPT. 12, 1920 AGE 42."[23]

John H. Pratt Jr.
Second Lieutenant, United States Army
Company G 47th Infantry Regiment 4th Division
Date of Action: August 7 - 9, 1918
Location: Near Bazoches, France
War Department General Orders 138 (1918)
Medal # 1076

CITATION
Second Lieutenant Pratt was untiring and fearless at all times in the performance of his duties as liaison officer. Under heavy fire he made three exceptionally hazardous trips with messages of vital importance when other means of communication had failed, volunteering for this service.

DETAILS
John Humphrey Pratt, Jr. (he was actually John Humphrey Pratt III) was born in New York City, New York on August 22, 1894, the son of John H. and Katherine Pratt. Before entering the Army, he attended Trinity College in Hartford, Connecticut. He entered Officers Training Camp at Plattsburg Barracks, New York on July 25, 1917, was commissioned a 2nd Lieutenant of Infantry on November 27, 1917 and was assigned to the 47th Infantry at Camp Greene, North Carolina. He sailed to France as a 2nd Lieutenant in Company G 47th Infantry 4th Division aboard the troopship *U.S.S. Princess Matoika* (seized German liner *Prinzess Alice*) on May 10, 1918. As his emergency contact, he listed his father in New York. Pratt served in the Aisne-Marne Offensive.

He was awarded the Distinguished Service Cross for his actions spanning three days in the Vesle Sector. On August 7, 1918, another attempt was made by the 4th Division to capture the village of Bazoches across the Vesle River from the village of St. Thibaut. The 47th Infantry led the way and managed to get about 350 soldiers across the river who consolidated their positions and dug in. German artillery and machine gun fire was intense everywhere on the battlefield. The village of St. Thibaut was especially hard hit by a bombardment of high explosive and gas artillery rounds. On August 8, elements of the Regiment continued to hold positions across the river and tried to move reinforcements through St. Thibaut to the river. Enemy artillery was even more active than the previous day and an estimated seven thousand shells fell upon the Americans.

The French 62nd Division on the left flank of the 4th Division failed to advance and the American attack stalled. The 47th Infantry spent August 9 patrolling and searching for and destroying German machine gun nests. During the three days of August 7-9, Pratt made three trips through heavy enemy fire to deliver important messages. Though he could have designated the duty to a subordinate, he volunteered to personally carry the messages, sparing the enlisted men the danger and placing that danger upon himself.

At some time after his award action, Pratt was promoted to 1st Lieutenant. He served in the Toulon Sector, the St. Mihiel Offensive, and the Meuse-Argonne Offensive. He served on occupation duty in Germany with the 4th Division and returned to the United States several months ahead of his Regiment as a 1st Lieutenant of the 47th Infantry and a member of a casual detachment of officers destined for discharge at Camp Dix, New Jersey aboard the troopship *U.S.S. Imperator* (surrendered

German liner) leaving Brest, France May 15, 1919 and arriving at Hoboken, New Jersey May 22, 1919. He was discharged at Camp Dix, New Jersey on May 24, 1919.

The 1930 Census showed Pratt living in Rye Village, Westchester County, New York where he was employed as an assistant production manager for a postage meter company and indicated he was a veteran of the World War. In 1933 he married Marjorie Allen. The 1940 Census showed him living with his wife in Darien, Fairfield County, Connecticut where he was employed as an Executive in the Wholesale Meter industry. When he registered for the draft in 1942, Pratt indicated he was still living in Darien and was employed by the Pitney Bowes Postage Meter Company in Stamford, Connecticut. Sometime after 1953, he divorced Marjorie and in August 1959 he married Gladys Leitch. He divorced Gladys in May 1968 and in December 1968 married Dorothy Ray. John H. Pratt Jr. died at the age of 77 on February 3, 1972 and is buried in Lakeview Cemetery, New Canaan, Fairfield County, Connecticut.

Fred N. Rapp

Corporal, United States Army
Army Service Number: 572451
Machine Gun Company 59th Infantry Regiment 4th Division
Date of Action: October 6, 1918
Location: In the Bois de Fays, France
War Department General Orders 44 (1919)
Medal # 6852

CITATION

While exposed to an exceptionally heavy barrage in the Bois de Fay, Corporal Rapp left his shelter and went to the aid of a seriously wounded comrade. He was killed by a fragment from a high explosive shell while in the performance of this gallant mission.

DETAILS

Fred Norton Rapp was born in Lorimer, Union County, Iowa on August 26, 1889, the son of Benjamin and Sarah Rapp. Before entering the Army, he was a farmer in Alexandria, Hanson

Michael D. Belis

County, South Dakota. He enlisted in the Regular Army at Bay Horse, Powder River County, Montana on March 6, 1918 and was assigned to the 59th Infantry at Camp Greene, North Carolina. He sailed to England as a Private in Machine Gun Company 59th Infantry 4th Division aboard the British troopship *RMS Aquitania* on May 7, 1918. As his emergency contact, he listed his mother in Shorty, Wyoming. From England he and his Company were sent over to France where they were issued their machine guns. Rapp was promoted to Corporal in France in July, 1918. He served in the Aisne-Marne Offensive, the Vesle Sector, the Toulon Sector and the St. Mihiel Offensive.

He was awarded the Distinguished Service Cross for his actions in the Meuse-Argonne Offensive in the wooded area known as the Bois de Fays. (See the entry for Joseph Bassi for a description of the battlefield conditions in the Bois de Fays.)

On October 6, 1918, the 4th Division had for the moment stopped all attacks and ordered its units in the Bois de Fays to hold and improve their positions in case of German counterattacks. During one particularly intense enemy artillery bombardment, Rapp disregarded his own personal safety by leaving his dug in position and ventured out into the barrage to aid a wounded soldier. He was killed by that enemy artillery in the performance of this action.

Rapp was buried in Plot 2 Grave # 42 in the American battlefield cemetery # 700 at the village of Brieulles-sur-Meuse.

His Distinguished Service Cross was presented to his mother Mrs. Sarah C. Rapp.

On May 14, 1919 Rapp was reinterred in Plot 3 Section 44 Grave # 123 in the Argonne American Cemetery # 1232. On June 28, 1921 his body was disinterred for preparation and shipment. His remains were returned to the United States aboard the United States Army Transport *U.S.A.T. Wheaton* leaving Antwerp, Belgium August 6, 1921 and arriving at Hoboken, New Jersey August 20, 1921. He was one of 5,759 American dead being brought home aboard the *Wheaton* on that journey. Fred N. Rapp was laid to his final rest in Ridge Cemetery, Fremont, Dodge County, Nebraska.

Michael D. Belis

Carl Rasmussen

Private, United States Army
Army Service Number: 2250434
Company B 39th Infantry Regiment 4th Division
Date of Action: September 27, 1918
Location: Near the Bois de Brieulles, France
War Department General Orders 98 (1919)
Medal # 1837

CITATION

Private Rasmussen, a company runner, volunteered and made two trips from the post of command of his own regiment to that of the regiment adjoining his own, passing each time more than a thousand yards under intense enemy machine gun fire.

DETAILS

Carl Rasmussen was born in Lisbon, Linn County, Iowa on October 11, 1894, the son of Jacob and Catherine Rasmussen. When he registered for the draft in June 1917, he indicated he was living in Edinburg, Texas where he was a farmer. Rasmussen was drafted into the Army at the camp at McAllen, Texas on September 19, 1917. He was assigned to Company B 39th Infantry on March 11, 1918 and was promoted to Private First Class on May 2, 1918. He sailed to France as a Private First Class in Company B 39th Infantry 4th Division aboard the Italian troopship *Dante Alegheiri* on May 10, 1918. As his emergency contact, he listed his father in Grand Rapids, Wisconsin. He served in the Aisne-Marne Offensive, the Vesle Sector, the Toulon Sector and the St. Mihiel Offensive.

He was awarded the Distinguished Service Cross for his actions in the Meuse-Argonne Offensive. On September 27, 1918, the second day of the Offensive, the 7th Brigade led the left side of the 4th Division's attack with the 39th Infantry on the left and the 47th Infantry on the right. The Germans had been surprised the first day of the Offensive but on the second day they were reorganized and offered stiff resistance to the American advance. The battlefield was alive with machine gun and artillery fire and those two Regiments of the 4th Division suffered a high rate of casualties that day. Rasmussen volunteered to maintain communications between his own 39th Infantry in the area near Hill 295 and the 47th Infantry on the right at the southern edge of the wooded area known as the Bois de Brieulles. He crossed the deadly battleground twice with important messages, both times traversing more than 1,000 yards under the heaviest of enemy machine gun fire.

Rasmussen served on occupation duty in Germany with the 4th Division. He returned to the United States as a Private First Class in Company B 39th Infantry 4th Division aboard the troopship *U.S.S. Leviathan* (seized German liner *Vaterland*) leaving Brest, France July 30, 1919 and arriving at Hoboken, New Jersey August 6, 1919. He was discharged at Camp Grant, Illinois on August 13, 1919.

Michael D. Belis

The 1930 Census showed Rasmussen living with his mother in Milwaukee, Wisconsin where he was employed as a plasterer and indicated he was a veteran of the World War. He was admitted to the U.S. National Home for Disabled Soldiers, Northwestern Branch in Milwaukee, Wisconsin in 1931 and was discharged from there in 1932. The 1940 Census showed him living by himself in Milwaukee where he was still employed as a plasterer. He was still living in Milwaukee when he registered for the draft in 1942. Carl Rasmussen died at the age of 52 on February 28, 1947 and is buried in Wanderer's Rest Cemetery (Lincoln Memorial Cemetery), Milwaukee, Milwaukee County, Wisconsin. His headstone application with the War Department was signed by his brother Hermann Rasmussen.

Lee M. Ray

Corporal, United States Army
Army Service Number: 556193
Headquarters Company 39th Infantry Regiment 4th Division
Date of Action: August 5, 1918
Location: Near St. Thibaut, France
War Department General Orders 44 (1919)
Medal # 3175

CITATION

Corporal Ray, clerk of headquarters, volunteered and delivered important operations messages to the French regiments attacking on the left flank of the 39th Infantry. He made his way for about one and one-half miles through heavy artillery, machine gun, and sniping fire parallel to the enemy's line, located the French headquarters, and delivered the message in time to stop flanking attacks by the enemy.

DETAILS

Lee Miller Ray was born in Philadelphia, Pennsylvania on May 6, 1894, the son of Chester and Cuba Ray. When he registered for the draft in June 1917, he indicated he was employed as the Treasurer for Brilliant Manufacturing Company in Philadelphia. He enlisted as a Private in Headquarters Company 39th Infantry (Regular Army) at Camp Greene, North Carolina on February 24, 1918. He was promoted to Corporal on March 18, 1918. Ray sailed to France as a Corporal in Headquarters Detachment 39th Infantry 4th Division aboard the Italian troopship *Duca D'Aosta* on May 10, 1918. As his emergency contact, he listed his mother Cuba in Philadelphia, Pennsylvania.

Ray was awarded the Distinguished Service Cross for his actions in the Aisne-Marne Offensive. At the time of his award actions, he was a Corporal and clerk in Regimental Headquarters Company. On August 5, 1918, the 39th Infantry spearheaded the left side of the attack against the village of Bazoches across the Vesle River from the village of St. Thibaut. The 39th Infantry was on the extreme left of the 4th Division advance with the French 62nd Division on their left flank. There was no liaison with the French Division and when German movements on the battlefield indicated an attack was forming against the French flank, Ray acted as a runner to deliver messages of warning to the French command.

Carrying messages was not the normal duty of a Headquarters clerk such as Ray, but since he was in a position where he monitored communications among all units on the battlefield, he understood the significance and danger of the enemy activities and thus he volunteered to bring the warning to the French regiments. The Germans were responding to the 39th Infantry attack at the river with heavy artillery concentrations all along the front in and around St. Thibaut and with incessant fire from their machine guns across the river and rifle fire from their trenches. Ray ventured through all of that to get to the French units. Realizing the importance of his messages, he pushed on for about a mile and a half through heavy enemy fire without stopping and managed to arrive with his warning in time to allow the French to prepare for and defeat the German attacks.

Ray served in the Vesle Sector and the Toulon Sector. He was promoted to Regimental Sergeant Major on September 8, 1918 and served in the St. Mihiel and Meuse-Argonne Offensives. He served on occupation duty in Germany with the 4th Division and was presented the Distinguished Service Cross by the Commanding General of the American Expeditionary Forces John J. Pershing in a 4th Division ceremony at Büchel, Germany on March 18, 1919.

Ray returned to the United States ahead of his Regiment as a Regimental Sergeant Major of Infantry and part of the St. Aignan Casual Company No. 3487 destined for Camp Mills, New York and special discharge aboard the transport *S.S. Colombia* leaving Marseille, France April 19, 1919 and arriving at Brooklyn, New York May 8, 1919. He was discharged on May 13, 1919.

The 1920 Census showed Ray living with his brother and his brother's family in Philadelphia, Pennsylvania where he had returned to his job as Treasurer of his family's sign manufacturing business (Brilliant Manufacturing Company). Lee M. Ray died at the age of 27 on December 1, 1921 and is buried in West Laurel Hill Cemetery, Bala Cynwyd, Montgomery County, Pennsylvania.

Michael D. Belis

James T. Rice

Private First Class, United States Army
Army Service Number: 570794
Company C 8th Field Signal Battalion 4th Division
Date of Action: September 29 and October 11 - 13, 1918
Location: Near the Bois de Fays, France and in the Bois de Malaumont, France
War Department General Orders 89 (1919)
Medal # 2169

CITATION

While at work with a group of men maintaining telephone communication, Private Rice went out under heavy fire and carried to shelter a comrade who had been wounded by a bursting shell, returning immediately and repairing breaks in the line. During the action in the Bois de Malaumont, he repeatedly exposed himself to heavy artillery and machine gun fire in order to maintain telephone lines for the infantry, displaying remarkable courage.

DETAILS

James Thomas Rice was born in Aden, Carter County, Kentucky on October 5, 1890, the son of Jacob and Mary Rice. He enlisted as a Private in the Army Signal Corps on January 27, 1910 at Columbus Barracks, Ohio and listed his previous occupation as farmer. He was discharged as a Private 1st Class on January 30, 1913 at Fort McDowell, California with a character rating of Good. When he registered for the draft in June 1917, he indicated his home of record as Seattle, Washington and that he was employed as a farmer in Alberta, Canada. He enlisted as a Private in the Regular Army at Fort Lawson, Washington on July 21, 1917. He was promoted to Private 1st Class on March 1, 1918. Rice sailed to France as a Private 1st Class in Company C 8th Field Signal Battalion 4th Division aboard the troopship *U.S.S. Von Steuben* (seized German liner *Kronprinz Wilhelm*) on May 26, 1918. As his emergency contact, he listed his father in Kehoe, Kentucky. He served in the Aisne-Marne Offensive, the Vesle Sector, the Toulon Sector and the St. Mihiel Offensive.

He was awarded the Distinguished Service Cross for his actions on two separate occasions in the Meuse-Argonne Offensive. The first date was on September 29, 1918 when the 4th Division again attacked in the direction of the wooded area known as the Bois de Fay. Heavy machine gun fire from that forest and heavy artillery fire from east of the Meuse River prevented the Americans from gaining the woods. Rice ventured out onto the battlefield under heavy fire and brought back a wounded soldier to safety. He then went back into the shell and machine gun fire swept area to continue repairing broken and destroyed field telephone lines.

The second date in his Citation was also in the Meuse-Argonne Offensive from October 11-13, 1918. The Division had finally occupied the Bois de Fay and were pushing out of it into the wooded areas known as the Bois de Malaumont and the Bois de Forêt. Once again Rice demonstrated his bravery by working on damaged lines across the battlefield, exposed to heavy fire and was an inspiration to his fellow soldiers.

Rice was the only soldier in the 8th Field Signal Battalion to be awarded the Distinguished Service Cross in World War One.

Rice was promoted to Corporal on May 1, 1919. He served

on occupation duty in Germany with the 4th Division and was promoted to Sergeant on July 16, 1919. He returned to the United States with Company C 8th Field Signal Battalion aboard the troop transport *U.S.S. Aeolus* (seized German liner *Grosser Kurfürst*) leaving Brest, France July 18, 1919 and arriving at Newport News, Virginia July 29, 1919. He was discharged on August 9, 1919.

On May 27, 1925 Rice married Juliaette Hudson. The 1930 Census showed him living with his wife, daughter and son in Dearborn, Wayne County, Michigan where he was employed as a streetcar conductor and indicated he was a veteran of the World War. The 1940 Census showed Rice still living in Dearborn with his wife, three daughters and four sons where he was still employed as a streetcar conductor. When he registered for the draft in 1942, his information was still the same. James T. Rice died at the age of 70 on December 18, 1960 and is buried in Roseland Park Cemetery, Berkley, Oakland County, Michigan.

Stefano Riggio

Private, United States Army
Army Service Number: 1688585
Company K 39th Infantry Regiment 4th Division
Date of Action: September 28, 1918
Location: Near Septsarges, France
War Department General Orders 46 (1919)
Medal # 7052

CITATION

While his company was halted by machine gun and sniper fire from the front and both flanks, Private Riggio moved forward to outflank the enemy sniping posts. He was wounded in the execution of his mission, but he managed to make his way back and reported the information he had obtained.

DETAILS

Stefano Riggio was born in Salemi, Sicily, Italy on October 2, 1889, the son of Stephano and Crocifissa Riggio. When he registered for the draft in June 1917, he declared himself to be an alien and not a citizen of the United States, that his home of record was Rockland, Maine, that he was employed by the Rockland Rockport Line Company in Rockland, Maine, and that he had three years prior military service as a Private in Italy (though he did not specify details on that service). Riggio entered the Army on April 2, 1918. He was assigned to Company K 39th Infantry at Camp Greene, North Carolina on May 1, 1918. He sailed to France as a Private in Company K 39th Infantry 4th Division aboard the troop transport *U.S.S.*

Michael D. Belis

Lenape on May 10, 1918. As his emergency contact, he listed a friend Tony Marmoni in Rockland, Maine. Riggio served in the Aisne-Marne Offensive, the Vesle Sector, the Toulon Sector, and the St. Mihiel Offensive.

He was awarded the Distinguished Service Cross for his actions in the Meuse-Argonne Offensive. On September 28, 1918, 2nd Battalion 39th Infantry led the left side of the advance with Riggio's 3rd Battalion 39th Infantry in support. The two Battalions moved out from Septsarges in the direction of Nantillois to the northwest. The 79th Division on the left had not moved as far as the 4th Division, which presented the 39th Infantry with a left flank exposed to enemy fire and counterattacks. When his Company was held up by German machine gun and small arms fire, Riggio went ahead alone under fire and scouted the enemy's positions. Although he was wounded, he completed his mission, returned, and informed his commander of the location of enemy machine gun nests and sniper positions, thus allowing his Company to be prepared when they next moved out.

Riggio was also awarded the French Croix De Guerre with gilt star. Because of his wounds, he returned to the United States several months ahead of his Regiment as a Private in Company K 39th Infantry and part of the Beau Desert Company A-83 of sick and wounded aboard the troop transport *U.S.S. Pastores,* leaving Bordeaux, France December 25, 1918 and arriving at Newport News, Virginia January 6, 1919. He was noted in the passenger list as having gunshot wounds to the buttock. He was discharged on January 31, 1919.

He married Virginia Micione Adamo on November 11, 1920 and they had two sons and a daughter. The city directories for Boston, Massachusetts showed Riggio living there in 1930 and 1931 and working as a baker. Riggio became a naturalized American citizen in Rockland, Maine on September 13, 1922. On April 15, 1931 his wife Virginia died. On July 3, 1937 he married Josephine LaRosa in Fitchburg, Massachusetts. The 1940 Census showed Riggio and Josephine with three sons and daughter living in Chelsea, Massachusetts where he was employed as a baker in a bread bakery. His registration for the draft in 1942 showed him and Josephine still in Chelsea where he worked for the North End Bakery Shop. Stefano Riggio died at the age of 77 on December 13, 1966 and is buried in Woodlawn Cemetery, Everett, Middlesex County, Massachusetts.

Michael D. Belis

Lowell H. Riley
Second Lieutenant, United States Army
Company F 58th Infantry Regiment 4th Division
Note: His burial card from the Register of Burials of Deceased American Soldiers, 1917 - 1922 Office of the Quartermaster General gives his Company as Company L. His grave marker gives his Company as Company H. The official history of the 58th Infantry in World War One gives his Company as Company F.
Date of Action: August 7, 1918
Location: At Ville Savoye, northeast of Chateau-Thierry, France
War Department General Orders 116 (1918)
Medal # No assigned number discovered

CITATION
Lieutenant Riley maintained an observing station for his battalion commander for two days although subjected during the whole of this time to intense artillery bombardment. He obtained valuable information as to the movements of the enemy which was used in directing artillery fire. While engaged in this very important and hazardous work, he was killed by shell fire.

DETAILS

Lowell Hobart Riley was born in Orange, Essex County, New Jersey on August 28, 1896, the son of Abram and Jessie Riley. He entered the Army in November 1917. He sailed to England as a 2nd Lieutenant of Infantry in the Officers Reserve Corps in Company B 58th Infantry 4th Division aboard the armed British troopship *Moldavia* on May 6, 1918. As his emergency contact, he listed his mother in Orange, New Jersey. On May 23, 1918 the *Moldavia* was torpedoed by a German submarine and sank with the loss of 56 men, 55 of which were Riley's fellow soldiers from Company B 58th Infantry The other soldier was from Company A 58th Infantry which had also sailed aboard the *Moldavia*. Riley was rescued by one of the escorting British destroyers and brought to England. From England he and the other survivors of his Company were sent over to France. At some time after reaching France, he was transferred to Company F 58th Infantry. Riley served in the Aisne-Marne Offensive.

He was awarded the Distinguished Service Cross for his actions in the Vesle Sector. On August 7, 1918, Riley and Captain Peter W. Ebbert also of the 58th Infantry were killed in action together at the village of Ville Savoye near the Vesle River and both were awarded the Distinguished Service Cross. The following passage from the official history of the 4th Division in World War One describes the event:

"In Villesavoye there stands the ruin of an old church. On the corner of this church at one time there was built a tower. On August 4th, the town was captured and occupied by the 58th Infantry. It was necessary that the enemy's movements across the Vesle be observed. The one position from which these observations could be made was the tower of the old church, and it was by far the most outstanding target for the Germans. First Lieutenant Peter W. Ebbert and Second Lieutenant Lowell H. Riley, 58th Infantry, volunteered to establish an observation station in this tower. For two days they occupied it and sent in very valuable information. During the whole of the time, they were subjected to heavy artillery fire. On the evening of August 7th, an enemy shell crashed into the tower and both officers were instantly killed."[24] (The passage gives Ebbert's rank incorrectly. The Army Register for 1918 shows Ebbert was promoted to Captain on June 12, 1918.)

Riley was buried in Grave # 10 in cemetery # C-64 at the village of Ville Savoye.

His Distinguished Service Cross was presented to his father Abram M. Riley.

At an unknown date Riley was reinterred in Plot 1 Section G Grave # 23 in the American cemetery # 617 at the village of Fismes. On May 24, 1921 his body was disinterred for preparation and shipment. His remains were returned to the United States aboard the United States Army Transport *U.S.A.T. Wheaton* leaving Antwerp, Belgium June 19, 1921 and arriving at Hoboken, New Jersey July 2, 1921. He was one of 5,827 American dead brought home by the *Wheaton* on that journey. Lowell H. Riley was laid to his final rest in Rosedale Cemetery, Orange, Essex County, New Jersey on July 30, 1921.

Charles C. Rismiller

Private, United States Army
Army Service Number: 571132
Medical Detachment 4th Engineer Regiment 4th Division
Date of Action: August 5, 1918
Location: Near St. Thibaut, France
War Department General Orders 5 (1920)
Medal # 7783

CITATION

Private Rismiller went forward exposed to intense rifle, machine gun, and artillery fire and assisted a seriously wounded comrade to a place of safety, thus saving his life. In the performance of this gallant act, Private Rismiller was mortally wounded.

DETAILS

Charles C. Rismiller was born in Leesport, Berks County, Pennsylvania on July 4, 1896, the son of Clayton and Maggie Rismiller. The 1910 Census showed him living with his grandfather (his mother's father) David H. Kline in Ontelaunee, Berks County, Pennsylvania. Rismiller enlisted in the Regular Army at Fort Slocum, New York on September 1, 1914. He sailed to France as a Private in Medical Detachment 4th Engineers 4th Division aboard the troopship *U.S.S. Martha Washington* (seized Austrian liner) on April 30, 1918. As his emergency contact, he listed his grandfather David H. Kline in Leesport, Pennsylvania.

Rismiller was awarded the Distinguished Service Cross for his actions in the Aisne-Marne Offensive. On August 5, 1918, elements of the 4th Division made a determined effort to cross the Vesle River near the village of St. Thibaut in order to attack the village of Bazoches on the north side of the river. The 4th Engineers worked under fire incessantly to build and emplace several footbridges for the Infantry and a couple of artillery bridges for the Artillery to use to cross the river and in so doing the Engineers suffered their highest rate of casualties during the war. Machine gun and rifle fire from Bazoches and artillery fire from the heights behind the village was intense and caused a high amount of killed and wounded among the Americans. On the south side of the river near the village of St. Thibaut, Rismiller charged out onto the battlefield under heavy fire and rescued a wounded soldier, removing him to safety and thereby saving the soldier's life. Rismiller was mortally wounded during this action. In the ebb and flow of movements on the battlefield that day his body was not recovered. Rismiller's status with the American Battle Monuments Commission is Missing In Action.

His Record of Service Form No. 724-6 indicates that the person notified of his death was his grandfather David H. Kline in Leesport, Pennsylvania.

His Distinguished Service Cross was presented to his mother Mrs. Maggie Yeager.

The name of C.C. Rismiller is inscribed on the Tablets of the Missing at the Aisne-Marne American Cemetery and Memorial, Belleau, Departement de l'Aisne, Picardie, France.

Edward D. Ritchie

Private, United States Army
Army Service Number: 562140
Company M 47th Infantry Regiment 4th Division
Date of Action: August 10, 1918
Location: Near St. Thibaut, France
War Department General Orders 46 (1919)
Medal # 1559

CITATION

While on an outpost near the Vesle River, Private Ritchie volunteered to accompany Corporal John S. Weimer in rescuing a wounded soldier who had been left by members of a patrol in a shell hole some distance to the front. Under fire from machine guns and snipers, Private Ritchie and Corporal Weimer proceeded to the shell hole and found the wounded man who was unable to walk. Suggesting that the three of them in a group would make a more conspicuous target for the enemy, Private Ritchie offered to run ahead to draw the enemy fire while his comrade assisted the wounded man. He made his way back to shelter under continuous machine gun and sniper fire while Corporal Weimer carried the wounded soldier to safety.

Note: The Citation for Edward D. Ritchie gives John S. Weimer's rank as Corporal. Weimer was actually a Private at the time of the award action and would not be promoted to Corporal until November 1, 1918.

DETAILS

Edward Douglas Ritchie was born in Dallas, Texas on November 20, 1891, the son of Amos and Margaret Ritchie. Before entering the Army, he was a stock farmer. He was drafted into the Army as a Private on February 27, 1918. Ritchie sailed to England as a Private in Company M 47th Infantry and part of Casual Detachment No. 1 of the 4th Division (made up of soldiers from various units in the Division) aboard the troop transport *U.S.S. Louisville* on May 19, 1918. As his emergency contact, he listed his stepfather John Wilson in Stratford, Texas. From England he was sent over to France where he joined his Company. Ritchie served in the Aisne-Marne Offensive.

He was awarded the Distinguished Service Cross for his actions in the Vesle Sector. On August 10, 1918, the 47th Infantry was established in defensive positions along the railroad south of the Vesle River near the village of St. Thibaut. For several days the Regiment had tried to take the village of Bazoches, which was directly across the river on the north side. Elements of the 47th Infantry had managed to cross the river and enter the village itself before being driven back across the river by German counterattacks. On August 10, the 47th Infantry dug in south of the river and consolidated its positions in anticipation of being relieved in place by the 77th Division the next night.

A wounded soldier from a patrol lay in a shell hole in advance of the front lines and Ritchie and Private John S. Weimer from the same squad as Ritchie decided to make an attempt to retrieve the soldier. Together the two Texans made their way across the battlefield under fire to the shell hole. Ritchie suggested he run alone in the direction of the American lines to provide the enemy with a target to shoot at while Weimer hoisted the wounded man onto his back and carried him to safety. While Ritchie conspicuously drew the Germans' fire, Weimer managed to get the wounded comrade back to where he could be attended to by American medics. Though cited in different General Orders, both Ritchie and Weimer were awarded the Distinguished Service Cross for their actions in saving the wounded soldier.

Ritchie served in the Vesle Sector, the Toulon Sector and the St. Mihiel Offensive. He was promoted to Private 1st Class on September 17, 1918. He served in the Meuse-Argonne Offensive. Ritchie was also awarded the French Croix De Guerre. He was one of 20 soldiers of the 4th Division and one of only 399 soldiers of the American Expeditionary Forces to be awarded the Italian War Merit Cross (Croce al Merito di Guerra). The Italian War Merit Cross was presented to him on June 11, 1919 in a 4th Division awards ceremony at Remagen, Germany.

He served on occupation duty in Germany with the 4th Division and returned to the United States as a Private 1st Class in Company M 47th Infantry and as part of Company I Third Army Composite Regiment aboard the troopship *U.S.S. Leviathan* (seized German liner *Vaterland*) leaving Brest, France September 1, 1919 and arriving at Hoboken, New Jersey September 8, 1919.

(See the entry for Joseph Bassi for an explanation of the Third Army Composite Regiment.)

Ritchie was discharged on September 29, 1919 and returned to Texas. At some time before 1923, he married Eskie King. They had a daughter in 1923 and a son in 1925. The 1930 Census showed Ritchie and his family living in Stratford, Sherman County, Texas where he was a farmer and indicated he was a veteran of the World War. The 1940 Census showed him and his family still in Stratford where he was employed as a junior clerk with the Soil Conservation Service. When he registered for the draft in 1942, he indicated that he was living in Stratford and working for the Department of Agriculture Soil Conservation Service in Dalhart, Texas. By 1960 he and Eskie were living in San Antonio, Texas. Edward D. Ritchie died at the age of 81 on September 2, 1974 and is buried in Stratford Cemetery, Stratford, Sherman County, Texas.

James H. Roberts

Sergeant, United States Army
Army Service Number: 557227
Company K 39th Infantry Regiment 4th Division
Date of Action: September 26 - 28, 1918
Location: Near Montfaucon, France
War Department General Orders 46 (1920)
Note: General Orders 46 (1920) rescinded a previous Citation for the Distinguished Service Cross issued to James Roberts [sic] in General Orders 98 (1919)
Medal # 6259

CITATION

Sergeant Roberts displayed marked courage and self-sacrifice when after being wounded in the arm he refused to leave the battlefield and continued to perform his duties as platoon sergeant until he was wounded in the knee two days later and had to be carried from the field.

Michael D. Belis

DETAILS

James Henry Roberts was born in Rouses Point, Clinton County, New York on September 25, 1886, the son of Peter and Louise Roberts. He served an enlistment as a Private in Battery B 3rd Field Artillery which included duty on Mexican Border Service. Before entering the Army, he worked as a trolley repairman for the Connecticut Railway Company in Hartford, Connecticut. He enlisted in the Regular Army as a Private on February 22, 1918 and was assigned to Company K 39th Infantry at Camp Greene, North Carolina. He was promoted to Private 1st Class on March 11, 1918 and to Corporal on March 18, 1918. On April 26, 1918 he was promoted to Sergeant. Roberts sailed to France as a Sergeant in Company K 39th Infantry 4th Division aboard the troop transport *U.S.S. Lenape* on May 10, 1918. As his emergency contact, he listed his father Peter in South Manchester, Connecticut. Roberts served in the Aisne-Marne Offensive, the Vesle Sector, the Toulon Sector, and the St. Mihiel Offensive.

He was awarded the Distinguished Service Cross for his actions spanning several days in the Meuse-Argonne Offensive. On September 26, 1918, the first day of the Offensive, Roberts' 3rd Battalion 39th Infantry spearheaded the left side of the 4th Division attack. Moving out from a position east of the village of Malancourt, the Regiment advanced in the direction of the village of Cuisy. After clearing Cuisy, the Battalion occupied the ridge north of the village of Septsarges to the northeast of the village of Montfaucon. At this location, Roberts and Company K repulsed three German counterattacks. Despite being wounded early in the action, Roberts remained with his platoon throughout the attack, inspiring his men with his courage.

On September 27, Roberts' Battalion again led the advance.

Intense enemy fire halted the Battalion not far beyond the Nantillois road and Roberts and his men dug in while 1st Battalion joined them in the line. The Germans concentrated a terrible artillery barrage on the two Battalions and forced them into a withdrawal to the reverse slope of Hill 295. Through all of the action for those two days, Roberts ignored his wound and continued to lead his men as their platoon Sergeant.

On September 28, the advance northward resumed with 2nd Battalion leading the way while Roberts and 3rd Battalion followed close behind in support. In this attack, Roberts was wounded again and had to be evacuated to the rear. He was transported to Base Hospital #48 at Mars-sur-Allier, Department of Nievre in central France where he eventually succumbed to his wounds and died on November 1, 1918. He also received a Citation Star.

In the official history of the 39th Infantry in World War One published in 1919 Roberts was one of 16 men in the 39th Infantry chosen by their fellow soldiers as the bravest men in the Regiment. He was described in that history with the following:

> "SERGEANT JAMES H. ROBERTS. Soon after the "jump off" on September 26th, a sniper got him in the left arm. Though ordered back by his Company Commander, Sergeant Roberts remained on duty, directing the attack of his platoon for two days, when he was wounded in the left knee. He again chose to stick with his badly shattered, but irresistibly successful comrades, but was overruled and carried to the rear. His conduct set such an example to his men that they were all delighted to see him awarded the D. S. C."[25]

Michael D. Belis

Roberts was buried in grave # B104 in the American cemetery # 85 at the village of Mars-sur-Allier.

His Distinguished Service Cross was presented to his father Peter Roberts.

On June 15, 1919 James H. Roberts was reinterred in grave # 291 in the American cemetery # 395 at the village of Nevers. He was reinterred and laid to his final rest on June 17, 1922 in Plot A Row 6 Grave 36 in the St. Mihiel American Cemetery, Thiaucourt, France.

In 1930 his mother Louise declined the offer to visit her son's grave in France under the U.S. Government program of Gold Star Pilgrimages. See the entry for Edward S. Blackman for an explanation of the Gold Star Pilgrimages.

Leo D. Roberts
Sergeant, United States Army
Army Service Number: 559000
Company A 11th Machine Gun Battalion 4th Division
Date of Action: October 12, 1918
Location: Near Nantillois, France
War Department General Orders 87 (1919)
Medal # 7220

CITATION
After the infantry had fallen back 200 meters under heavy fire, Sergeant Roberts stayed at his one remaining machine gun and operated it until the infantry had re-established its position, capturing a German machine gun and three prisoners.

DETAILS
Leo Devaux Roberts was born in Harper, Logan County, Ohio on February 26, 1894, the son of Clarence and Minora "Minnie" Roberts. On April 25, 1917, he enlisted as a Private in Company D 30th Infantry (Regular Army) at the camp at Syracuse, New York. When elements of the 30th Infantry were used to create the 39th Infantry, Roberts was transferred to Company D 39th Infantry on June 1, 1917 and thus became an original member of the 39th Infantry. On October 1, 1917 he was promoted to Corporal. He went with the 39th Infantry to Camp Greene, North Carolina on October 30, 1917. At some time before sailing overseas, Roberts was transferred to Company A 11th Machine Gun Battalion. He sailed to France as a Corporal in Company A 11th Machine Gun Battalion 4th Division

aboard the troopship *U.S.S. Rijndam* (former Dutch liner) on May 10, 1918. As his emergency contact, he listed his mother Minnie in Bellefontaine, Ohio.

Roberts was promoted to Sergeant in France on June 4, 1918. He served in the Aisne-Marne Offensive, the Vesle Sector, the Toulon Sector, and the St. Mihiel Offensive.

He was awarded the Distinguished Service Cross for his actions in the Meuse-Argonne Offensive. By October 12, 1918 elements of the 4th Division had pushed the Germans out of the wooded area known as the Bois de Fays and had occupied the wooded area known as the Bois de Forêt in an area north/northeast of the village of Nantillois. Roberts' Company was operating in support of the 39th Infantry at that time. During the morning, several enemy counterattacks in the Bois de Forêt were defeated. At half past noon a heavy barrage was laid on the woods by the Germans, causing the Americans to pull back. Roberts and his platoon leader, Second Lieutenant Orval Kline, remained with their machine gun in an exposed forward position and held the ground until the Infantry could establish new defensive positions further back. Not only did they prevent the enemy from counterattacking, but they maneuvered their gun about the battlefield and captured a German machine gun and several prisoners. Their actions allowed the Infantry to withdraw in an orderly fashion and establish a new defensive perimeter further back. Although cited in different General Orders, both Roberts and Kline received the Distinguished Service Cross for the action. Roberts also received a Citation Star.

Roberts returned to the United States several months ahead of his Battalion as a Sergeant of Infantry and part of Le Mans Casual Company #1208 aboard the Brazilian troop transport *S.S. Sobral* leaving Brest, France February 15, 1919 and arriving at Brooklyn, New York March 1, 1919. He was discharged on March 15, 1919.

On May 29, 1920 he married Helen Zanetta Rose. The 1930 Census showed Roberts, his wife, their two daughters and son living in Columbus, Ohio where he was employed as a mattress maker at a wholesale mattress factory and indicated he was a veteran of the World War. The 1940 Census showed Roberts and his family still in Columbus where he was a finisher at the mattress factory. At some time after 1940, he married his second wife Julia. Leo D. Roberts died of cancer at the age of 64 on January 17, 1959 and is buried in Resthaven Park East Cemetery, Phoenix, Maricopa County, Arizona.

Michael D. Belis

Raymond D. Robertson
Sergeant, United States Army
Army Service Number: 569255
Company F 4th Engineer Regiment 4th Division
Date of Action: August 5, 1918
Location: West of Fismes, France
War Department General Orders 15 (1919)
Medal # 1081

CITATION

Sergeant Robertson was a member of a small detachment of engineers which went out in advance of the front line of the infantry through an enemy barrage from 77-mm and one-pounder guns to construct a footbridge over the River Vesle. As soon as their operations were discovered machine gun fire was opened upon them but undaunted, the party continued at work, removing the German wire entanglements and successfully completing the bridge which was of great value in subsequent operations.

DETAILS

Raymond Douglas Robertson was born in Valley Ford, Sonoma County, California on October 18, 1895, the son of John and Mary Robertson. Before the war, he was an engineer with a lumber company in California. When war was declared, he was in school in Oakland, California pursuing studies in civil engineering. He quit school and enlisted in the 4th Engineers (Regular Army) on June 7, 1917. He went with his Regiment to Camp Greene, North Carolina where it became part of the 4th

Division. He sailed to France as a Corporal with Company F 4th Engineers 4th Division aboard the troopship *U.S.S. Martha Washington* (seized Austrian liner) on April 30, 1918. As his emergency contact, he listed his sister Rose Linebaugh in Two Rock, California.

Robertson was awarded the Distinguished Service Cross for his actions in the Aisne-Marne Offensive. After several days of trying to cross the Vesle River under fire and only managing to put across small numbers of men, on August 5, 1918 the 4th Division made a determined effort to get across the river. The 4th Engineers, including Robertson, worked under fire round the clock to build and emplace footbridges which the Infantry could utilize to make a mad dash across the river. Robertson's award actions occurred when he was part of a detachment that worked to emplace bridges across the river west of the village of Fismes near the village of Ville Savoye. On August 5, that detachment particularly distinguished themselves by remaining at their work under the heaviest of fire and completing a bridge for the Infantry.

(See the entry for William B. Beach for an overview of the kinds of activities carried out by the soldiers of the 4th Engineers involved in building the bridges across the Vesle River.)

Robertson served in the Vesle Sector, the Toulon Sector, the St. Mihiel Offensive, and the Meuse-Argonne Offensive. At some time after his award actions, he was promoted to Sergeant 1st Class. He served on occupation duty in Germany with the 4th Division where in February and March 1919 he was Inspector of Roads at Adenau, Germany. Robertson's Distinguished Service Cross was pinned on him by the Commanding General of the American Expeditionary Forces John J. Pershing in a 4th Division ceremony at Büchel, Germany on March 18, 1919.

Robertson was also awarded the Belgian Knight of the Order of Leopold II (Chevalier de L'Ordre Leopold II) as one of two soldiers in the 4th Division to receive this decoration during World War One. (The other soldier was Captain Murray R. MacKall, also from the 4th Engineers and also a Distinguished Service Cross recipient.) Robertson was one of only 47 soldiers of the American Expeditionary Forces to receive the Belgian Order of Leopold II.

Robertson returned to the United States as a Sergeant 1st Class with Company F 4th Engineers aboard the troopship *U.S.S. Von Steuben* (seized German liner *Kronprinz Wilhelm*) leaving Brest, France July 21, 1919 and arriving at Hoboken, New Jersey July 29, 1919. He was discharged on August 8, 1919.

He married Leila MaMarr in 1923 and went to work as a laborer on a stock farm. The 1930 Census showed him and his wife living in Salt Point, Sonoma County, California where he was still employed as a laborer on a stock farm and indicated he was a veteran of the World War. Starting in 1931, he worked in Alaska for over seven years in a government program to distribute food to the native population. In October 1938, he broke his leg in an accident and died three months later of complications resulting from his injury. Raymond D. Robertson died at the age of 43 on January 2, 1939 and is buried in Spring Hill Cemetery, Sebastopol, Sonoma County, California.

Alvan C. Sandeford

Major, United States Army
13th Field Artillery Regiment 4th Division
Date of Action: August 8-17, 1918
Location: Near Chery-Chartreuve, France
War Department General Orders 47 (1921)
Medal # 7898

CITATION

Twice gassed, Major Sandeford declined to be evacuated and continued in active command of his battalion. Having been advised and knowing that failure to be evacuated would probably result in his death, he nevertheless continued until he fell from his saddle in a state of total collapse. His fortitude and spirit of self-sacrifice were conspicuous.

DETAILS

Alvan Crosby "Sandy" Sandeford was born in Midville, Georgia on October 31, 1888, the son of John and Beatrice Sandeford. He entered the U.S. Military Academy on June 15, 1907 and graduated 63 out of a class of 82 on June 13, 1911. He was commissioned a 2nd Lieutenant of Infantry and served in the 8th Infantry in the Philippines in engagements against hostile Moros. In May 1915 he was assigned to the 22nd Infantry and served with the Regiment on Mexican Border Service. He was promoted to 1st Lieutenant on July 1, 1916 and was transferred to the Field Artillery on January 13, 1917. Sandeford was promoted to Captain of Field Artillery on May 15, 1917 and joined the 13th Field Artillery at Camp Stuart, Texas in July 1917. He went with the Regiment to Camp Greene, North Carolina in December 1917 where it became part of the 4th Division.

He sailed to France as a Captain in the 13th Field Artillery and part of the advance detachment of the 4th Division aboard the troop transport *U.S.S. Kroonland* on April 30, 1918. As his emergency contact, he listed his father in Midville, Georgia. In France on June 1, 1918 Sandeford was promoted to the wartime temporary rank of Major. He served in the Aisne-Marne Offensive.

He was awarded the Distinguished Service Cross for his actions spanning nine days in the Vesle Sector. From August 7-11, 1918, the 13th Field Artillery supported the operations of the 4th Division at the Vesle River. The Infantry units of the 4th Division were relieved in place along the Vesle River by elements of the 77th Division on the night of August 11, but the artillery units of the 4th Division were held in the front lines until the artillery of the 77th Division could be brought up, which did not fully take place until August 17. From August 12-17, the 13th Field Artillery supported the operations of the 77th Division.

Sandeford and the 155mm guns of the 13th Field Artillery

were situated near the village of Chery-Chartreuve, south of the Vesle River. German aircraft had control of the skies in the area and kept their artillery informed of the locations of the American artillery. The only safety for Sandeford and his men was in frequent shifting of their positions, which involved moving the guns by horse teams. Even so, the enemy artillery continued to find them and fire high explosive and gas shells at the American batteries. Though he was suffering from being wounded twice by gas, Sandeford continued to command the repositioning and supervise the firing of his guns until August 17 when he collapsed from exhaustion and the effects of the gas and fell unconscious from his horse. He also received a Citation Star.

Sandeford spent the rest of August and much of September in hospital at Orleans, France and rejoined his Regiment in time to take part in the Meuse-Argonne Offensive. He was awarded the French Legion of Honor. He was promoted to the wartime temporary rank of Lieutenant Colonel on November 7, 1918 with date of rank back to July 30, 1918. Sandeford left France in November 1918, returned to the United States and was assigned to the Personnel Branch of the General Staff in Washington, D.C. He reverted back to his permanent rank of Captain on March 15, 1920 and was promoted to Major on July 1, 1920. He served various assignments in the General Staff and Field Artillery branch. After 1932, he was awarded the Purple Heart with oak leaf cluster for the wounds he received in World War One. On August 1, 1935, he was promoted to Lieutenant Colonel. He was appointed Assistant Chief of Staff for Intelligence Fifth Corps Area.

On September 1, 1938, while driving from Ohio to Fort Knox, Kentucky, Sandeford attempted to pass a slow moving truck. The truck turned abruptly into his path and to avoid hitting it Sandeford crashed into a ditch. Though his passenger received minor injuries, Sandeford was severly hurt in the crash. He was transported to a hospital in Columbus, Ohio where he died the same day from his injuries. He was never married and was 49 years old at the time of his death. Alvan C. Sandeford is buried in Arlington National Cemetery.

Otto A. Schwanke

Private First Class, United States Army
Army Service Number: 2024343
Company B 47th Infantry Regiment 4th Division
Date of Action: August 1, 1918
Location: At Sergy, France
War Department General Orders 21 (1919)
Medal # 1072

CITATION

Private First Class Schwanke displayed the greatest devotion to duty, loyalty, and courage by repeatedly volunteering, night and day, to carry messages under the heaviest machine gun and shell fire from his battalion commander to the company commanders thereby maintaining efficient liaison at all times.

DETAILS

Otto Albert Schwanke was born in Rockland, La Crosse County, Wisconsin on April 16, 1894, the son of William (Wilhelm) and Ella Schwanke. Before entering the Army, he was a farmer in Potter, Wisconsin. He entered the Army on October 4, 1917. Schwanke sailed to France as a Private in Company B 47th Infantry 4th Division aboard the troopship *U.S.S. Princess Matoika* (seized German liner *Prinzess Alice*) on May 10, 1918. As his emergency contact, he listed his father in Brill, Wisconsin. At some time after reaching France, he was promoted to Private 1st Class.

Schwanke was awarded the Distinguished Service Cross for his actions in the Aisne-Marne Offensive. During the battle for Sergy, 1st Battalion 47th Infantry was temporarily attached to the 42nd Division. Schwanke operated as a runner carrying messages from 1st Battalion Headquarters to all four Companies in the Battalion. The battlefield at Sergy was alive with German artillery and small arms fire and Schwanke braved it over and over to deliver vital communications, thereby insuring the American units operated efficiently and in coordination with one another. Schwanke's 1st Battalion and 3rd Battalion 47th Infantry who fought alongside each other paid a heavy price at Sergy, losing a combined 27 officers and 462 enlisted men killed, wounded, or gassed and six enlisted men missing in action.

Schwanke served in the Vesle Sector, the Toulon Sector, the St. Mihiel Offensive and the Meuse-Argonne Offensive. At some time after his award action, he was promoted to Sergeant. He was awarded the French Croix De Guerre. He served on occupation duty in Germany with the 4th Division and returned to the United States with Company B 47th Infantry aboard the troopship *U.S.S. Mobile* (seized German liner *Cleveland*) leaving Brest, France July 16, 1919 and arriving at Hoboken, New Jersey July 27, 1919. He was discharged on August 4, 1919.

The 1920 Census showed him living with his parents in Rantoul, Calumet County, Wisconsin where he was employed as a farmer. On November 18, 1920, he married Wilhelmina Bleck and in 1923 they had a son. The 1930 Census showed Schwanke, his wife, and son living in Manitowoc, Manitowoc County, Wisconsin where he was employed as a truck driver for a bakery and indicated he was a veteran of the World War. The 1940 Census showed him, his wife and son living in Newton, Manitowoc County, Wisconsin where he was employed as a farmer. Otto A. Schwanke died at the age of 79 on January 16, 1974 and is buried in Grace Lutheran Cemetery, Cato, Manitowoc County, Wisconsin.

Louis Scionti

Sergeant, United States Army
Army Service Number: 558045
Company F 47th Infantry Regiment 4th Division
Date of Action: August 9, 1918
Location: Near Bazoches, France
War Department General Orders 46 (1919)
Medal # 2030

CITATION

Responding to a call for volunteers to destroy a hostile machine gun Sergeant Scionti with two other soldiers boldly went forward through machine gun fire and accomplished this mission.

DETAILS

Louis (Luciano) Scionti was born in Messina, Italy on October 25, 1898, the son of Joseph (Giuseppe) and Grace (Grazia) Scionti. He immigrated to the United States aboard the Italian passenger/cargo ship *S.S. San Giovanni* sailing from Palermo, Italy May 8, 1909 and arriving at New York City May 24, 1909. Scionti entered the Army on June 5, 1917 and listed his home of record as Boston, Massachusetts. He sailed to France as a Private 1st Class in Company F 47th Infantry 4th Division aboard the troopship *U.S.S. Princess Matoika* (seized German liner *Prinzess Alice*) on May 10, 1918. As his emergency contact, he listed his father Joseph in Boston, Massachusetts.

At some time in France, Scionti was promoted to Sergeant, though it is unclear if his promotion came before or after his award action. Scionti served in the Aisne-Marne Offensive.

He was awarded the Distinguished Service Cross for his actions in the Vesle Sector. By August 9, 1918, Scionti's Battalion had crossed the Vesle River and was engaged in battle before the village of Bazoches. The area in front of the village was full of German machine gun and sniper positions and the day was largely spent by the 47th Infantry hunting for and eliminating those positions. Scionti along with Cook Henry Garst of Company H and Bugler Richard Marcella of Company F braved machine gun fire to assault and destroy an enemy machine gun position. Garst and Marcella also received the Distinguished Service Cross for the action. Scionti was also awarded a Citation Star and the French Croix De Guerre.

He served in the Toulon Sector, the St. Mihiel Offensive, and the Meuse-Argonne Offensive. At some time after being promoted to Sergeant, he was reduced to the rank of Private. Scionti served on occupation duty in Germany with the 4th Division and returned to the United States as a Private in Company F 47th Infantry and part of Company I Third Army Composite Regiment aboard the troopship *U.S.S. Leviathan* (seized German liner *Vaterland*) leaving Brest, France September 1, 1919 and arriving at Hoboken, New Jersey September 8, 1919.

(See the entry for Joseph Bassi for an explanation of the Third Army Composite Regiment.)

Scionti was discharged on September 27, 1919.

The 1920 Census showed him living with his parents in Boston, Massachusetts where he was employed as a salesman for a retail grocery. On May 26, 1920 Scionti became a naturalized American citizen. He was married sometime between 1920 and 1930. The 1930 Census showed him living with his wife Mary and their son and daughter in Revere, Suffolk County, Massachusetts where he was employed as a house painter and indicated he was a veteran of the World War. The 1940 Census showed him, his wife and son and daughter living in Boston where he was employed as a Boss Painter at the Navy Yard. One source lists Louis Scionti dying at the age of 86 on March 10, 1985 in Randolph, Norfolk County, Massachusetts.

William A. Shea

Sergeant, United States Army
Army Service Number: 556292
Machine Gun Company 39th Infantry Regiment 4th Division
Date of Action: September 26, 1918
Location: Near Cuisy, France
War Department General Orders 46 (1919)
Medal # 7075

CITATION

Although painfully wounded by machine gun fire, Sergeant Shea placed himself in an exposed position between two machine guns and by the use of his glasses directed the fire of a heavy machine gun barrage on the enemy. He remained in this exposed position for two hours and his were the only guns which remained in action under the sweeping fire of the enemy.

DETAILS

William Anson Shea was born in Holyoke, Hampden County, Massachusetts on July 17, 1889. He enlisted in the Army as a Private in Company 147 Coast Artillery Corps on December 1, 1910 and was discharged as a Private with a character rating of Excellent on December 1, 1913 at Fort Winfield Scott, California. He enlisted as a Private in the Regular Army on April 15, 1916 at Columbus Barracks, Ohio and listed his home of record as Niagara Falls, New York. He served with the 10th Recruiting Company at Columbus Barracks. He was promoted to Corporal on July 23, 1916 and to Sergeant on August 29, 1917.

Shea was transferred to Machine Gun Company 39th In-

Michael D. Belis

fantry on October 16, 1917. He was reduced to the rank of Corporal on November 1, 1917 at Camp Greene, North Carolina and promoted back to the rank of Sergeant on March 9, 1918. He sailed to France as a Sergeant in Machine Gun Company 39th Infantry 4th Division aboard the Italian troopship *Duca D'Aosta* on May 10, 1918. As his emergency contact, he listed his mother Matilda in Hamilton, Ontario, Canada. Shea served in the Aisne-Marne Offensive, the Vesle Sector, the Toulon Sector, and the St. Mihiel Offensive.

He was awarded the Distinguished Service Cross for his actions in the Meuse-Argonne Offensive. On September 26, 1918, the first day of the Offensive, the 39th Infantry advanced from east of the village of Malancourt in the direction of the village of Cuisy. Shea and his machine guns supported the attack of the 3rd Battalion 39th Infantry against Cuisy. Although he was shot through both buttocks, Shea remained in an exposed position under fire for two hours using his binoculars in order to direct the fire of the two guns of his section upon the German lines. His were the only machine guns of his Company which remained in action covering the Infantry's attack and under Shea's expert calculations and dynamic leadership his gunners lay down a withering barrage of indirect suppressive fire on the enemy all the while under heavy fire themselves.

He was one of 20 soldiers of the 4th Division and one of only 399 soldiers of the American Expeditionary Forces to be awarded the Italian War Merit Cross (Croce al Merito di Guerra). He also received a Citation Star.

Because of his wounds he was evacuated back to the United States several months ahead of his Regiment as a Sergeant of Machine Gun Company 39th Infantry and part of the Sick and Wounded Convalescent Detachment Ambulatory Surgical

Base Hospital #119 aboard the troop transport *U.S.S. Manchuria* leaving St. Nazaire, France January 10, 1919 and arriving at Hoboken, New Jersey January 22, 1919. Shea was discharged on June 21, 1919 with a 2% disability rating.

When he registered for the draft in 1942, Shea indicated his home of record as Detroit, Michigan and that he was unemployed. No further details for William A. Shea could be found.

Michael D. Belis

Anthony F. Shedlock
Sergeant, United States Army
Army Service Number: 559887
Company H 58th Infantry Regiment 4th Division
Date of Action: August 6, 1918
Location: Near Ville Savoye, France
War Department General Orders 27 (1920)
Medal # 6027

CITATION
Sergeant Shedlock, when the officers of the company became casualties took command, reorganized the scattered groups into a platoon and personally led them across the Vesle River in the face of heavy machine gun fire and drove the enemy from their position on the railroad embankment 500 yards beyond the river. He defended his position under the heavy fire and attacks of the enemy.

DETAILS

Anthony Frank Shedlock was born in Chesterfield, Pennsylvania on February 20, 1898, the son of Anthony J. (Antonius) and Mary (Maria) Shedlock. He enlisted in Company H 58th Infantry (Regular Army) on August 2, 1917 at Gettysburg, Pennsylvania as the Regiment was in the process of being formed. He moved with the Regiment to Camp Greene, North Carolina and was promoted to Private First Class on January 5, 1918. On April 1, 1918 he was promoted to Corporal. Shedlock sailed to England as a Corporal in Company H 58th Infantry 4th Division aboard the British troopship *S.S. Rhesus* on May 7, 1918. As his emergency contact, he listed his father Anthony in Utahville, Pennsylvania. From England he and his Company were sent over to France. Shedlock served in the Aisne-Marne Offensive.

He was awarded the Distinguished Service Cross for his actions in the Vesle Sector. Though his Citation states he was a Sergeant, Shedlock was actually a Corporal at the time of his award actions and was not promoted to Sergeant until three days later. On August 6, 1918, the 1st Battalion of the 58th Infantry was designated as the attack Battalion in the continuing effort to cross the Vesle River and move the right side of the 4th Division advance northward. Shedlock's Company as part of 2nd Battalion 58th Infantry was working in support of 1st Battalion 58th Infantry. All four Companies of Shedlock's Battalion had been moved up from their holding positions near the village of Ville Savoye to a location close to the south bank of the river, ready to cross the river and assist 1st Battalion should the need arise.

As 1st Battalion got across the river and established a foothold, German artillery and machine gun fire caused casualties in the American positions on both sides of the river. When all of the officers in his Company and many of the enlisted men became casualties, Shedlock organized what was left of the Company and led them over the improvised footbridges and crossed the river. He then led his men in an assault against an enemy position along the railroad embankment several hundred yards beyond the river, drove the Germans out, occupied and defended the position until relieved by elements of the 59th Infantry the next day.

Shedlock was promoted to Sergeant on August 9, 1918. He served in the Toulon Sector, the St. Mihiel Offensive, and the Meuse-Argonne Offensive. He was promoted to 1st Sergeant on November 10, 1918. He served on occupation duty in Germany with the 4th Division and returned to the United States with Company H 58th Infantry aboard the troopship *U.S.S. Mount Vernon* (captured German liner *Kronprinzessin Cecilie*) leaving Brest, France July 24, 1919 and arriving at Hoboken, New Jersey August 1, 1919. He was discharged on August 7, 1919 at Camp Dix, New Jersey.

The 1920 Census showed Shedlock living with his father, sister, and brothers in Bigler, Clearfield County, Pennsylvania where he was employed as a coal miner. The 1930 Census showed him living in East Cleveland, Cuyahoga County, Ohio where he was employed as a wood worker for an automobile body company and indicated he was a veteran of the World War. The 1940 Census showed him still living in Cleveland where he was employed as a metal worker in the auto industry. At some time after 1940, he moved to Florida. Anthony F. Shedlock died at the age of 93 on January 13, 1991 and is buried in Woodlawn Memory Gardens, Saint Petersburg, Pinellas County, Florida.

Albert B. Simpson
First Lieutenant, United States Army
11th Machine Gun Battalion 4th Division
Date of Action: September 27-28, 1918
Location: Near Nantillois, France
War Department General Orders 27 (1919)
Medal # No assigned number discovered

CITATION
Though he was wounded, Lieutenant Simpson remained with his Company and by skillful arrangement of his machine gun covered a retirement of the 39th Infantry. The next day he was again wounded and although urged by the surgeon to go to the rear, this gallant officer replied that there was too much work yet to be done at the front. He left to rejoin his command and had gone about half the distance when he was killed by a high explosive shell.

DETAILS
Albert Bartow Simpson was born in Eelbeck, Georgia on March

10, 1890, the son of Robert and Kate Simpson. Before the war Simpson worked as a salesman for the Haverty Furniture Company in Atlanta, Georgia. He attended Officers Training Camp and was commissioned a 2nd Lieutenant of Infantry in the Officers Reserve Corps at Fort Oglethorpe, Georgia on November 27, 1917. He was assigned to the 39th Infantry at Camp Greene, North Carolina. At some time before going overseas he was transferred to the 11th Machine Gun Battalion. Simpson sailed to France as a 2nd Lieutenant in 11th Machine Gun Battalion 4th Division and part of the advance detachment of the 4th Division aboard the troop transport *U.S.S. Kroonland* on April 30, 1918. As his emergency contact, he listed his father Robert in Waverly Hall, Georgia.

Simpson served in the Aisne-Marne Offensive, the Vesle Sector, the Toulon Sector, and the St. Mihiel Offensive. He was promoted to 1st Lieutenant on September 21, 1918.

He was awarded the Distinguished Service Cross for his actions on two days in the Meuse-Argonne Offensive. At the start of the Offensive, Simpson's 11th Machine Gun Battalion was operating in direct support of the 39th Infantry. On September 26, 1918, the first day of the Offensive, the 39th Infantry captured the villages of Cuisy and Septsarges southeast of the village of Nantillois. The machine guns of the 11th Machine Gun Battalion were instrumental that day in suppressing the enemy fire from Hill 315 during the attack. On September 27, the attack was resumed with the 39th Infantry leading the advance across the open ground between the village of Nantillois and the wooded area known as the Bois de Brieulles.

Simpson and his machine guns were right alongside 3rd Battalion 39th Infantry with his guns protecting the left flank. As soon as the advance began, the Germans opened up an intense barrage of high explosive and gas shells from their artillery. They also opened up heavy fire from machine guns and trench mortars in the woods around and on Hill 266. By about ten o'clock in the morning, 3rd Battalion was disorganized from taking heavy casualties and in a piecemeal fashion began pulling back. Simpson was wounded but remained with his machine gun in a forward position to cover the retreat of the Infantry. He stayed in his position, laying down a heavy fire until eight o'clock that night when he was ordered to join the withdrawn units that had moved back and dug in on Hill 295.

The next day, September 28 the attack was resumed with Simpson and his gun again supporting the lead element which that day was 2nd Battalion 39th Infantry. As the 79th Division came up on the left to join in the attack against Nantillois, Simpson was wounded again. He went to the forward aid station and had his wound dressed where the surgeon recommended he go to the rear so he could be sent to a hospital. Anxious to rejoin the fight however, Simpson started for the front lines but only got halfway there before being killed by a German artillery round.

Simpson was buried in Plot 2 Grave # 42 in the American cemetery # 702 at the village of Septsarges.

His Distinguished Service Cross was presented to his father Robert N. Simpson.

On May 20, 1919 Simpson was reinterred in Plot 4 Section 46 Grave # 186 in the Argonne American Cemetery # 1232. On July 8, 1921 his body was disinterred for preparation and shipment. His remains were returned to the United States aboard the United States Army Transport *U.S.A.T. Wheaton* leaving Antwerp, Belgium August 6, 1921 and arriving at Hoboken, New Jersey August 20, 1921. He was one of 5,759 American

Michael D. Belis

dead being brought home aboard the *Wheaton* on that journey. Albert B. Simpson was laid to his final rest in Waverly Hall Cemetery, Waverly Hall, Harris County, Georgia.

Ralph Slate

Captain, United States Army
Company I 39th Infantry Regiment 4th Division
Date of Action: September 27, 1918
Location: Near Bois de Septsarges, France
War Department General Orders 81 (1919)
Medal # 3365

CITATION

After being wounded in a previous action, Captain Slate led his command in the face of unusual machine gun fire, repeatedly exposing himself to prevent his units from becoming scattered and strengthening and holding his line until again severely wounded.

DETAILS

Ralph Slate was born in Grand Rapids, Kent County, Michigan on January 4, 1892. He enlisted as a Private in Company H 12th Infantry (Regular Army) on December 2, 1914 and was promoted to Corporal and then Sergeant. He was a Sergeant in Company H 62nd Infantry when that unit was formed from elements of the 12th Infantry in 1917. Slate was discharged on July 4, 1917 to accept a commission as a 2nd Lieutenant of Infantry on July 5, 1917 with date of rank back to June 14, 1917. He was immediately promoted to 1st Lieutenant and assigned to the 39th Infantry. Slate sailed to France as a 1st Lieutenant and Executive Officer of Company I 39th Infantry 4th Division aboard the French troopship *Espagne* on May 8, 1918. As his emergency contact, he listed his sister Mrs. Mary P. Matteson in Kalamazoo, Michigan. After reaching France he was promoted to the war time temporary rank of Captain with date of rank back to January 16, 1918.

Slate commanded Company I 39th Infantry in the Aisne-Marne Offensive. While leading an advance through dense barbed wire entanglements near the Vesle River, he was wounded on August 5, 1918, but refused to leave his command. Instead, he brought his Company to the bank of the river and directed his soldiers to dig in and establish a strong defensive position from which any possible counterattacks by the enemy could be repulsed.

Slate was awarded the Distinguished Service Cross for his actions in the Meuse-Argonne Offensive. On September 27, 1918, Slate and Company I as part of 3rd Battalion 39th Infantry were in the assaulting line of the Regiment's attack. The previous day the Battalion had occupied a ridge running east and west one kilometer north of the village of Septsarges and withstood three German counterattacks. On September 27, the Battalion left the ridge, attacked Hill 266, and moved through the wooded area known as the Bois de Septsarges. Upon reaching the Nantillois road, enemy artillery and machine gun fire became so extraordinarily intense it broke down the advance, caused the Americans to become disorganized, and they started to fall back. Slate moved about through the heavy fire, keeping his men from scattering and directed them into holding a strong defensive line. He was wounded by enemy fire while doing so but refused to be evacuated until he was certain his Company had been organized into strong defensive positions.

He was twice awarded the French Croix De Guerre, once with bronze star and a second time with gold star. He served on occupation duty in Germany with the 4th Division and was promoted to the temporary rank of Major on May 6, 1919. He returned to the United States as a Major in command of 3rd Battalion 39th Infantry aboard the troopship *U.S.S. Leviathan* (seized German liner *Vaterland*) leaving Brest, France July 30, 1919 and arriving at Hoboken, New Jersey August 6, 1919. Slate remained in the Army.

The January 1920 Census showed him as a Major at Camp Dodge, Iowa. On March 10, 1920, he was discharged from his temporary rank of Major and reverted back to his permanent rank of 1st Lieutenant. He was promoted to Captain on July 1, 1920. After 1932, he was awarded the Purple Heart with oak leaf cluster for the wounds he received in World War One. Slate was promoted to Major on October 1, 1934 and to Lieutenant Colonel on July 1, 1940. On February 1, 1942, he was promoted to the temporary rank of Colonel. During World War Two he served as Provost Marshall at the New York Port of Embarkation and as a G3 officer in the General Staff Corps in the United States. He retired from the Army with the rank of Colonel on July 31, 1946. Ralph Slate died at the age of 93 in 1985 and is buried in Masonic Cemetery, Des Moines, Polk County, Iowa.

Ford D. Smith

Corporal, United States Army
Army Service Number: 568844
Company D 4th Engineer Regiment 4th Division
Date of Action: August 11, 1918
Location: Near Ville Savoye, France
War Department General Orders 71 (1919)
Medal # 2925

CITATION

Leaving a sheltered position, Corporal Smith exposed himself to an intense artillery barrage to rescue a wounded officer. He carried him across the Vesle River to where he could obtain aid in taking him to a dressing station. He displayed utter disregard of personal danger while under heavy fire.

DETAILS

Ford Dewey Smith was born in Wyandotte, Wayne County, Michigan on November 2, 1897, the son of William and Hattie Smith. He entered the Army on May 1, 1917. Smith sailed to France as a Private 1st Class in Company D 4th Engineers 4th Division aboard the troopship *U.S.S. Martha Washington* (seized Austrian liner) on April 30, 1918. As his emergency contact, he listed his mother Hattie in Antioch, California. Smith served in the Aisne-Marne Offensive.

He was awarded the Distinguished Service Cross for his actions in the Vesle Sector. On August 11, 1918, Smith was on the north side of the Velse River and had helped to emplace a footbridge for the 59th Infantry to use to cross the river. The day was marked by heavy artillery firing from both the Americans and the Germans which lasted nearly the entire day. Smith ventured out into the bombardment, picked up a wounded officer, and under fire carried him across the river to American positions on the south side of the river where he could get soldiers to help him bring the officer to an aid station near the village of Ville Savoye.

Smith served in the Toulon Sector, the St. Mihiel Offensive, and the Meuse-Argonne Offensive. He served on occupation duty in Germany with the 4th Division and returned to the United States as a Musician 1st Class in Headquarters Company 4th Engineers aboard the troopship *U.S.S. Von Steuben* (seized German liner *Kronprinz Wilhelm*) leaving Brest, France July 21, 1919 and arriving at Hoboken, New Jersey July 29, 1919. He was discharged soon after returning.

The 1920 Census showed him living with his parents in Contra Costa County, California where he was employed as a house carpenter. He married Zella Ayers. In 1923 they had a son who died in 1927. The 1930 Census showed Smith and Zella living in Contra Costa County, California where he was employed as a Gauger for an oil company and indicated he was a veteran of the World War. The 1940 Census showed him and Zella living in Avenal, Kings County, California where he was employed as a Truck Driver in the oil fields. When he registered for the draft in 1942, he indicated he was living in Avenal and was employed by Standard Oil Company of California. Ford D. Smith died at the age of 93 on December 16, 1990 and is buried in Chapel of the Chimes Columbarium and Mausoleum, Oakland, Alameda County, California.

Joe Smith

Private, United States Army
Army Service Number: 556655
Company C 39th Infantry Regiment 4th Division
Date of Action: October 10 - 13, 1918
Location: Near Bois de Fays, France
War Department General Orders 64 (1919)
Medal # 3509

CITATION

Acting as battalion runner, Private Smith repeatedly carried messages over a route swept by machine gun and artillery fire. It was necessary to send runners night and day in order to maintain communication with the front lines. He volunteered out of his turn for this dangerous but all-important work.

DETAILS

Joe Barton Smith was born in Montgomery County, Arkansas on October 1, 1894, the son of Bascum and Annie Smith. Before entering the Army, he was a farmer in Porter, Oklahoma. He entered the Army on February 21, 1918. Smith sailed to France as a Private in Company C 39th Infantry 4th Division aboard the Italian troopship *Dante Alegheiri* on May 10, 1918. As his emergency contact, he listed his brother Dee Smith in Center Point, Arkansas. Smith served in the Aisne-Marne Offensive, the Vesle Sector, the Toulon Sector, and the St. Mihiel Offensive.

He was awarded the Distinguished Service Cross for his actions spanning three days in the Meuse-Argonne Offensive.

From October 10 - 13, 1918, the 39th Infantry attacked out of the wooded area known as the Bois de Fays and headed northeast through the wooded area of the Bois de Malaumont and into the wooded area known as the Bois de Forêt. The enemy positions in the Bois de Forêt were taken by a bayonet charge. Paths through the woods were blocked by trees felled by the Germans and the undergrowth was wired together with barbed wire. Enemy machine guns were everywhere and took a high toll on the Americans as they struggled through the entanglements in the forests.

The dangerous job of runner in the 39th Infantry became even more deadly during those three days of October 10 – 13 in the Bois de Malaumont and the Bois de Forêt as fifteen runners in the Regiment were wounded and one was killed. The job was rotated by turns so that exhausted runners could rest before venturing out across the battleground again. Through it all, Smith continued to volunteer out of his turn and braved the enemy fire over and over to carry messages and insure communications between his Battalion and other units remained unbroken. He also received a Citation Star.

At some time after his award action, Smith was promoted to Private 1st Class. He was one of 20 soldiers of the 4th Division and one of only 399 soldiers of the American Expeditionary Forces to be awarded the Italian War Merit Cross (Croce al Merito di Guerra).

Smith served on occupation duty in Germany with the 4th Division and returned to the United States with Company C 39th Infantry aboard the troopship *U.S.S. Leviathan* (seized German liner *Vaterland*) leaving Brest, France July 30, 1919 and arriving at Hoboken, New Jersey August 6, 1919. The date of his discharge could not be found.

Michael D. Belis

When he registered for the draft in 1942, he indicated he was living and working in Firebaugh, Fresno County, California. Joe Barton Smith died at the age of 82 on February 21, 1977 and is buried in Liberty Veterans Cemetery, Fresno, Fresno County, California.

Raymond R. Smith

Corporal, United States Army
Army Service Number: 559162
Company C 11th Machine Gun Battalion 4th Division
Date of Action: October 12, 1918
Location: Northeast of Cunel, France
War Department General Orders 21 (1919)
Medal # 1065

CITATION

During a heavy bombardment after a shell had struck his machine gun, knocking it and his squad completely out of action, Corporal Smith assembled three men from another squad and obtaining another gun again took up a position on the line and remained throughout the action, as the front was at that time thinly held and in constant danger of counterattack; the prompt initiative and splendid courage on the part of this soldier not only inspired and encouraged his men but aided materially in the success of the action.

DETAILS

Raymond R. Smith (His middle name is spelled in different accounts as Reader, Rader or Raeder.) was born in Burlington, Iowa on April 1, 1897, the son of Jonah and Elizabeth Smith. He entered the Army on April 24, 1917. He married Lola Prim in December 1917. Their one-year-old daughter died a few weeks before Smith sailed overseas. Smith sailed to France as a Corporal in Company C 11th Machine Gun Battalion 4th Division aboard the troop transport *U.S.S. Rijndam* (former

Michael D. Belis

Dutch liner) on May 10, 1918. As his emergency contact, he listed his wife Mrs. Raymond R. Smith in Charlotte, North Carolina. While he was in France his second daughter was born back home. He served in the Aisne-Marne Offensive, the Vesle Sector, the Toulon Sector and the St. Mihiel Offensive.

Smith was awarded the Distinguished Service Cross for his actions in the Meuse-Argonne Offensive. By October 12, 1918, the 7th Brigade of the 4th Division which included the 11th Machine Gun Battalion had established a foothold in the western edge of the wooded area known as the Bois de Forêt near the village of Cunel. At eleven o'clock that morning, a German counterattack was repulsed and at half past noon the enemy fired an artillery barrage that killed or wounded two hundred Americans in the front lines. Smith's machine gun and his crew were knocked out of action by that barrage, but he procured another gun, gathered several gunners from his Company to constitute another crew, and continued to fire upon the enemy and hold his position at the forward area. His courageous action held the front line, prevented enemy counterattacks from forming, and was an inspiring example of bravery and dedication to duty for his fellow soldiers.

At some time after his award action, Smith was promoted to Sergeant. He was awarded the French Croix De Guerre. He served on occupation duty in Germany with the 4th Division and returned to the United States as a Sergeant with Company C 11th Machine Gun Battalion aboard the troop transport *U.S.S. Tiger* leaving Brest, France July 17, 1919 and arriving at Hoboken, New Jersey July 29, 1919. He remained in the Army and was discharged on September 29, 1921.

Smith had a son born with Lola in 1920. He divorced his wife Lola and in June 1920 he married Nettie McDaneld. He had a second son born with Nettie. He apparently went back into the Army sometime after 1921. The 1930 Census showed Smith and his wife Nettie living in Denver, Colorado where he was employed as a Recruiting Officer for the Army and indicated he was a veteran of the World War. Raymond R. Smith died at the age of 54 on December 5, 1951 in Los Angeles County, California.

Michael D. Belis

Rutherford H. Spessard

Major, United States Army
Commanding Officer 1st Battalion 58th Infantry Regiment 4th Division
Date of Action: August 6 and October 2, 1918
Location: Near Ville Savoye, France and near Bois de Fays, France
War Department General Orders 98 (1919)
Medal # No assigned number discovered

CITATION

During the crossing of the Vesle River, Major Rutherford H. Spessard (then Captain), when his battalion commander was killed, immediately assumed command of the battalion without orders and led them across the Vesle River against strongly fortified enemy positions, displaying absolute disregard for his personal danger. On 2 October in the vicinity of the Bois de Fays, Major Spessard exposed himself to intense enemy artillery and machine gun fire while making observations and directing the movement of his men. He established his battalion headquarters a short distance to the rear of his lines in a position continually subjected to severe enemy artillery fire.

DETAILS

Rutherford Houston Spessard was born in New Castle, Craig County, Virginia on January 28, 1896, the son of Nathaniel and Alice Spessard. He graduated from the Virginia Military Institute (VMI) in 1915. Before the war he was a teacher at Marion Military Institute in Alabama. He was commissioned a Captain in the Officers Reserve Corps on August 15, 1917. Spessard was detailed to the School at the Officers Training Camp at Fort Oglethorpe, Georgia as an instructor, taught one class there and asked for duty with a unit going overseas. He was assigned to Company D 58th Infantry. On March 15, 1918, he married Matilda Hayesworth. Spessard sailed to England as a Captain and Commanding Officer of Company D 58th Infantry 4th Division aboard the British transport *City of Brisbane* on May 7, 1918. As his emergency contact, he listed his wife Matilda in Uniontown, Alabama. From England he and his Company were sent over to France.

He was awarded the Distinguished Service Cross for his actions on two separate dates. The first date mentioned in his Citation was in the Aisne-Marne Offensive. On August 4, 1918, Spessard was commanding Company D 1st Battalion 58th Infantry in the attack toward the Vesle River, coming from a direction southwest of the village of Ville Savoye. As his Battalion moved out across an open area, it was caught in a German artillery barrage. The Battalion Commander Major Samuel H. Houston was killed, the Battalion Executive Officer Captain Evan M. Sherrill was wounded, and Spessard as a 22-year-old

Michael D. Belis

Captain, without waiting for orders to do so, assumed command of the Battalion. He led his men to the river where it took two days under heavy fire to finally get across and establish a foothold on the other side. Spessard and his Battalion were the first soldiers of the 58th Infantry to make it across the Vesle River.

Spessard continued to command 1st Battalion in the Vesle Sector, the Toulon Sector, and the St. Mihiel Offensive. The second date mentioned in his Citation was in the Meuse-Argonne Offensive. By October 2, 1918, Spessard had been promoted to Major. On that date the 58th Infantry was stalled in its attack into the wooded area known as the Bois de Fays by heavy enemy resistance. After moving through the nearby wooded area of the Bois des Ogons and realizing there was no support on either flank, Spessard had to pull his Battalion back south of those woods and dig in. During this phase of the battle, Spessard was conspicuous in his command by bravely moving about in the open, constantly under fire, to direct the activities of his Battalion. He set up his Battalion headquarters close to the front line where he could better command his troops even though the location brought his headquarters under enemy fire.

Spessard was shifted to command of 3rd Battalion after its commander had been wounded on October 4, 1918. He was awarded the French Croix De Guerre. He served on occupation duty in Germany with the 4th Division where he was assigned as the Provost Marshal of the city of Trier. Spessard returned to the United States as Commanding Officer of 3rd Battalion 58th Infantry 4th Division aboard the troopship *U.S.S. Mount Vernon* (captured German liner *Kronprinzessin Cecilie*) leaving Brest, France July 24, 1919 and arriving at Hoboken, New Jersey August 1, 1919. He was discharged soon after returning. During his service he had attained the rank of Major and commanded two different Battalions of the 58th Infantry in combat and was still only 23 years old.

By 1920 Spessard was back at Marion Military Institute in Alabama as a Professor. The 1920 Census showed him and his wife Matilda living in Marion, Alabama where he was employed as a college professor. He became the Commandant of Marion Military Institute. In 1920 he and Matilda had a son. For a while Spessard was Superintendent of Virginia State Penitentiary and then went to work for his family's pulpwood manufacturing business in Richmond, Virginia, the business being called N.E. Spessard & Sons.

The 1930 Census showed Spessard, his wife, and son living in Richmond with his occupation as Broker of Pulpwood and indicated he was a veteran of the World War. He took over the family business in 1934 when his father died. The 1940 Census showed him and his family living in Tuckahoe, Henrico County, Virginia with his occupation indicated as President of a Pulpwood business. Rutherford H. Spessard committed suicide by jumping off a bridge and drowning in Chesterfield County, Virginia at the age of 59 on November 8, 1955 and is buried in Hollywood Cemetery, Richmond, Richmond City, Virginia.

Michael D. Belis

Dave W. Stearns

Corporal, United States Army
Army Service Number: 568697
Company C 4th Engineer Regiment 4th Division
Date of Action: August 6, 1918
Location: Near St. Thibaut, France
War Department General Orders 46 (1919)
Medal 1492

CITATION

Corporal Stearns was a member of a platoon ordered to precede the infantry, to construct footbridges across the Vesle River. Enemy sniper, machine gun, and artillery fire was so intense that four attempts of his platoon failed. Acting upon his own initiative, Corporal Stearns made his way along the river in the face of the deadly fire and for one hour reconnoitered the enemy's positions, reporting back to his commanding officer with information of the greatest value.

DETAILS

David Whitaker "Dave" Stearns was born in Waldport, Lincoln County, Oregon on October 10, 1892, the son of Joseph and Isabella Stearns. Before entering the Army, he was a logger for Larkin Green Logging Company in Oregon. He entered the Army on September 18, 1917. Stearns sailed to France as a Private 1st Class in Company C 4th Engineers 4th Division aboard the troopship *U.S.S. Martha Washington* (seized Austrian liner) on April 30, 1918. As his emergency contact, he listed his father Joseph in Portland, Oregon. At some time after reaching France, he was promoted to Corporal. Stearns served in the Aisne-Marne Offensive.

He was awarded the Distinguished Service Cross for his actions in the Vesle Sector.

(See the entry for William B. Beach for an overview of the kinds of activities carried out by the soldiers of the 4th Engineers involved in building the bridges across the Vesle River.)

In early August 1918, the 4th Engineers had been working for several days to construct and emplace bridges which the Infantry could use to cross the Vesle River. There was no fixed front line along the river and enemy positions were scattered everywhere, to include machine gun nests and sniper posts. Roving German patrols happened upon parties of Engineers setting up footbridges and exchanges of gunfire were numerous. One of these patrols surprised Stearns and his fellow Engineers while they were assembling logs into a footbridge and the result was a short firefight. Before the Germans retired, one of them took a shot at Stearns with a flare pistol but missed.

On August 6, after four attempts by his platoon to construct a bridge across the river near the village of St. Thibaut had failed due to enemy fire, Stearns took it upon himself to scout along the river for a suitable location to build a footbridge. For an hour, alone and at the risk of being detected by the enemy, he searched out and noted enemy locations along and across the river and was able to report them when he returned to his Company. His report was of immense value in helping his commander to determine a relatively defensible place at the river where they could construct a bridge for the Infantry.

At some time after his award action, Stearns was promoted to Sergeant. He returned to the United States as a Sergeant of Company C 4th Engineers and part of a detachment of sick and wounded from various units in St. Nazaire Convalescent Detachment No. 234 aboard the troop transport *U.S.S. Dekalb* (seized German liner *Prinz Eitel Friedrich*) leaving St. Nazaire, France June 12, 1919 and arriving at Newport News, Virginia June 23, 1919. He was discharged on July 7, 1919.

The 1920 Census showed him living with his parents in Portland, Oregon where he was employed as a laborer in a flour mill. He married Violet Berntsen in about 1924 and they had a daughter in 1925 and another in 1926. The 1930 Census showed Stearns living with his wife and two daughters in Coquille, Coos County, Oregon where he was employed as a laborer for Coos County and indicated he was a veteran of the World War. When he registered for the draft in 1942, he indicated he was living in Portland, Oregon with his wife and was unemployed. His wife Violet died in 1978. David W. Stearns died at the age of 92 on March 18, 1985 and is buried in Willamette National Cemetery, Portland, Multnomah County, Oregon.

Wallis H. Sturtevant

Corporal, United States Army
Army Service Number: 1684164
Company D 12th Machine Gun Battalion 4th Division
Date of Action: August 10, 1918
Location: Near Chéry-Chartreuve, France
Note: Though his Citation indicates the location as being near the village of Chéry-Chartreuve the actual area was 10 miles or more north of Chéry-Chartreuve across the Vesle River from and nearer to the village of Ville Savoye.
War Department General Orders 59 (1919)
Medal # 7664

CITATION

Corporal Sturtevant voluntarily ran through a terrific shell fire into a burning ammunition dump and rescued a badly wounded and burned comrade. The ammunition was exploded a few seconds after this heroic act was performed.

DETAILS

Wallis Hall Sturtevant was born in Greenfield, Franklin County, Massachusetts on April 22, 1891, the son of Harry and Edith Sturtevant. Before entering the Army, he was a teacher at Fitchburg Normal School in the Massachusetts Public School system in Fitchburg, Massachusetts. He attended Officers Training Camp at Plattsburg Barracks, New York but did not graduate. He entered the Army as a Private on March 29, 1918.

Sturtevant sailed to England as a Private in Company D 12th Machine Gun Battalion 4th Division aboard the British

Michael D. Belis

troopship *RMS Aquitania* on May 7, 1918. As his emergency contact, he listed his father Harry in Fitchburg, Massachusetts. From England he and his Company were sent over to France where they were issued their machine guns. At some time after reaching France, he was promoted to Corporal. Sturtevant served in the Aisne-Marne Offensive.

He was awarded the Distinguished Service Cross for his actions in the Vesle Sector. On August 7-10, 1918, his Company was working in support of 1st Battalion 59th Infantry on the north side of the Vesle River directly across the river from the village of Ville Savoye. The 59th Infantry advanced along the Rouen-Rheims Road and attacked enemy positions in the woods around the Château du Diable. Sturtevant's Company was one of the elements of the 12th Machine Gun Battalion that moved with the Infantry and provided direct support with their machine guns.

As the enemy retreated, they left behind several ammunition dumps. The Germans fired artillery in the area of the dumps, hoping to blow them up and prevent the Americans from using them. Several of the soldiers from his Company were moving down the road next to one of the dumps when it was struck and exploded, wounding the men, and setting fire to the brush in the area around the dump. While incoming enemy artillery was impacting around him, Sturtevant charged into the burning brush, picked up a wounded soldier who was unable to move by himself, and carried the man to safety moments before the dump exploded. Though cited in different orders, this appears to be the same action for which William H. McGinnis was also awarded the Distinguished Service Cross. Both soldiers were from Company D 12th Machine Gun Battalion and each rescued a fellow soldier from that burning ammunition dump that day.

Sturtevant served in the Toulon Sector, the St. Mihiel Offensive, and the Meuse-Argonne Offensive.

On October 2, 1918, while his Company was near the wooded area known as the Bois de Fays and preparing to resume the attack toward the Meuse River, Sturtevant was severely wounded by enemy artillery. Shrapnel hit him in the neck by his ear and tore a large gash in his leg. He was removed to a Field Hospital and then to a Base Hospital. His leg wound was so severe that amputation was considered but the doctors were able to save his leg without amputating. He returned to the United States several months ahead of his Company as a Corporal in Company D 12th Machine Gun Battalion and part of a detachment of sick and wounded in Convalescent Detachment No. 25 Ambulatory Surgical from Base Hospital # 119 aboard the troopship *U.S.S. Rijndam* (former Dutch liner) leaving St. Nazaire, France January 19, 1919 and arriving at Newport News, Virginia February 9, 1919. The passenger list indicated he had a wound to the thigh. Upon arrival Sturtevant was admitted to the Base Hospital at Camp Devens, Massachusetts. He was discharged on May 10, 1919.

Sturtevant married Beulah Harvey on November 9, 1919. The 1920 Census showed him living in Boston, Massachusetts where he was unemployed. In 1921 he and Beulah had a son. The 1925 New York Census showed Sturtevant living with his wife and son in the Bronx in New York City where he was employed as an artist. The 1930 Census showed him living with his wife and son in Montclair, Essex County, New Jersey where he was employed as an Art Director at an Advertising Agency and indicated he was a veteran of the World War.

Michael D. Belis

When he registered for the draft in 1942, Sturtevant indicated he lived in Louisville, Jefferson County, Kentucky while Beulah lived in New Jersey. On that registration he stated that he was employed by the U.S. Army War Department at the Federal Building in Louisville and also indicated he had a scar from a wound on the left side of his neck behind the ear. In 1958, he and Beulah were living in Springfield, Massachusetts where he was employed as a draftsman. His wife Beulah died in 1978. Wallis H. Sturtevant died at the age of 98 on June 17, 1989 and is buried in Oak Grove Cemetery, Springfield, Hampden County, Massachusetts.

Clark O. Tayntor

First Lieutenant, United States Army
Company M 47th Infantry Regiment 4th Division
Date of Action: July 29 - 30, 1918
Location: At Sergy, France
War Department General Orders 66 (1919)
Medal # 3753

CITATION

Disregarding two wounds from shell fire which he had suffered, Lieutenant Tayntor continued in the advance with his platoon, keeping his men well organized, directing the consolidation of the line throughout the night and refusing medical attention until all the wounded men in his platoon had received treatment.

DETAILS

Clark Olds Tayntor was born in Barre, Washington County, Vermont on October 10, 1891, the son of Eugene and Inez Tayntor. He graduated from Yale University in 1915. When he registered for the draft in May 1917, he indicated he was a Student at Harvard Law School. He was commissioned a 2nd Lieutenant of Infantry in the Regular Army on October 26, 1917 and assigned to Company M 47th Infantry at the camp at Syracuse, New York. Tayntor moved with the Regiment to Camp Greene, North Carolina where it became a part of the 4th Division. He was promoted to the wartime temporary rank of 1st Lieutenant on March 25, 1918. He sailed to France as a 1st Lieutenant in Company M 47th Infantry 4th Division aboard the Italian troopship *Caserta* on May 10, 1918. As his emergency contact, he listed his mother in Erie, Pennsylvania.

Tayntor was awarded the Distinguished Service Cross for his actions on two days in the Aisne-Marne Offensive. The Allied advance in the Offensive was pushing the Germans back with a speed they had not expected, and the enemy decided to make a delaying stand in the valley of the Ourcq River with the village of Sergy as the central point. The 42nd Division attack was stopped in this location by the 4th Prussian Guards and the Division was taking severe losses. The 1st and 3rd Battalions of the 47th Infantry were temporarily attached to the 42nd Division to aid in the assault on Sergy. Tayntor's 3rd Battalion 47th Infantry was attached to the 168th Infantry of the 42nd Division on the right side of the attack.

On July 29, 1918, 3rd Battalion 47th Infantry attacked the village and lost three successive Battalion Commanders wounded and a fourth killed. Major Gulielmus Heidt was wounded before the advance got to the village. Major James Cole was wounded as soon as the Battalion entered the village. Captain Louis Roberts was wounded in the village almost immediately after assuming command. Tayntor's Company Commander Captain Ross Snyder then took command of the Battalion and was killed a few hours later.

Tayntor was wounded twice by shrapnel in the attack on Sergy but after witnessing the loss of so many senior officers, he refused to be evacuated to the aid station and remained in command of his platoon. He was determined not to leave his men leaderless and directed his soldiers forward in the attack during the day and established them in their defensive position for the night. Even though he was severely wounded, he did not allow the medics to look after him until the next morning when he was sure all the wounded enlisted men in his platoon had received medical aid before him.

Tayntor's further service in 1918 could not be determined but he apparently was in a hospital for some time as his service record lists no more campaigns after Aisne-Marne. He was transferred to Headquarters 89th Division on February 7, 1919, and was promoted to the permanent Regular Army rank of 1st Lieutenant on February 14, 1919. He was assigned to the Civil Affairs Office of 3rd Army and served on occupation duty in Germany with that organization from May 12 to September 5, 1919. He was then stationed in Paris, France with the 29th Infantry until November 14, 1919. Tayntor returned to the United States as a 1st Lieutenant of the Motor Transport Corps in a casual detachment of officers aboard the troopship *U.S.S. George Washington* (seized German liner) leaving Brest, France November 19, 1919 and arriving at Hoboken, New Jersey November 25, 1919. He resigned his commission on January 19, 1920 at Camp Benning, Georgia.

Tayntor returned to Harvard Law School and received his Law degree in 1922. He was married to Mary Wheeler after 1920. Their daughter was born in 1928. The 1930 Census showed Tayntor living with his wife and daughter at his wife's parents' home in Erie, Pennsylvania where Tayntor was employed as a lawyer and indicated he was a veteran of the World War. His wife Mary died of tuberculosis in 1936. The 1940 Census showed Tayntor living with his daughter in Erie where he was still a lawyer. Clark O. Tayntor died of cirrhosis of the liver at the age of 49 on March 23, 1941 and is buried in Erie Cemetery, Erie, Erie County, Pennsylvania. He was a member of the New Jersey Society of the Sons of the American Revolution.

Michael D. Belis

Henry Tudury

Private, United States Army
Army Service Number: 562712
Company C 12th Machine Gun Battalion 4th Division
Date of Action: July 18 - 20, 1918
Location: Near Courchamps, France
War Department General Orders 46 (1919)
Medal # 3344

CITATION

Engaged as runner, Private Tudbury made repeated trips through intense shelling and machine gun fire. On July 18 he was gassed but bravely continued with his heroic work until he fell exhausted on the 20th.

DETAILS

Henry Jetton Tudury was born in Bay Saint Louis, Hancock County, Mississippi on November 22, 1885, the son of Peter (Pierre) and Ida Tudury. Before entering the Army, he was a baker. He enlisted as a Private in the 4th Infantry (Regular Army) on April 24, 1917. He was assigned to the rank/position of Cook on June 27, 1917. When elements of the 4th Infantry were used to create the 59th Infantry he thus became one of the original members of the 59th Infantry at the camp at Gettysburg, Pennsylvania in July 1917. He was promoted to Sergeant on September 14, 1917. Tudury went with the 59th Infantry to Camp Greene, North Carolina on November 8, 1917 and in the latter part of that month was transferred to Company C 12th Machine Gun Battalion where he became Mess Sergeant of the Company.

He went absent without leave to Mississippi around Christmas time and married Zelma Bermond on December 27, 1917. When he returned to his Company, he was reduced to the rank of Private. Tudury sailed to England as a Private in Company C 12th Machine Gun Battalion 4th Division aboard the British troopship *RMS Aquitania* on May 7, 1918. As his emergency contact, he listed his father Peter in Bay St. Louis, Mississippi. From England he and his Company were sent over to France where they were issued their machine guns.

Tudury was awarded the Distinguished Service Cross for his actions spanning three days in the Aisne-Marne Offensive. During the advance, Tudury and his Company were supporting 1st Battalion 58th Infantry which was attached to the 133rd Infantry Regiment of the French 164th Division. On July 18, 1918, the first day of the Offensive, the 1st Battalion 58th Infantry assisted in the capture of the village of Hautevesnes and then supported the French as they attacked and occupied the village of Courchamps. The next day, July 19, the attack was resumed but was stopped on a ridge line about 800 meters east of Courchamps and the French and Americans dug in for the night. On July 20, the Battalion advanced as far as the wooded area known as the Bois Pétret but was driven back to the ridge line east of Courchamps by a German counterattack and then later retook the ground it had lost.

During those three days, Tudury was the personal runner for his Company Commander, Captain Harold C. Hoopes. On July 18, Tudury was knocked down by concussion from a close artillery round and wounded by gas but continued carrying messages. He was gassed once more yet still remained on his missions. On July 20, he was again struck down by the explosion of a gas shell which hit close to him and for a time rendered him unable to move. Once he recovered his senses, he was found to be injured by the effects of the gas and too exhausted to resume his duties as a runner. He was carried from the battlefield and spent a couple of weeks in Base Hospital No. 46 before rejoining his Company.

Tudury served in the St. Mihiel Offensive. He was promoted to Corporal on September 23, 1918 and served in the Meuse-Argonne Offensive where he was placed in charge of his Company's runners. With Tudury leading the group, the total of runners for his Company came to twenty. When his Battalion's participation in the Offensive ended, Tudury was one of only three out of those twenty runners who made it through the Meuse-Argonne without being killed or wounded.

During the Offensive an enemy three-inch artillery round glanced off his helmet without exploding, leaving him dazed for about five minutes yet unwounded except for a lump on his head and a considerable dent in his helmet. At some time after September 1918, his rank was changed to the rank/position of Mechanic. He served on occupation duty in Germany with the 4th Division. On March 18, 1919 in a 4th Division award ceremony at Büchel, Germany Tudury was presented with the Distinguished Service Cross by the Commanding General of the American Expeditionary Forces John J. Pershing. In a ceremony on July 18, 1919 at Brest, France the Commanding General of the French Army Field Marshall Philippe Pétain awarded Tudury the French Croix De Guerre.

Tudury returned to the United States as a Mechanic of Infantry with Company C 12th Machine Gun Battalion aboard the troopship *U.S.S. Von Steuben* (seized German liner *Kronprinz Wilhelm*) leaving Brest, France July 21, 1919 and arriving at Hoboken, New Jersey July 29, 1919. He was discharged on August 9, 1919.

The 1920 Census showed Tudury and his wife living in New Orleans, Louisiana where he was employed as a baker. The 1930 Census showed him living with his wife and four daughters in Bay St. Louis, Mississippi where he was employed as a baker and indicated he was a veteran of the World War. After 1932, he was awarded the Purple Heart for the wounds he received from gas in World War One. The 1940 Census showed him and his family still in Bay St. Louis where he was still a baker. Henry Tudury died of cancer at the age of 66 on May 21, 1952 and is buried in Saint Mary's Cemetery, Bay Saint Louis, Hancock County, Mississippi.

No soldier from Mississippi received the Medal of Honor in World War One. Twenty-five soldiers from Mississippi received the Distinguished Service Cross in World War One. With his Distinguished Service Cross, Purple Heart, World War One Victory Medal with battle clasps, Army of Occupation in Germany Medal and French Croix De Guerre, Henry Tudury is considered the most decorated soldier from the State of Mississippi in World War One. Among several of his mementos on display at the Mississippi Armed Forces Museum at Camp Shelby, Mississippi is the scarred and dented helmet he wore throughout the war.

Michael D. Belis

Henry D. Turner

Sergeant, United States Army
Army Service Number: 2101130
Company B 10th Machine Gun Battalion 4th Division
Date of Action: July 23, August 9, September 29 and October 6, 1918
Location: Near Le Vallee, France; near St. Thibaut, France; near the Bois de Septsarges, France and near the Bois de Fays, France
War Department General Orders 66 (1919)
Medal # 3752

CITATION

Sergeant Turner, a runner, repeatedly went out under shell and machine gun fire to maintain liaison between units, frequently volunteering for especially hazardous missions. After other runners had been killed, he rendered valuable service by repeatedly crossing dangerous areas in order to maintain communications.

DETAILS

Henry David Turner was born in Burnt Prairie, White County, Illinois on May 22, 1893, the son of Henry K. and Margaret Turner. Before entering the Army, he was a clerk in an Abstract Office in Fairfield, Illinois. He sailed to France as a Private in Company B 10th Machine Gun Battalion 4th Division aboard the French troopship *Rochambeau* on May 7, 1918. As his emergency contact, he listed his father in Fairfield, Illinois.

There are four specific dates in his Citation on which his actions as a runner were singled out and combined to merit the award of the Distinguished Service Cross. The first date in his Citation is July 23, 1918 in the Aisne-Marne Offensive.

On that date, his Battalion was attached to the French 164th Division and was still operating in support of the French when all other Infantry and Machine Gun units of the 4th Division which had been attached to the French had been detached and sent to the rear for a rest two days earlier. By July 23, the 10th Machine Gun Battalion had advanced with the French as far as the wooded area known as the Bois du Roi, west of the Château-Thierry-Soissons Road before finally being pulled out of the front lines and being reattached to the 4th Division.

The second date in his Citation is August 9, 1918 in the Vesle Sector. On that date, the 10th Machine Gun Battalion was assigned to Division reserve in the area between the villages of St. Thibaut and Ville Savoye at the Vesle River. August 9 was a day in which no attack was made, instead patrols were sent out to determine German defensive positions and locate and destroy enemy machine gun nests.

Turner then served in the Toulon Sector and the St. Mihiel Offensive.

The third date in his Citation is September 29, 1918, the fourth day of the Meuse-Argonne Offensive. On that date, the 10th Machine Gun Battalion was acting as part of the Brigade Reserve for the 7th Brigade of the 4th Division. The Brigade was pulled back out of the front attacking line and held in reserve in the wooded area known as the Bois de Septsarges.

The fourth date in his Citation is October 6, 1918, also in the Meuse-Argonne Offensive. On this date the 10th Machine Gun Battalion was working with the 8th Brigade of the 4th Division and had supported the attack of the 58th Infantry into the wooded area known as the Bois de Fays. A foothold had been gained in the woods and on October 6 the attack had been suspended while all units consolidated and improved their positions and held the line against German patrols and counterattacks.

Turner carried out the dangerous job of runner through the above four dates and continued to volunteer for assignments even after other runners were killed in their attempts to deliver messages. His actions helped to ensure communications were maintained so that his unit could coordinate its operations with other units which it supported.

At some time in France, Turner was promoted to Sergeant though it is not clear if he was promoted before or after his award actions. He served on occupation duty in Germany with the 4th Division and returned to the United States as a Sergeant with Company B 10th Machine Gun Battalion aboard the troopship *U.S.S. Mobile* (seized German liner *Cleveland*) leaving Brest, France July 16, 1919 and arriving at Hoboken, New Jersey July 27, 1919. The date of his discharge could not be found.

The 1920 Census showed Turner living with his parents in Grover, Wayne County, Illinois where he was indicated as being a law student. On May 20, 1922 he married Stella Young. The 1930 Census showed him living with his wife and two sons in Douglas, Effingham County, Illinois where he was employed as a postal railway clerk and indicated he was a veteran of the World War. The 1940 Census showed him living with his wife, two sons and sister-in-law in Hudson, McLean County, Illinois where he was employed as a postal railway clerk for the Alton Railroad. When he registered for the draft in 1942, he indicated he was living in Hudson, McLean County, Illinois and was employed by the U.S. Post Office Department in Chicago, Illinois on the Alton Railroad. He was back in uniform during World War Two with the rank of Captain as an Army Postal Officer from May 15, 1942 to October 5, 1945. His wife Stella died in 1970 and he married Ethel Bosher on January 3, 1981. Henry D. Turner died at the age of 93 on March 13, 1987 and is buried in Hudson Cemetery, Hudson, McLean County, Illinois.

Michael D. Belis

Thomas Vander Veen

Private First Class, United States Army
Army Service Number: 573830
Company C 11th Machine Gun Battalion 4th Division
Date of Action: October 10 - 13, 1918
Location: Near Nantillois, France
War Department General Orders 87 (1919)
Medal # 1519

CITATION

As company liaison agent, Private First Class Vander Veen maintained continual contact between his company commander and the battalion Post Commander, repeatedly exposing himself to artillery, machine gun, and sniper's fire to deliver important messages. On one occasion it was necessary for him to pass through the German and our own barrages, but he accomplished this mission fearlessly, showing marked personal bravery.

DETAILS

(In half the records found for him his last name is spelled Vanderveen.)

Thomas Vander Veen was born in Oenkerk, Netherlands on October 20, 1888, the son of Henry Gubben Vander Veen. He immigrated to the United States aboard the *S.S. Potsdam* in 1912. He entered the Army on October 3, 1917 and at that time listed his home of record as San Fernando, Los Angeles County, California. He sailed to France as a Private in Company C 11th Machine Gun Battalion 4th Division aboard the troopship *U.S.S. Rijndam* (former Dutch liner) on May 10, 1918. As his emergency contact, he listed his father Henry in Burum, Holland.

He served in the Aisne-Marne Offensive, the Vesle Sector, the Toulon Sector, and the St. Mihiel Offensive. At some time in France, he was promoted to Private 1st Class.

Vander Veen was awarded the Distinguished Service Cross for his actions spanning three days in the Meuse-Argonne Offensive. From October 10 - 13, 1918 he performed the dangerous job of Company runner, delivering messages from his Company in the wooded area known as the Bois de Forêt to 11th Machine Gun Battalion Headquarters near the village of Nantillois south of the wooded areas known as the Bois de Fays and the Bois de Brieulles. He constantly crossed open areas subjected to every kind of fire the enemy could throw at him. On one of his missions, he had to venture through not only German artillery fire but American artillery fire as well. Because of his courage to repeatedly traverse the battlefield and risk his life under the most dangerous and deadly conditions, his Company was able to keep fully coordinated in its movements with its orders from Headquarters.

Michael D. Belis

Vander Veen served on occupation duty in Germany with the 4th Division and returned to the United States with Company C 11th Machine Gun Battalion aboard the troop transport *U.S.S. Tiger* leaving Brest, France, July 17, 1919 and arriving at Hoboken, New Jersey July 29, 1919. He was discharged on August 18, 1919.

Vander Veen became a naturalized American citizen on September 16, 1919. When he registered for the draft in 1942, he gave his residence as Delavan, Walworth County, Wisconsin and indicated he was employed as a farmer on his own farm. Thomas Vander Veen died at the age of 91 on May 29, 1980 and is buried in Spring Grove Cemetery, Delavan, Walworth County, Wisconsin.

John C. Vann

Second Lieutenant, United States Army
Company H 47th Infantry Regiment 4th Division
Date of Action: August 7, 1918
Location: Near Bazoches, France
War Department General Orders 46 (1919)
Medal # 2029

CITATION

Lieutenant Vann concealed the fact that he was wounded and led the advance platoon of his company to their objectives despite heavy losses. He remained with his command, displaying the highest leadership and courage until he was wounded a second time.

DETAILS

John Cannon Vann was born in Valdosta, Georgia on June 13, 1893, the son of Arthur and Mary Vann. Before entering the Army, he was a Travel Agent in Jacksonville, Florida. He attended Officers Training Camp at Fort Oglethorpe, Georgia, was commissioned a 2nd Lieutenant of Infantry in the Officers Reserve Corps on November 27, 1917 and assigned to the 47th Infantry at Camp Greene, North Carolina. Vann sailed to France as a 2nd Lieutenant in Company H 47th Infantry 4th Division aboard the troopship *U.S.S. Princess Matoika* (seized German liner *Prinzess Alice*) on May 10, 1918. As his emergency contact, he listed his mother in Columbus, Georgia. Vann served in the Aisne-Marne Offensive.

He was awarded the Distinguished Service Cross for his

Michael D. Belis

actions in the Vesle Sector. In early August 1918, the 4th Division attempted to move their advance across the Vesle River. On August 7, 2nd Battalion 47th Infantry succeeded in getting about 350 men across the river south of the village of Bazoches. Vann led his platoon as the spearhead of the attack with his ultimate objective being the Rouen-Rheims National Highway running along an east-west line north of Bazoches. At sometime during the advance and the crossing of the river, Vann was slightly wounded but kept the men of his platoon from learning about it and led them to the Highway, his platoon taking heavy casualties during their attack. After reaching the objective, he directed his men in holding their position until ordered to pull back to the south and change the direction of their advance. He skillfully executed the movement and led his platoon in a further attack until he was wounded again, this time severely and had to relinquish his command.

For his bravery and leadership, he was also awarded a Citation Star and the French Croix De Guerre with gilt star. He must have spent considerable time in hospital as his Form No. 84c-9 A.G.O. Statement of Service Card shows no further campaigns or sectors. Vann was promoted to 1st Lieutenant on October 28, 1918. He returned to the United States as a 1st Lieutenant of Infantry assigned to Headquarters District, Paris aboard the troopship U.S.S. Imperator (surrendered German liner) leaving Brest, France July 7, 1919 and arriving at Hoboken, New Jersey July 13, 1919. He was discharged on July 31, 1919.

He was married in about 1925 and had a daughter in 1926. By 1935 he was living in Houston, Texas. The 1940 Census showed Vann living with his wife and daughter in Houston, Texas where he was the owner of an oil drilling business. John

C. Vann died of malignant hypertension at the age of 51 on March 30, 1945 and is buried in Forest Park Cemetery, Houston, Harris County, Texas.

Michael D. Belis

Paul Von Krebs
First Sergeant, United States Army
Army Service Number: 558501
Company M 47th Infantry Regiment 4th Division
Date of Action: July 29 - 30, 1918
Location: At Sergy, France
War Department General Orders 55 (1920)
Medal # 6285

CITATION
Sergeant Von Krebs displayed exceptional bravery in voluntarily carrying wounded men to safety across shell-swept areas. Later he took charge of two platoons, whose officers had become casualties, and reorganized them. Strengthening these with stragglers from other organizations, he led them all into the attack at a critical moment.

DETAILS
Paul Von Krebs was born in Berlin, Germany on December 5, 1875, the son of Richard Von Krebs. He immigrated to the United States in either 1889 or 1898. (According to the 1900 Census he arrived in the U.S. in 1889. According to the 1910 Census he arrived in the U.S. in 1898. Sometime between 1900 and 1910 he became a naturalized American citizen.) He enlisted in Company L 27th United States Volunteer Infantry in 1899 and served in the Philippine Insurrection. At the time of that enlistment, Von Krebs listed his civilian occupation as upholsterer. He was discharged from the Volunteers on April 1, 1901 and enlisted in Company E 15th Infantry (Regular

Army) on December 20, 1901. He was discharged in 1904, returned to his trade as an upholsterer and in 1906 enlisted in the Army Ordnance Department for duty at the Benicia Arsenal in California. Upon his discharge in 1909 he reenlisted, this time in Company M 9th Infantry and served in the Philippines. He was discharged in 1912 as a Corporal.

The dates of further enlistments for Von Krebs could not be found. At the time of his service with the 4th Division his home of record was listed as Franklin Park, New Jersey. He sailed to France as 1st Sergeant in Company M 47th Infantry 4th Division aboard the Italian troopship *Caserta* on May 10, 1918. As his emergency contact, he listed his sister Mrs. Gertrude Welker in New York City.

Von Krebs was awarded the Distinguished Service Cross for his actions on two days during the battle for Sergy in the Aisne-Marne Offensive.

(See the entry for Thomas W. Kearns for an overview of the conditions on the battlefield at Sergy.)

During the battle for Sergy, Von Krebs carried wounded soldiers to safety through enemy artillery fire. His Company was part of 3rd Battalion 47th Infantry and suffered heavy casualties during the fighting in and around Sergy. At one point, Von Krebs took over two platoons who lost their officers and assembled them together, along with other soldiers who had become stragglers and led his makeshift unit in an attack upon the Germans.

Von Krebs is listed in the official history of the 47th Infantry in World War One as being gassed and returned to duty (dates not given) and also of dying of wounds (date not given). In the Register of Burials of Deceased American Soldiers, 1917 – 1922, Office of the Quartermaster General he is recorded as dying of Septicemia on November 4, 1918. One account states he died in Base Hospital No. 15 at Chaumont, Haute-Marne, Champagne-Ardenne, France.

Von Krebs was buried in Plot J Grave # 419 in the American cemetery # 10 at the village of Chaumont.

His Distinguished Service Cross was presented to his stepmother Mrs. Leonie Von Krebs.

On February 13, 1922, Von Krebs' body was disinterred for preparation and shipment. His remains were returned to the United States aboard the United States Army Transport *U.S.A.T. Cantigny* leaving Antwerp, Belgium April 8, 1922 and arriving at Brooklyn, New York April 21, 1922. He was one of 213 American dead brought home on the *Cantigny* on that journey. His remains were shipped to Henry Stolzenberger, Undertaker, Bronx, New York who received them on April 26, 1922. No further details for Paul Von Krebs could be found.

Emmett W. Waltman

Corporal, United States Army
Army Service Number: 569248
Company F 4th Engineer Regiment 4th Division
Date of Action: August 5, 1918
Location: West of Fismes, France
War Department General Orders 145 (1918)
Medal # 1087

CITATION

Corporal Waltman was a member of a small detachment of engineers which went out in advance of the front line of the infantry through an enemy barrage from 77-mm and one-pounder guns to construct a footbridge over the River Vesle. As soon as their operations were discovered, machine gun fire was opened up on them, but undaunted, the party continued at work, removing the German wire entanglements and completing a bridge which was of great value in subsequent operations.

DETAILS

Emmett Wesley Waltman was born in Rockford, Spokane, Washington on August 25, 1894, the son of Victor and Amelia "Ella" Waltman. He entered the Army on May 10, 1917. Waltman sailed to France as a Corporal in Company F 4th Engineers 4th Division aboard the troopship *U.S.S. Martha Washington* (seized Austrian liner) on April 30, 1918. As his emergency contact, he listed his mother Ella in Kellogg, Indiana.

Waltman was awarded the Distinguished Service Cross for his actions in the Aisne-Marne Offensive. On August 5, 1918,

Michael D. Belis

Waltman and his Engineer Company were on the extreme right of the 4th Division attack at the Vesle River west of the village of Fismes and near the village of Ville Savoye. After several days of trying to cross the Vesle River under fire and only managing to put across small numbers of men, on August 5 the 4th Division made a determined effort to get across the river. The 4th Engineers, including Waltman, worked under fire incessantly to build and emplace footbridges which the Infantry could utilize to make a mad dash across the river. On August 5 he was part of a small detachment of Engineers who particularly distinguished themselves by remaining at their work under the heaviest of fire and completing a bridge for the Infantry.

(See the entry for William B. Beach for an overview of the kinds of activities carried out by the soldiers of the 4th Engineers involved in building the bridges across the Vesle River.)

Waltman served in the Vesle Sector, the Toulon Sector, the St. Mihiel Offensive, and the Meuse-Argonne Offensive.

Waltman also received a Citation Star. He was awarded the French Croix De Guerre. He was one of three soldiers in the 4th Division to be awarded the Belgian Military Decoration (Décoration Militaire) during World War One. All three soldiers were Corporals from Company F 4th Engineers and all received the Belgian Military Decoration and the Distinguished Service Cross for their actions at the Vesle River on August 5, 1918. Waltman was one of only 20 soldiers of the American Expeditionary Forces to receive the Belgian Military Decoration.

At some time after his award action, Waltman was promoted to Sergeant. He served on occupation duty in Germany with the 4th Division and returned to the United States as a Sergeant in Headquarters Company 4th Engineers aboard the troopship *U.S.S. Von Steuben* (seized German liner *Kronprinz Wilhelm*) leaving Brest, France July 21, 1919 and arriving at Hoboken, New Jersey July 29, 1919. The date of his discharge could not be found.

The 1920 Census showed Waltman living with his parents in Kellogg, Shoshone County, Idaho where he was employed as a Mining Engineer at a lead mine. On July 20, 1922, he married Mary Erfle and in 1924 their daughter was born. The 1930 Census showed Waltman living with his wife and daughter in Kellogg where he was employed as a Laborer at a lead smelter and indicated he was a veteran of the World War. The 1940 Census showed him living with his wife and two daughters in Kellogg where he was employed as a Maintenance Superintendent at a zinc smelter. Emmett W. Waltman died at the age of 71 on May 2, 1966 and is buried in Forest Cemetery, Coeur d'Alene, Kootenai County, Idaho.

Michael D. Belis

Arthur H. Warfield

Sergeant, United States Army
Army Service Number: 557740
Company B 47th Infantry Regiment 4th Division
Date of Action: August 1, 1918
Location: At Sergy, France
War Department General Orders 145 (1918)
Medal # 1079

CITATION

Sergeant Warfield displayed exceptional courage and loyalty by remaining in active command of his section after being wounded twice.

DETAILS

Arthur Henry Warfield Jr. was born in Conway, Franklin County, Massachusetts on October 10, 1887, the son of Arthur Henry and Frances Warfield. Before entering the Army, he was a farmer on his father's farm in West Brookfield, Massachusetts. He entered the Army on June 13, 1917. Warfield sailed to France as a Corporal in Company B 47th Infantry 4th Division aboard the troopship *U.S.S. Princess Matoika* (seized German liner *Prinzess Alice*) on May 10, 1918. As his emergency contact, he listed his father in West Brookfield, Massachusetts. At some time after reaching France, he was promoted to Sergeant.

Warfield was awarded the Distinguished Service Cross for his actions in the Aisne-Marne Offensive. During the battle for the village of Sergy two Battalions of the 47th Infantry were attached to the 42nd Division. Warfield's 1st Battalion 47th

Infantry was attached to the 167th Infantry Regiment of the 42nd Division. Altogether, the village of Sergy was won and lost nine times before the Americans finally captured and held it permanently. Once committed to the battle on July 29, 1918, Warfield and Company B 47th Infantry were constantly at the front of the action, engaged in hand-to-hand combat in taking the slopes of the high ground outside the village and fighting in the village itself. On August 1, 1918, Warfield stayed with his squad although he had been wounded twice and inspired his men with his bravery and leadership. He was also awarded the French Croix De Guerre.

He served on occupation duty in Germany with the 4th Division and returned to the United States with Company B 47th Infantry aboard the troopship *U.S.S. Mobile* (seized German liner *Cleveland*) leaving Brest, France July 16, 1919 and arriving at Hoboken, New Jersey July 27, 1919. He was discharged in August 1919.

The 1920 Census showed Warfield living with his parents in West Brookfield, Worcester County, Massachusetts where he was employed as a farmer on his father's poultry farm. In 1925, he married Lillian Miner and in 1926 they had a daughter. The 1930 Census showed him living with his wife, daughter, stepson, and mother-in-law in West Brookfield where he was employed as a farm laborer and indicated he was a veteran of the World War. The 1940 Census showed him and his family still in West Brookfield where he was still a farmer. When he registered for the draft in 1942, he indicated he was self-employed and had scars from gunshot wounds to his arm and body. Arthur H. Warfield died at the age of 76 on July 18, 1964 and is buried in Pine Grove Cemetery, West Brookfield, Worcester County, Massachusetts.

Michael D. Belis

Joseph Waskiewic

Private, United States Army
Army Service Number: 559066
Company A 11th Machine Gun Battalion 4th Division
Date Of Action: October 9 - 13, 1918
Location: Near Bois De Brieulles, France
War Department General Orders 87 (1919)
Medal # 1423

CITATION

As a runner between company and battalion headquarters, Private Waskiewic crossed heavily shelled areas to deliver important messages. Wounded when crossing an open space, subjected to artillery and machine gun fire, he refused to be evacuated, but continued the performance of his duties.

DETAILS

Joseph Waskiewic was born in Thorndike, Minnesota on April 24, 1898. Upon entering the Army, he listed his home of record as New Bedford, Massachusetts. He entered the Army on September 6, 1917. Waskiewic sailed to France as a Private in Company A 11th Machine Gun Battalion 4th Division aboard the troopship *U.S.S. Rijndam* (former Dutch liner) on May 10, 1918. As his emergency contact, he listed his father Joseph in New Bedford, Massachusetts. Waskiewic served in the Aisne-Marne Offensive, the Vesle Sector, the Toulon Sector, and the St. Mihiel Offensive.

He was awarded the Distinguished Service Cross for his actions spanning several days in the Meuse-Argonne Offensive. From October 9–13, 1918, the 7th Brigade of the 4th Division which included the 11th Machine Gun Battalion made daily attacks on a line from the northwestern edge of the wooded area known as the Bois de Brieulles to the eastern edge of the wooded area known as the Bois de Fays. The attack of October 9 was made through a horror on the battlefield of heavy German artillery barrages, most of which were gas. The attack was begun late in the afternoon and continued into the night. Gas masks had to be worn and in the wooded areas men stumbled over their dead and wounded fellow soldiers as they advanced in the darkness through the tangled undergrowth of the forest. Waskiewic was wounded when crossing the open area between the woods yet continued his duties as a runner until his unit was relieved from the front lines on October 13.

At some time after his award action, Waskiewic was promoted to Private 1st Class. He served on occupation duty in Germany with the 4th Division and returned to the United States with Company A 11th Machine Gun Battalion aboard the troop transport *U.S.S. Tiger* leaving Brest, France July 17, 1919 and arriving at Hoboken, New Jersey July 29, 1919. He was discharged on August 4, 1919.

No further details for Joseph Waskiewic could be found. There is a Joseph Walter Waskiewicz buried at Pine Grove Cemetery, New Bedford, Bristol County, Massachusetts. His grave marker indicates he was born on April 24, 1898 and died on December 16, 1981 and that he was a Private 1st Class in the U.S. Army in World War One. City directories show that he and his wife Josephine lived in New Bedford in the 1920's, 1930's, and 1940's and moved to Amherst, Massachusetts in about 1950. Though confirmation for such could not be found, it is believed this is the same soldier who received the Distinguished Service Cross under the name of Joseph Waskiewic.

Michael D. Belis

Harrison B. Webster

Major, United States Army
47th Infantry Regiment 4th Division
Date of Action: September 26 - October 12, 1918
Location: Near Bois de Brieulles, France
War Department General Orders 74 (1919)
Medal # 7108

CITATION

After seeing that his personnel were functioning properly, Major Webster went fearlessly to positions in the front lines. When stretcher bearers were unable to handle the large number of casualties, he personally took a light German wagon to the front lines and gathered the wounded. His personal bravery was an inspiration to his men throughout his service. He was killed by shell fire on October 12, 1918.

DETAILS

Harrison Briggs Webster was born in Boston, Massachusetts on January 26, 1884, the son of Andrew and Elizabeth Webster. He graduated from Harvard University in 1905 and from Harvard Medical School in 1909. On May 1, 1913, he married Margaret Gleason. Their son was born in June 1914 and their first daughter was born in June 1916. Webster was commissioned a 1st Lieutenant in the Officers Reserve Corps (Medical Department) on May 25, 1917. He was promoted to Captain on September 23, 1917 and in March 1918 was promoted to Major. He was assigned to the 4th Division prior to sailing overseas. Webster sailed to England as a Major in Field Hospital 19 Sanitary Train 4th Division aboard the British/Canadian troop transport *Tunisian* on May 27, 1918. As his emergency contact, he listed his wife in Northampton, Massachusetts. From England he and his unit were sent over to France. His second daughter was born in June 1918, the same month he was arriving in France.

Sometime after reaching France, Webster was temporarily attached to the 47th Infantry. He served in the Aisne-Marne Offensive where on the night of July 30, 1918 even though he was the senior surgeon in the Regiment, he personally drove a motorized ambulance to the fighting at the village of Sergy and evacuated wounded while under German artillery fire. In August 1918 Webster was officially assigned to the 47th Infantry as their Regimental Surgeon. He served in the Vesle Sector, the Toulon Sector, and the St. Mihiel Offensive.

He was awarded the Distinguished Service Cross for his actions spanning two weeks in the Meuse-Argonne Offensive. During the days of September 26 - October 13, 1918, Webster's bravery under fire while directing and working in his forward field hospital near the front lines was an inspiration to the men under his command and was instrumental in keeping the unit working smoothly, thus saving many lives among the Infantry casualties. After making sure his field hospital was working efficiently, he went to the front lines to view the conditions concerning the wounded. When he saw that stretcher bearers were unable to keep up with the high rate of casualties, Webster brought a captured German wagon to the front lines and personally picked up wounded soldiers, getting them off the battlefield, and back to medical aid as quickly as possible. He was killed by enemy artillery while bringing wounded back with his wagon near the wooded area known as the Bois de Brieulles south of the village of Brieulles-sur-Meuse near the Meuse River. His date of death is given in his Citation as October 12, 1918, however in the records of the American Battle Monuments Commission and in the Register of Burials of Deceased American Soldiers, 1917 – 1922, Office of the Quartermaster General his date of death is indicated as October 13, 1918.

Webster was the highest ranking Medical officer of the 4th Division to be killed in action during the war. He also received a Citation Star.

He was buried in Plot 4 Grave # 160 in the American cemetery # 702 at the village of Septsarges.

His Distinguished Service Cross was presented to his widow Mrs. Harrison B. Webster.

On May 21, 1919, Harrison B. Webster was reinterred in Plot 4 Section 46 Grave # 180 in the Argonne American Cemetery # 1232. He was reinterred and laid to his final rest on November 25, 1921 in Plot B Row 37 Grave 16 in the Meuse-Argonne American Cemetery, Romagne-sous-Montfaucon, France.

At some time between 1930 and 1933, Mrs. A. G. Webster,

the mother of Major Harrison B. Webster visited her son's grave in France under the Gold Star Pilgrimage program. See the entry for Edward S. Blackman for an explanation of the Gold Star Pilgrimages.

Michael D. Belis

John Samuel Weimer

Private, United States Army
Army Service Number: 2225018
Company M 47th Infantry Regiment 4th Division
Date of Action: August 10, 1918
Location: Near the Vesle River, France
War Department General Orders 81 (1919)
Medal # 7910

CITATION

While on outpost duty, Private Weimer learned that a soldier from another organization was lying wounded in a shell hole 200 yards away. With another member of his squad, Private Weimer voluntarily went through machine gun and sniper fire and carried the wounded man to shelter.

DETAILS

John Samuel Weimer was born in Waxahachie, Ellis County, Texas on October 26, 1888, the son of John and Margurette Weimer. Before entering the Army, he was a school-teacher in Mount Pleasant, Titus County, Texas. He was drafted into the Army as a Private on October 9, 1917 and was assigned to Company M 47th Infantry on March 29, 1918. Weimer sailed to France as a Private in Company M 47th Infantry 4th Division aboard the Italian troopship *Caserta* on May 10, 1918. As his emergency contact, he listed his father in Mt. Pleasant, Texas. Weimer served in the Aisne-Marne Offensive.

He was awarded the Distinguished Service Cross for his actions in the Vesle Sector. On August 10, 1918, the 47th

Infantry was established in defensive positions along the railroad south of the Vesle River near the village of St. Thibaut. For several days the Regiment had tried to take the village of Bazoches, which was directly across the river. Elements of the 47th Infantry had managed to cross the river and enter the village itself before being driven back across the river by German counterattacks. On August 10, the 47th Infantry dug in south of the river and consolidated its positions in anticipation of being relieved in place by the 77th Division the next night.

A wounded soldier from a patrol lay in a shell hole in advance of the front lines and Weimer and Private Edward D. Ritchie from the same squad as Weimer decided to make an attempt to retrieve the soldier. Together, the two Texans made their way across the battlefield under fire to the shell hole. Ritchie suggested he run alone in the direction of the American lines to provide the enemy with a target to shoot at while Weimer hoisted the wounded man onto his back and carried him to safety. As Ritchie conspicuously drew the Germans' fire, Weimer managed to get the wounded comrade back to where he could be attended to by American medics. Both Weimer and Ritchie were awarded the Distinguished Service Cross for their actions in saving the wounded soldier.

Weimer was promoted to Private First Class on September 17, 1918, and to Corporal on November 1, 1918. The official history of the 47th Infantry in World War One indicates that Weimer was wounded but does not give a date. He was also awarded a Citation Star. He was one of 20 soldiers of the 4th Division and one of only 399 soldiers of the American Expeditionary Forces to be awarded the Italian War Merit Cross (Croce al Merito di Guerra). The Italian War Merit Cross was presented to Weimer on June 11, 1919 in a 4th Division awards ceremony at Remagen, Germany.

He served on occupation duty in Germany with the 4th Division and returned to the United States as a Corporal in Company M 47th Infantry and as part of Company I Third Army Composite Regiment aboard the troopship *U.S.S. Leviathan* (seized German liner *Vaterland*) leaving Brest, France September 1, 1919 and arriving at Hoboken, New Jersey September 8, 1919.

(See the entry for Joseph Bassi for an explanation of the Third Army Composite Regiment.)

Weimer was discharged on September 29, 1919.

In 1919, Weimer married Lucille Lemmon. They had a daughter in 1921, another daughter in 1923, another daughter in 1924, and another daughter in 1926. The 1930 Census showed Weimer living with his family in Titus County, Texas where he was employed as a Teacher in Public School and indicated he was a veteran of the World War. The 1940 Census showed Weimer living with his family in Titus County where he was employed as a farmer on his own farm. His wife died in 1953. John Samuel Weimer died of cancer at the age of 68 on January 17, 1957 and is buried in Laurel Land Memorial Park, Fort Worth, Tarrant County, Texas.

CARROLL B. WEST

Sergeant, United States Army
Army Service Number: 573878
Company B 12th Machine Gun Battalion 4th Division
Date of Action: September 30 and October 2, 1918
Note: Though his Citation gives a time frame of September 30 - October 2, 1918 the action described in his Citation more closely fits the days of September 29 - October 2.
Location: Near the Bois des Ogons, France
War Department General Orders 66 (1919)
Medal # 7027

CITATION

Sergeant West displayed exceptional courage and leadership in leading the section forward and maintaining fire on the enemy from an advanced position in the wood, successfully covering the withdrawal of the infantry to a more secure position. This gallant soldier was killed two days later while he was successfully directing his section in breaking up an enemy counterattack.

DETAILS

Carroll Benjamin West was born in Lake Mills, Jefferson County, Wisconsin on April 28, 1895, the son of Allen and Hattie West. He grew up in rural Wisconsin and attended Milton College. Before entering the Army, he was employed as the General Secretary of the Dakota Wesleyan University Y.M.C.A. in Mitchell, South Dakota. West sailed to England as a Private in Company B 12th Machine Gun Battalion 4th Division aboard the British troopship *RMS Aquitania* on May 7, 1918. As his emergency contact, he listed his father Allen in Milton Junction, Wisconsin. From England he and his Company were sent over to France where they were issued their machine guns.

West served in the Aisne-Marne Offensive and the Vesle Sector. He was promoted to Sergeant on August 21, 1918 and placed in charge of a machine gun section consisting of 18 men and two machine guns. He served in the Toulon Sector and the St. Mihiel Offensive.

He was awarded the Distinguished Service Cross for his actions spanning several days in the Meuse-Argonne Offensive. West's Company was part of the 8th Brigade of the 4th Division and on September 29 was supporting 1st Battalion 58th Infantry in its attack toward the wooded area known as the Bois de Fays. Heavy machine gun fire from the Bois de Fays and from the wooded area known as the Bois des Ogons to the west and also artillery fire from east of the Meuse River prevented the Infantry from securing the Bois de Fays. Company C of 1st Battalion 58th Infantry fought its way through the Bois des Ogons but with no friendly troops on its flanks had to pull back to a more protected location south of the woods. West and his machine gun section advanced with the Infantry, established an exposed forward position in the southeastern part of the Bois des Ogons and covered the movements of Company C, including the Company's withdrawal. West and his guns succeeded in stopping two enemy counterattacks that day. The next three days found 1st Battalion 58th Infantry dug in and holding the line, repelling repeated enemy counterattacks with West and his guns supporting the Battalion in the defense. West was wounded during one of those enemy counterattacks and died of his wounds on October 2, 1918.

West was buried in Row 15 Grave # 56 in the French Military Cemetery (designated cemetery # 290 by the U.S. Graves Registration Services) at the village of Froidos.

His Distinguished Service Cross was presented to his father Allen B. West.

Carroll B. West was reinterred and laid to his final rest on October 18, 1921 in Plot G Row 30 Grave 35 in the Meuse-Argonne American Cemetery and Memorial, Romagne-sous Montfaucon, Departement de la Meuse, Lorraine, France.

Mrs. Hattie E. West, the mother of Sergeant Carroll B. West, sailed aboard the troopship *U.S.S. America* (seized German liner *Amerika*) on May 13, 1931 and visited her son's grave in France under the Gold Star Pilgrimage program. See the entry for Edward S. Blackman for an explanation of the Gold Star Pilgrimages.

Michael D. Belis

Stephen J. Weston

Sergeant, United States Army
Army Service Number: 558269
Company I 47th Infantry Regiment 4th Division
Date of Action: September 28, 1918
Location: Near Bois de Brieulles, France
War Department General Orders 46 (1919)
Medal # 3223

CITATION

Sergeant Weston charged an enemy machine gun which was inflicting heavy losses upon our troops and delaying the advance. He wounded the gunner and captured the gun, thereby enabling our advance to continue.

DETAILS

(Stephen J. Weston's actual name was Stephen J. Whiston. He served in the Army under the last name of Weston and his award Citation for the Distinguished Service Cross is under the name of Weston. In the official histories of the 4th Division and of the 47th Infantry in World War One his name is recorded as Weston. In the U.S. Army Transport Service Passenger lists his name is given as Weston. However, before and after his military service he went by the last name of Whiston. Therefore, most records found for him outside of the Army are under the name of Whiston. When he filled out his State of Connecticut Military Service Record in 1933, he indicated his last name as Whiston with Weston coming before it in parentheses and signed his name the same way.

For the sake of continuity in this profile, his name will be given as Weston.)

Stephen Joseph Weston was born in Waterbury, New Haven County, Connecticut on November 18, 1897, the son of John and Hanna (Johanna) Whiston. Before entering the Army, he served in the British Merchant Marine. He enlisted as a Private in Company I 9th Infantry (Regular Army) on June 6, 1917. Weston was promoted to Corporal on October 9, 1917. When elements of the 9th Infantry were used to form the 47th Infantry, he was transferred to Company I 47th Infantry at Camp Greene, North Carolina on October 30, 1917 and thus became one of the first members of the 47th Infantry. He was with the Regiment when it became part of the 4th Division. On April 15, 1918, he was promoted to Sergeant. Weston sailed to France as a Sergeant in Company I 47th Infantry 4th Division aboard the Italian troopship *Caserta* on May 10, 1918. As his emergency contact, he listed his mother in Waterbury, Connecticut.

Weston served in the Aisne-Marne Offensive. He received a Citation Star for his actions in the Vesle Sector near the village of St. Thibaut on August 8, 1918 while leading a platoon of his Company in an attack against a strongly defended German position near the Vesle River. He received a second Citation Star for his actions the next day, August 9, when after crossing the river he made a reconnaissance alone and under fire of the German lines near the village of Bazoches, obtaining valuable information about the enemy defenses which aided in the subsequent attack against the village.

Weston served in the Toulon Sector and the St. Mihiel Offensive.

He was awarded the Distinguished Service Cross for his actions in the Meuse-Argonne Offensive. On September 28,

1918, the 47th Infantry had advanced into most of the wooded area known as the Bois de Brieulles which was northeast of the village of Nantillois and south of the village of Brieulles-sur-Meuse. Weston's Company fought through the tangled underbrush of the woods in a painstaking process of destroying German machine gun positions one at a time. Weston single-handedly charged an enemy machine gun which had stopped the advance by causing numerous casualties among the Americans, wounded the gunner, and captured the gun, thus eliminating the position and allowing his Company to continue the attack.

On October 12, 1918, the 3rd Division began the process of relieving the 4th Division in place. As units of the 4th Division were relieved and sent to the rear, the 47th Infantry was held in the front lines in the Bois de Brieulles and was the last of the 4th Division Infantry units to be relieved. Enemy artillery contantly harrassed Weston and his men and on October 15 he was wounded by gas and also suffered the effects of shell shock from a near miss by a German artillery round. He refused medical treatment and remained with his men until they were finally relieved on the night of October 18-19, 1918.

Weston served on occupation duty in Germany with the 4th Division and was promoted to 1st Sergeant on June 21, 1919. He returned to the United States as the 1st Sergeant of Company E 47th Infantry aboard the troopship *U.S.S. Mobile* (seized German liner *Cleveland)* leaving Brest, France July 16, 1919 and arriving at Hoboken, New Jersey July 27, 1919. He was discharged on August 1, 1919.

Sometime after returning home Weston was awarded the French Croix De Guerre for his actions near the village of St. Thibaut, France on August 5, 1918 when he evacuated wounded French soldiers while under artillery and machine gun fire. The medal was presented to him by E. Dumont the French Ambassador to the United States in Washington, D.C.

The 1920 Census showed Weston living with his parents in Waterbury, New Haven County, Connecticut where he was employed as a Machine Hand at a Brass Mill. On December 15, 1923, he married Rosalia McGrail. Their daughter was born in March 1925. City directories from 1926 through 1939 showed Weston still living in Waterbury, Connecticut. Weston and his wife had a son born in May 1929 and a second son born in February 1933. His 1933 dated State of Connecticut Military Service Record indicated he had suffered from gas and psychoneurosis during the war and had to give up his trade of Brass Rolling due to illness. The form also stated he had received treatment for his condition at the U.S. Veterans Hospital in Newington, Connecticut. The 1940 Census showed him living with his wife, daughter, and two sons in Meriden, New Haven County, Connecticut with no employment or occupation for him listed.

Weston died at the age of 92 on November 7, 1990. The Connecticut Death Index indicated he was a Disabled Vet. Stephen J. Weston is buried in Arlington National Cemetery (under the name of Stephen J. Whiston).

Michael D. Belis

Columbus Whipple
Private, United States Army
Army Service Number: 1630549
Company H 47th Infantry Regiment 4th Division
Date of Action: August 7, 1918
Location: Near Bazoches, France
War Department General Orders 147 (1918)
Medal # 1144

CITATION
Private Whipple crossed the Vesle River in the face of enemy fire and rescued a drowning comrade in the deep, swift current of the stream.

DETAILS
Columbus Whipple was born in Show Low, Navajo County, Arizona on February 29, 1896, the son of Edson and Rowena Whipple. Before entering the Army, he was a farmer in Arizona. He entered the Army on October 2, 1917. Whipple sailed to England as a Private of Infantry in the Camp Kearney June Automatic Replacement Draft Company 13 of Infantry aboard the British troopship *Lapland* on June 28, 1918. As his emergency contact, he listed his father Edson in Show Low, Arizona. From England he was sent over to France where he was eventually assigned to Company H 47th Infantry 4th Division.

Whipple was awarded the Distinguished Service Cross for his actions in the Vesle Sector. On August 7, 1918, 2nd Battalion 47th Infantry which included Whipple's Company H made a strong effort to cross the Vesle River and attack the village

Soldiers Steadfast and Loyal

of Bazoches. The Battalion suffered heavy casualties in an enemy artillery barrage in the morning and another in the afternoon. By five o'clock in the afternoon, nearly all of Companies G and H and one platoon of Company F had crossed the river and dug in. The men crossed on trees felled at the river, on a footbridge built by the Engineers which had been destroyed by German artillery only to be rebuilt by the Engineers, and some got across by swimming and wading. For most of its length the river was about chest deep with a mostly medium current and about thirty-five to forty-five feet wide. At various places however, the river narrowed, deepened, and the current was quite swift. Whipple saw a fellow soldier struggling in the water as he tried to swim across and who was about to drown. Despite a heavy volume of machine gun, artillery, and mortar fire being laid upon the Americans by the enemy, Whipple rushed out into the river under that fire, saved his comrade from drowning, and brought him to safety.

Whipple served in the Toulon Sector, the St. Mihiel Offensive, and the Meuse-Argonne Offensive. He received the Verdun Medal from the French Government. He served on occupation duty in Germany with the 4th Division and returned to the United States as a Cook in Company H 47th Infantry and part of Company I Third Army Composite Regiment aboard the troopship *U.S.S. Leviathan* (seized German liner *Vaterland*) leaving Brest, France September 1, 1919 and arriving at Hoboken, New Jersey September 8, 1919.

(See the entry for Joseph Bassi for an explanation of the Third Army Composite Regiment.)

Whipple was discharged on September 29, 1919.

The January 1920 Census showed Whipple living with his parents in Show Low, Arizona and his occupation indicated as

Army Cook. (No record could be found to show any subsequent enlistment for Whipple after September 29, 1919 and it is believed this entry of Army Cook on the Census probably should have read Unemployed Army Cook.) On September 29, 1922, he married Bessie Warner. The 1930 Census showed him living with his wife, daughter, and son in Ogden, Weber County, Utah where he was employed as a Rail Road Mail Clerk for the Federal Government and indicated he was a veteran of the World War.

In 1931, Whipple and Bessie had another daughter who died the same day she was born. His wife Bessie died in December 1939. The 1940 Census showed Whipple living with his children, his brother, and his brother's family in Ogden where he was still a railroad mail clerk for the Federal Government. He married Dorothea Heiner in July 1941. He was a member of the Legion of Valor. Columbus Whipple died at the age of 94 on March 10, 1990 and is buried in Aultorest Memorial Park, Ogden, Weber County, Utah. His obituary mentioned that he was a member of the "Eight Years Without a Birthday Club" due to his being born on February 29.

Gilbert W. Wilcox

Private First Class, United States Army
Army Service Number: 569486
Company D 4th Engineer Regiment 4th Division
Date of Action: August 11, 1918
Location: On the Vesle River near Ville Savoye, France
War Department General Orders 128 (1918)
Medal # 888

CITATION

Private First Class Wilcox volunteered to go into Ville Savoye at a time when it was under a heavy bombardment to rescue a wounded officer.

DETAILS

Gilbert Wallace Wilcox was born in Dartmouth, Nova Scotia, Canada on January 22, 1893, the son of Nathan and Isabell Wilcox. His family immigrated to the United States in May 1899. He entered the Army on May 12, 1917 and listed his home of record as Linton, Oregon. He sailed to France as a Private in Headquarters Detachment 4th Engineers 4th Division aboard the troopship *U.S.S. Martha Washington* (seized Austrian liner) on April 30, 1918. As his emergency contact, he listed his mother in Linton, Oregon. At some time after reaching France, Wilcox was transferred to Company D 4th Engineers. He served in the Aisne-Marne Offensive.

Wilcox was awarded the Distinguished Service Cross for his actions in the Vesle Sector. On August 11, 1918 Wilcox rescued 2nd Lieutenant Frank B. Cook of Company D 4th Engineers from the village of Ville Savoye which at the time was under a heavy artillery barrage. Wilcox was accompanied by Corporal Tom F. Barto, also of Company D 4th Engineers. The two soldiers charged into the deadly artillery fire to get their Lieutenant who lay seriously wounded in the shell torn village.

Barto was killed during the action, but Wilcox managed to bring out Lieutenant Cook who recovered from his wounds and returned to the United States with his Regiment when it sailed from France in the summer of 1919. Barto was awarded the Distinguished Service Cross for his bravery and sacrificing his life in the rescue. Wilcox was awarded the Distinguished Service Cross for his bravery in rescuing Lieutenant Cook under heavy fire. Cook was awarded the Distinguished Service Cross for his own actions at the Vesle River that day before going into the village and getting wounded in the horrific bombardment.

Wilcox served in the Toulon Sector, the St. Mihiel Offensive, and the Meuse-Argonne Offensive. At some time, he was transferred back to Headquarters Company 4th Engineers and promoted to the rank/position of Wagoner. He received his Distinguished Service Cross from Major General Mark L. Hersey, Commander of the 4th Division, in a ceremony in France on November 7, 1918.

Wilcox served on occupation duty in Germany with the 4th Division and returned to the United States as a Wagoner with Headquarters Company 4th Engineers aboard the troopship *U.S.S. Von Steuben* (seized German liner *Kronprinz Wilhelm*) leaving Brest, France July 21, 1919 and arriving at Hoboken, New Jersey July 29, 1919. He was discharged on August 9, 1919.

The 1920 Census showed Wilcox living with his parents in Portland, Multnomah County, Oregon where he was employed as a Mechanic for Standard Oil Company. On September 7,

1920, he married Lydia Hoffman. They had a daughter born in July 1921. On July 21, 1921, Wilcox became a naturalized citizen of the United States. At that time, he listed his occupation as Hoseman for the Portland, Oregon Fire Department. The 1930 Census showed Wilcox living with his family in Portland, Oregon where he was employed as a Fireman with the Portland Fire Department and indicated he was a veteran of the World War. The 1940 Census showed him living with his wife and daughter in Hazelia, Clackamas County, Oregon where he was employed as a Fireman with the Hazelia Fire Department. When he registered for the draft in 1942, he and his family were living in Portland, Oregon where he was employed by the Portland City Hall. Gilbert W. Wilcox died at the age of 72 on February 8, 1965 and is buried in Willamette National Cemetery, Portland, Multnomah County, Oregon.

Michael D. Belis

Merle R. Winsor
Corporal, United States Army
Army Service Number: 567353
Company D 12th Machine Gun Battalion 4th Division
Date of Action: July 19, 1918
Location: Near Hautevesnes, France
War Department General Orders 50 (1919)
Medal # 3363

CITATION
Although severely wounded by a flanking machine gun fire, Corporal Winsor remained with his gun crew in an exposed position and under sweeping artillery and machine gun fire. He received aid from members of his company and remained on duty with the platoon until the company had withdrawn and he had been ordered to the aid station.

DETAILS
Merle Robinson Winsor was born in West Falmouth, Barnstable County, Massachusetts on February 11, 1894, the son of Arthur and Louise Winsor. Before entering the Army, he was a Machinist at Sterling Motor Company in Brockton, Massachusetts. He entered the Army on August 30, 1917. Winsor sailed to England as a Private 1st Class in Company D 12th Machine Gun Battalion 4th Division aboard the British troopship *RMS Aquitania* on May 7, 1918. As his emergency contact, he listed his father Arthur in Campello, Massachusetts. From England he and his Company were sent over to France where they were issued their machine guns.

Soldiers Steadfast and Loyal

At some time after reaching France, he was promoted to Corporal.

Winsor was awarded the Distinguished Service Cross for his actions in the Aisne-Marne Offensive. During the Offensive Winsor and Company D 12th Machine Gun Battalion were acting in support of 1st Battalion 59th Infantry 4th Division which was temporarily attached to the French 164th Division. On July 18, 1918, the first day of the Offensive, the Americans had advanced to a location east of the village of Hautevesnes and were in between the village of Courchamps and the small, wooded area known as the Bois de l'Orme where they dug in for the night.

On July 19, 1st Battalion supported by Winsor's Company moved out of its positions and attacked eastward toward the wooded area known as la Remise. The Infantry moved up and down gentle slopes and through fields of wheat, bayonets fixed, and determined to reach the German trenches which lay in front of la Remise. They were stopped on a small rise short of their objective by intense artillery and machine gun fire. The Battalion Commmander was wounded, and seven other officers were killed or wounded. The soldiers of 1st Battalion began to pull back about 300 yards to the Courchamps-Priez Road where they dug in and made a defensive line. Though wounded and with his wounds temporarily treated by his men, Winsor remained in an exposed position and directed the fire of his crew and their gun to cover the withdrawl of the Infantry. Only after his Company was withdrawn and he was ordered to do so did he report to the aid station to have his wounds properly treated.

He was also awarded the French Croix De Guerre.

At some time after his award action, Winsor was promoted

to Sergeant. He served on occupation duty in Germany with the 4th Division and returned to the United States as a Sergeant in Company D 12th Machine Gun Battalion aboard the troopship *U.S.S. Von Steuben* (seized German liner *Kronprinz Wilhelm*) leaving Brest, France July 21, 1919 and arriving at Hoboken, New Jersey July 29, 1919. He was discharged on August 4, 1919.

A record from the town of West Bridgewater, Massachusetts showed that Winsor was a member of Company F 14th Infantry Massachusetts State Guard in 1919. The 1920 Census showed him living with his parents in Brockton, Plymouth County, Massachusetts where he was employed as an Edge Trimmer in a Shoe Factory. He married Amy Keith sometime between 1920 and 1928. The 1940 Census showed Winsor living in East Bridgewater, Plymouth County, Massachusetts with his wife Amy and their son and four daughters where he was employed as a Timekeeper. Merle R. Winsor died at the age of 54 on June 19, 1948 and is buried in Central Cemetery, East Bridgewater, Plymouth County, Massachusetts.

William J. Wood

Sergeant, United States Army
Army Service Number: 568737
Company D 4th Engineer Regiment 4th Division
Date of Action: August 11, 1918
Location: Near Ville Savoye, France
War Department General Orders 147 (1918)
Medal # 1089

CITATION

Although his eyes had been burned by gas, Sergeant Wood volunteered for duty and assisted in the construction of an artillery bridge across the Vesle River under constant machine gun and artillery fire, setting a conspicuous example of personal bravery and devotion to duty.

DETAILS

William J. Wood was born in Hesler, Owen County, Kentucky on April 21, 1878, the son of Sidney and Nancy Wood. He entered the Army on June 1, 1917. Wood sailed to France as a Sergeant in Company D 4th Engineers aboard the troopship *U.S.S. Martha Washington* (seized Austrian liner) on April 30, 1918. As his emergency contact, he listed his sister Mrs. Katherine K. Vanderdenter in New Castle, Indiana. Wood served in the Aisne-Marne Offensive.

He was awarded the Distinguished Service Cross for his actions in the Vesle Sector. On August 11, 1918, a defensive line across the Vesle River from the village of Ville Savoye was being held by Infantry elements of the 4th Division along the railroad embankment which was just north of the river and which ran east to west. The 4th Division was due to be relieved that night by the 77th Division but the 4th Artillery Brigade of the 4th Division was going to remain in the line and was not due to be withdrawn until the night of August 17. Wood and a detachment from his Company were tasked with building a bridge which the artillery could use to move their 75mm guns across the river. Although he was wounded in the eyes by gas, Wood volunteered for the job and worked under constant enemy fire until the job was finished, providing his fellow soldiers with an example of courage and an attitude of strong dedication to duty. He was one of four enlisted men from Company D 4th Engineers who, along with their officer, were awarded the Distinguished Service Cross for their actions during the building of that artillery bridge across the Vesle River under fire on August 11, 1918.

Wood was also awarded the French Croix De Guerre.

At some time after his award action, he was reduced to the rank of Private. He served on occupation duty in Germany with the 4th Division and returned to the United States as a Private with Company D 4th Engineers aboard the troopship *U.S.S. Von Steuben* (seized German liner *Kronprinz Wilhelm*) leaving Brest, France July 21, 1919 and arriving at Hoboken, New Jersey July 29, 1919. He was discharged on August 6, 1919.

The 1930 Census showed Wood living in Chicago, Illinois with his sister and her family where he was unemployed and indicated he was a veteran of the World War. When he registered for the draft in 1942, he indicated he was living in Chicago and unemployed. William J. Wood died at the age of 68 on October 10, 1946 and is buried in Irving Park Cemetery, Chicago, Cook County, Illinois.

Robert L. Worden

Wagoner, United States Army
Army Service Number: 934300
21st Ambulance Company 4th Sanitary Train 4th Division
Date of Action: August 7, 1918
Location: At Ville Savoye, France
War Department General Orders 99 (1918)
Medal # 136

CITATION

While driving an ambulance through the town, Wagoner Worden heard cries for help. Voluntarily and under heavy shell and machine gun fire he climbed a tower in which he found two officers and a corporal severely wounded. He rendered first aid and assisted in carrying the wounded men to a place of safety.

DETAILS

Robert LeRoy Worden was born in Wellington, Sumner County, Kansas on September 17, 1892, the son of James and Lizzie Worden. Before entering the Army, he was an Installation Mechanic for United Metal Weather Strip Company in Wichita, Kansas. He entered the Army on August 8, 1917, and received his intitial training at Fort Riley, Kansas. Worden sailed to England as a Wagoner in Ambulance Company Number 21 Medical Department Sanitary Train 4th Division aboard the British/Canadian troop transport *S.S. Scotian* on May 27, 1918. As his emergency contact, he listed his father James in Wichita, Kansas. (His younger brother Austin sailed with him on the *Scotian*. Austin served alongside Robert also as a Wagoner in the 21st Ambulance Company.) From England he and his Company were sent over to France. Worden served in the Aisne-Marne Offensive.

He was awarded the Distinguished Service Cross for his actions in the Vesle Sector. On August 7, 1918, the 4th Division made a strong effort to cross the Vesle River with the 7th Brigade attacking near the village of St. Thibaut on the left and the 8th Brigade attacking near the village of Ville Savoye on the right. The Germans were resisting the attack with heavy concentrations of artillery fire from across the river and machine gun fire from their positions on both sides of the river. While driving his ambulance through Ville Savoye, Worden responded to cries for help from several wounded Americans in an observation tower and brought them down from their exposed position, dressed their wounds, and got them to safety, all the while under heavy artillery and machine gun fire from the enemy.

Robert L. Worden was the only soldier from the 21st Ambulance Company to receive the Distinguished Service Cross in World War One.

Worden served in the Toulon Sector and the St. Mihiel Offensive. In the Meuse-Argonne Offensive he was wounded in the head by shrapnel near the village of Cuisy on October 5, 1918. He served on occupation duty in Germany with the 4th Division and returned to the United States as a Wagoner with Ambulance Company 21 4th Sanitary Train aboard the troop transport *U.S.S. Minnesotan* leaving Brest, France July 23, 1919 and arriving at Philadelphia, Pennsylvania August 3, 1919. (His brother Austin returned on the same ship.) Worden was discharged on August 9, 1919.

On November 3, 1919, he married Elizabeth Sailer. The 1920

Census showed him living with his wife in Wichita, Sedgwick County, Kansas where he was employed as an Auto Mechanic. In 1922 he and his wife had a son. The 1930 Census showed him living with his family in Wichita where he was employed as an Electric Welder in his own shop and indicated he was a veteran of the World War. He had a second son in 1930. The 1940 Census showed him living with his family in Wichita where he was still employed as a welder in his own welding shop. When he registered for the draft in 1942, he indicated he was then employed by the Beech Aircraft Corporation in Wichita and that he had a scar from a gunshot wound on top of his head. Robert L. Worden died at the age of 68 on May 4, 1961 and is buried in Wichita Park Cemetery and Mausoleum, Wichita, Sedgwick County, Kansas.

APPENDIX

Distinguished Service Cross Assigned Medal Numbers for the 4th Division in World War One

136 Robert L. Worden 21st Ambulance Company
604 William Shemin 47th Infantry
888 Gilbert W. Wilcox 4th Engineers
1013 Gustav J. Braun 47th Infantry
1060 Edward R. Lawless 39th Infantry
1065 Raymond R. Smith 11th Machine Gun Battalion
1066 Earl M. McKinley 11th Machine Gun Battalion
1067 Homer S. Jarvis 11th Machine Gun Battalion
1071 Max S. Koss 47th Infantry
1072 Otto A. Schwanke 47th Infantry
1073 Charles F. Carbaugh 47th Infantry
1076 John H. Pratt Jr. 47th Infantry
1077 George N. Brigham 47th Infantry
1079 Arthur H. Warfield 47th Infantry
1080 Charles T. Dunbar 4th Engineers
1081 Raymond D. Robertson 4th Engineers
1082 James P. Growdon 4th Engineers
1083 Roy Harris 4th Engineers

Michael D. Belis

1085 Murray R. MacKall 4th Engineers
1086 Frank Jaworski 4th Engineers
1087 Emmett W. Waltman 4th Engineers
1088 Frank B. Cook Jr. 4th Engineers
1089 William J. Wood 4th Engineers
1090 Arthur J. Goetsch 4th Engineers
1091 Charles Glenn 4th Engineers
1092 Francis K. Newcomer 4th Engineers
1120 John J. Madore 47th Infantry
1144 Columbus Whipple 47th Infantry
1190 Thomas W. Kearns 47th Infantry
1303 William B. Beach 4th Engineers
1327 Leonard E. Guy 58th Infantry
1328 Joseph Longowski 59th Infantry
1329 Arthur H.G. Mallet French Army attached to 47th Infantry
1423 Joseph Waskiewic 11th Machine Gun Battalion
1492 Dave W. Stearns 4th Engineers
1519 Thomas Vander Veen 11th Machine Gun Battalion
1559 Edward D. Ritchie 47th Infantry
1589 Andrew Edmiston 39th Infantry
1669 Benjamin A. Poore 7th Infantry Brigade
1780 Arthur I. Clark 39th Infantry
1793 William B. Hook 4th Engineers
1796 Ernest R. Potter 39th Infantry
1832 Reuben L. Johnson 47th Infantry
1837 Carl Rasmussen 39th Infantry
1846 Isaac Gataino 47th Infantry
1847 Fred E. Billman 47th Infantry
1918 John C. Persons 47th Infantry
1973 James Manning 4th Engineers
2026 Luther E. Lindahl 47th Infantry

2028 Harold W. Enright 47th Infantry
2029 John C. Vann 47th Infantry
2030 Louis Scionti 47th Infantry
2033 Albert L.J. Ihrke 47th Infantry
2154 Harold V. Beal 13th Field Artillery
2156 James P. Behan 13th Field Artillery
2169 James T. Rice 8th Field Signal Battalion
2257 Reuben L. George 59th Infantry
2265 Pietro Formica 59th Infantry
2274 Charles H. Epler 59th Infantry
2287 Elwyn L. Berwick 13th Field Artillery
2293 Joseph H. Carvo 47th Infantry
2788 William Herren 58th Infantry
2925 Ford D. Smith 4th Engineers
2933 Albert Dietz 59th Infantry
3175 Lee M. Ray 39th Infantry
3176 Samuel P. Adkisson 39th Infantry
3177 Frank C. Bolles 39th Infantry
3178 Paul J. Pappas 39th Infantry
3223 Stephen J. Weston 47th Infantry
3308 Frank B. Gresham 39th Infantry
3310 Roy E. Mathews 58th Infantry
3311 Clyde H. Lindsey 59th Infantry
3312 Earl W. Curtis 59th Infantry
3313 William J. Cahill 59th Infantry
3335 Robert W. Norton 39th Infantry
3344 Henry Tudury 12th Machine Gun Battalion
3363 Merle R. Winsor 12th Machine Gun Battalion
3365 Ralph Slate 39th Infantry
3435 Joseph Bassi 59th Infantry
3509 Joe Smith 39th Infantry

Michael D. Belis

3543 Orval Kline 11th Machine Gun Battalion
3593 Etienne Escudier French Army attached to 59th Infantry
3751 Clinton Day 58th Infantry
3752 Henry D. Turner 10th Machine Gun Battalion
3753 Clark O. Tayntor 47th Infantry
3770 James Conway 58th Infantry
4158 Edward S. Blackman 59th Infantry *
5517 Robert G. Marshall 58th Infantry
5559 James J. Pirtle 59th Infantry
5666 William L. Hunter 58th Infantry
5670 Thomas D. Drake 59th Infantry
5684 Harold Bedolfe 4th Engineers
5689 Marion H. Cardwell 58th Infantry
5728 Emil J. Eklund 58th Infantry
5772 Samuel H. Hanna 12th Machine Gun Battalion
5787 Charles Kelly 12th Machine Gun Battalion
5801 John Legnosky 58th Infantry
5869 Frank J. Downs 58th Infantry
6001 Fred E. Cullen 12th Machine Gun Battalion
6019 Ernest Fosnes 59th Infantry
6027 Anthony F. Shedlock 58th Infantry
6087 Jacob Kreis 47th Infantry
6111 Lawrence Boop 59th Infantry
6121 Walter Currie 59th Infantry
6122 Martin Nelson 58th Infantry
6153 August Aibner 58th Infantry
6177 Oscar W. Peterson 59th Infantry
6247 Forrest L. Martz 58th Infantry
6250 Fred A. Lieuallen 47th Infantry
6259 James H. Roberts 39th Infantry
6284 Manton C. Mitchell 39th Infantry

Soldiers Steadfast and Loyal

6285 Paul Von Krebs 47th Infantry
6305 Arthur M. Hamilton 58th Infantry
6321 Earl R. Fretz 12th Machine Gun Battalion
6358 James B. Carpenter 47th Infantry
6384 Herbert A. Cohn 39th Infantry
6616 George C. McCelvey 47th Infantry
6684 Henry J. Garst 47th Infantry
6701 Donald A. Pegg 12th Machine Gun Battalion
6702 Harold Lamonte Hall 59th Infantry
6704 Arthur M. Miller 47th Infantry
6730 Howard C. McCall 59th Infantry
6769 Raymond Buma 39th Infantry
6785 Richard Marcella 47th Infantry
6852 Fred N. Rapp 59th Infantry
6904 Samuel H. Houston 58th Infantry
6968 John H. Norton 47th Infantry
6982 Charles E. Deleuw 4th Engineers
6986 Mathias W. Haney 39th Infantry
6997 William H. McGinnis 12th Machine Gun Battalion
7027 Carroll B. West 12th Machine Gun Battalion
7052 Stefano Riggio 39th Infantry
7058 John W. Norton 39th Infantry
7075 William A. Shea 39th Infantry
7108 Harrison B. Webster 47th Infantry
7109 Charles J. Love 59th Infantry
7135 Joe Limon 47th Infantry
7159 Morton Osborn 47th Infantry
7167 Edward McAndrew 12th Machine Gun Battalion
7175 Walter H. Detrow 47th Infantry
7198 Alberis Callewaert 58th Infantry
7220 Leo D. Roberts 11th Machine Gun Battalion

Michael D. Belis

7246 Robert A. Madden 47th Infantry
7348 Edward D. Haskew 33rd Ambulance Company
7363 Arthur Paulson 59th Infantry
7380 Joseph Dilworth 39th Infantry
7467 Olaf Brekke 58th Infantry
7473 Charles B. Duncan 77th Field Artillery
7475 Cornelius T. Glynn 59th Infantry
7480 Hans E. Morgan 47th Infantry
7552 Glenn M. Grove 11th Machine Gun Battalion
7555 William H. Hammond 39th Infantry
7562 Cecil N. Martin 47th Infantry
7564 Guy W. Boardman 59th Infantry
7617 Homer J. Bleau 59th Infantry
7620 Arnot L. McArthy 59th Infantry
7664 Wallis H. Sturtevant 12th Machine Gun Battalion
7667 Henry Howard 39th Infantry
7681 Joseph McCollum 10th Machine Gun Battalion
7783 Charles C. Rismiller 4th Engineers
7788 Charles H. Evans 39th Infantry
7898 Alvan C. Sandeford 13th Field Artillery
7910 John S. Weimer 47th Infantry
7911 Lester C. Dill 47th Infantry
7947 James W. Hanbery 59th Infantry
7989 Joseph T. Clement 39th Infantry

* No records exist for medal numbers 3800-5500. Assignment of the number 4158 to Edward S. Blackman is based on a medal found with that number officially stamped on it and his name engraved on the back.

NO ASSIGNED MEDAL NUMBERS DISCOVERED

Tom F. Barto 4th Engineers
George Brown 59th Infantry
Willis M. Campbell 59th Infantry
Claude E. Cherry 11th Machine Gun Battalion
Peter W. Ebbert 58th Infantry
Edwin A. Elliott 39th Infantry
Daris V. Ford 4th Engineers
David S. Grant 39th Infantry
Morrison Hayes 12th Machine Gun Battalion
Ralph E. Ladue 11th Machine Gun Battalion
Henry F. Martin 47th Infantry
Arno S. McClellan 47th Infantry
Jean L. Meurisse French Army attached to 58th Infantry
Robert H. Murdoch 47th Infantry
Cornelius J. O'Brien 4th Engineers
James K. Parsons 39th Infantry
Richard G. Plumley 39th Infantry
Lowell H. Riley 58th Infantry
Albert B. Simpson 11th Machine Gun Battalion
Rutherford H. Spessard 58th Infantry

ENDNOTES

1. *THE FOURTH DIVISION Its Services and Achievements In The World War* by Christian A. Bach and Henry Noble Hall 1920 pp187.

2. *Columbia to the Rhine: being a brief history of the Fourth Engineers, and their trip from the Columbia River, in the State of Washington, U.S.A., to the Rhine River in Germany Written and Illustrated by Men of the Regiment* Westdeutsche Grossdruckerei G.M.B.H. Wald-Germany 1919 pp94-95.

3. Ibid pp113.

4. Ibid pp106.

5. *The History of the 39th U.S. Infantry During the World War* by Richard Barrett Cole and Barnard Eberlin, New York, Press of J.D. McGuire 1919 pp66.

6. *THE FORTY-SEVENTH INFANTRY A HISTORY 1917-1918-1919* by James E. Pollard, Seeman & Peters Saginaw, Michigan 1919 pp112.

7. *The History of the 39th U.S. Infantry During the World War* by Richard Barrett Cole and Barnard Eberlin, New York, Press of J.D. McGuire 1919 pp130.

Michael D. Belis

8. 77th Artillery Regiment Association website: http://www.77fa.org/index.html.

9. *THE FOURTH DIVISION Its Services and Achievements In The World War* by Christian A. Bach and Henry Noble Hall 1920 pp121.

10. *Los Angeles California Examiner* March 30, 1920.

11. *THE FOURTH DIVISION Its Services and Achievements in The World War* by Christian A. Bach and Henry Noble Hall 1920 pp123.

12. *Memoirs of The Harvard Dead in The War Against Germany; Volume V* M.A. DeWolfe Howe, Cambridge: Harvard University Press 1924 pp279.

13. *SOLDIERS ALL; PORTRAITS AND SKETCHES OF THE MEN OF THE A.E.F.* Joseph Cummings Chase, New York, George H. Doran company 1920 pp425.

14. *THE FORTY-SEVENTH INFANTRY A HISTORY 1917-1918-1919* James E. Pollard, Seeman & Peters Saginaw, Michigan 1919 pp107.

15. THE FORTY-SEVENTH INFANTRY A HISTORY 1917-1918-1919 James E. Pollard, Seeman & Peters Saginaw, Michigan 1919 pp107. (In the book this is incorrectly listed as the Citation for his Distinguished Service Cross).

16. *Lists of Incoming Passengers, 1917-1938.* Textual records. 360 Boxes. NAI: 6234465. Records of the Office of the Quartermaster General, 1774-1985, Record Group 92. The National Archives at College Park, Maryland. Passenger list for U.S.S. Pastores leaving St. Nazaire, France June 17, 1919 and

arriving at Hoboken, New Jersey June 26, 1919. Accessed through Ancestry.com.

17. *Pennsylvania, WWI Veterans Service and Compensation Files, 1917-1919, 1934-1948* [database on-line]. Provo, UT, USA: Ancestry.com Operations, Inc., 2015. Original data: World War I Veterans Service and Compensation File, 1934–1948. RG 19, Series 19.91. Pennsylvania Historical and Museum Commission, Harrisburg Pennsylvania. From a two-page report endorsed on an unrecorded date in 1920 and included in McCall's service record as part of his Pennsylvania, WWI Veterans Service and Compensation Files. Accessed through Ancestry.com.

18. Ibid. From an article published in an unrecorded Philadelphia, Pennsylvania newspaper from 1918 included in McCall's service record as part of his Pennsylvania, WWI Veterans Service and Compensation Files. Accessed through Ancestry.com.

19. From an article written by one of McCall's 2nd Lieutenants in *The Pennsylvania Gazette* (date unrecorded). Taken from an entry on the Find A Grave memorial page for Howard Clifton McCall: https://www.findagrave.com/memorial/138657075/howard-clifton-mccall.

20. *The History of the 39th U.S. Infantry During The World War* by Richard Barrett Cole and Barnard Eberlin, New York, Press of J.D. McGuire 1919 pp44.

21. Ibid pp134.

22. Ibid pp130.

Michael D. Belis

23. Find A Grave website: https://www.findagrave.com/memorial/146939586

24. *THE FOURTH DIVISION Its Services and Achievements In The World War* by Christian A. Bach and Henry Noble Hall 1920 pp121.

25. *The History of the 39th U.S. Infantry During the World War* by Richard Barrett Cole and Barnard Eberlin, New York, Press of J.D. McGuire 1919 pp134.

SOURCES

BOOKS, PERIODICALS, PUBLICATIONS

The Agromek Volume Seventeen 1919 Edwards & Broughton Printing Company, Raleigh, N.C. 1919

A Marine Tells It To You Frederic May Wise & Meigs Oliver Frost, J.H. Sears & Company, Inc., New York 1929

AMERICAN DECORATIONS A List of Awards of the Congressional Medal of Honor the Distinguished-Service Cross and the Distinguished-Service Medal Awarded Under Authority of the Congress of the United States 1862-1926 Compiled in the Office of The Adjutant General of the Army and published by order of the Secretary of War United States Government Printing Office, Washington, D.C. 1927

AMERICAN DECORATIONS A List of Awards of the Congressional Medal of Honor the Distinguished-Service Cross the Distinguished-Service Medal the Soldier's Medal and the Distinguished-Flying Cross Awarded Under Authority of the Congress of the United States January 1, 1937-June 30, 1937 Compiled in the Office of The Adjutant General of the Army and published by order of the Secretary of War Supplement I, United States Government Printing Office, Washington, D.C. 1937

Michael D. Belis

AMERICAN DECORATIONS A List of Awards of the Congressional Medal of Honor the Distinguished-Service Cross the Distinguished-Service Medal the Soldier's Medal and the Distinguished-Flying Cross Awarded Under Authority of the Congress of the United States July 1, 1937-June 30, 1938 Compiled in the Office of The Adjutant General of the Army and published by order of the Secretary of War Supplement II, United States Government Printing Office, Washington, D.C. 1939

AMERICAN DECORATIONS A List of Awards of the Congressional Medal of Honor the Distinguished-Service Cross the Distinguished-Service Medal the Soldier's Medal and the Distinguished-Flying Cross Awarded Under Authority of the Congress of the United States July 1, 1938-June 30, 1939 Compiled in the Office of The Adjutant General of the Army and published by order of the Secretary of War Supplement III, United States Government Printing Office, Washington, D.C. 1940

AMERICAN DECORATIONS A List of Awards of the Congressional Medal of Honor the Distinguished-Service Cross the Distinguished-Service Medal the Soldier's Medal and the Distinguished-Flying Cross Awarded Under Authority of the Congress of the United States July 1, 1939-June 30, 1940 Compiled in the Office of The Adjutant General of the Army and published by order of the Secretary of War Supplement IV, United States Government Printing Office, Washington, D.C. 1940

AMERICAN DECORATIONS A List of Awards of the Congressional Medal of Honor the Distinguished-Service Cross the Distinguished-Service Medal the Soldier's Medal and the Distinguished-Flying Cross Awarded Under Authority of the Congress of the United States July 1, 1940-June 30, 1941 Compiled in the Office of The Adjutant General of the Army and published by order of the Secretary of War Supplement V, United States Government Printing Office, Washington, D.C. 1941

Soldiers Steadfast and Loyal

Army Serial Numbers - A Neglected World War I Research Tool Gary A. Mitchell, The Medal Collector Vol. 35 No. 10 October 1984 pp 18-22

The Awards of the AEF - Part II, The Purple Heart Gary A. Mitchell, The Medal Collector Vol. 36 No. 7 July 1985 pp 4-6

The Awards of the AEF - Part III, Significant Decorations And Other Awards Gary A. Mitchell, The Medal Collector Vol. 36 No. 8-9 August/September 1985 pp 6-8

The Awards of the AEF - Part IV, Foreign Awards Gary A. Mitchell, The Medal Collector Vol. 36 No. 10 October 1985 pp 22-26

A Withering Fire: American Machine Gun Battalions in World War I George T. Raach, BookLocker.com, Inc. Bradenton, Florida 2015

BATTLE PARTICIPATION OF ORGANIZATIONS OF THE AMERICAN EXPEDITIONARY FORCES IN FRANCE, BELGIUM AND ITALY 1917-1918 War Office United States of America, United States Government Printing Office, Washington 1920

Biographical Directory of the United States Congress, 1774-2005 Washington, D.C.: Government Printing Office, 2005

BRIEF HISTORIES OF DIVISIONS, U.S. ARMY 1917-1918 Prepared in the Historial Branch, War Plans Division, General Staff June 1921 Army Command & General Staff College, Combined Arms Research Library, Fort Leavenworth, Kansas 1921

Columbia to the Rhine: Being A Brief History of the Fourth Engineers, and their trip from the Columbia River, in the State of Washington, U.S.A., to the

Michael D. Belis

Rhine River in Germany Written and Illustrated by Men of the Regiment Westdeutsche Grossdruckerei G.M.B.H. Wald-Germany 1919

CONGRESSIONAL MEDAL OF HONOR THE DISTINGUISHED SERVICE CROSS AND THE DISTINGUISHED SERVICE MEDAL Issued by the War Department Since April 6, 1917 Up to and including General Orders, No. 126, War Department, November 11, 1919 The Adjutant General, Washington, D.C. 1919

Disposition Codes Department of the Army Office of the Quartermaster General Memorial Division Registration Branch Quartermaster Manual QMC 16-3 August 1947 as amended. Washington, D.C. 1947

The Doughboy The Infantry School, Camp Benning, Georgia 1922

The Doughboy The Infantry School, Fort Benning, Georgia 1924

The Fifty-Eighth Infantry in THE WORLD WAR George L. Morrow, 58th Infantry History Association 1919

Foreign Decorations and Translated Citations General Order No. 1, The Adjutant General's Office, January 1, 1923 Commonwealth of Pennsylvania, Harrisburg, Pennsylvania J.L.L. Kuhn Printer 1923

FOR VALOR Distinguished Service Cross Issues 1918 – 1925 Volume I Gary A. Mitchell, Planchet Press, Ft. Meyer, VA 1989

FOR VALOR Distinguished Service Cross Issues 1918 – 1925 Volume II Gary A. Mitchell, Planchet Press, Ft. Meyer, VA 1987

Soldiers Steadfast and Loyal

FOR VALOR Distinguished Service Cross Issues 1918 – 1925 Volume III Gary A. Mitchell, Planchet Press, Ft. Meyer, VA

THE FORTY-SEVENTH INFANTRY A HISTORY 1917-1918-1919 James E. Pollard, Seeman & Peters Saginaw, Michigan 1919

THE FOURTH DIVISION Its Services and Achievements In The World War Christian A. Bach and Henry Noble Hall, Issued by the Division 1920

4th Division Summary of Operations in the World War American Battle Monuments Commission, United States Government Printing Office, Washington, D.C. 1944

General Orders United States War Department, Adjutant General's Office 1920

General Orders United States War Department, Adjutant General's Office 1922

The Georgia State Memorial Book Bert E. Boss, American Memorial Publishing Company, 1921

Harvard College Class of 1914 Secretary's Second Report June, 1917 Plimpton Press Norwood, Massachusetts 1917

Heroes All!: A Compendium of the Names and Official Citations of the Soldiers and Citizens of the United States and of Her Allies who Were Decorated by the American Government for Exceptional Heroism and Conspicuous Service Above and Beyond the Call of Duty in the War with Germany, 1917-1919 Harry R. Stringer, Fassett Publishing Company 1919

Michael D. Belis

History of Erie County, Pennsylvania Volume II John Elmer Reed, Historical Publishing Company, Topeka, Kansas 1925

History of the 40th Infantry Division (Mech) 1917 - 1997 CALIFORNIA MILITARY DEPARTMENT HISTORY PROGRAM 1997/2020

A History of Fort Benjamin Harrison 1903-1982 Stephen E. Bower, Command History Office US Army Soldier Support Center, Fort Benjamin Harrison, Indiana June 1984

HISTORY OF THE 7TH INFANTRY BRIGADE DURING THE WORLD WAR 1918 Printed by M.D. Schauberg, Cologne, Germany 1919

The History of the 39th U.S. Infantry During The World War Richard Barrett Cole and Barnard Eberlin, New York, Press of J.D. McGuire 1919

INDIANA HISTORICAL COLLECTIONS VOLUME XVIII Indiana Book of Merit: Official Individual Decorations and Commendations Awarded to Indiana Men and Women for Services in the World War Harry A. Rider, Indianapolis: The Historical Bureau, Indiana Library and Historical Dept., 1932

Indiana World War Records GOLD STAR HONOR ROLL A Record of Indiana Men and Women who died in the service of the United States and the Allied Nations in the World War 1914-1918 Indiana Historical Commission, Indianapolis, Indiana 1921

INFANTRY IN BATTLE The Infantry Journal Incorporated, U.S. Army, Washington, D.C. Printed by Garrett & Massie, Richmond, Virginia 1939

Soldiers Steadfast and Loyal

INFANTRY REGIMENTS OF THE U.S. ARMY James A. Sawicki, Wyvern Publications, Dumfries, Virginia 1981

The KANZA Volume VI 1915 State Manual Training Normal Pittsburg, Kansas 1915

The Katyn Forest Massacre: Hearings Before the Select Committee to Conduct an Investigation of the Facts, Evidence and Circumstances of the Katyn Forest Massacre, Eighty-second Congress U.S. Government Printing Office, 1952 pp 1880

Maryland in the World War 1917-1919 Military and Naval Service Records Volume I Maryland War Records Commission Baltimore, Maryland 1933

Memoirs of The Harvard Dead In The War Against Germany; Volume V M.A. DeWolfe Howe, Cambridge: Harvard University Press 1924

Michigan in the World War Charles H. Landrum, George N. Fuller The Michigan Historical Commission 1924

MONOGRAPHS OF THE WORLD WAR Infantry School, Fort Benning, Georgia 1923

Official Army Register The Adjutant General's Office, Washington D.C. (Published yearly except for 1917 - volumes consulted were 1886-1972)

Official Bulletin Committee on Public Information, Washington, D.C. 1917

The Official Roster of Ohio Soldiers, Sailors and Marines in the World War 1917-18 (Volumes 1-23) Columbus, Ohio: The F.J. Heer Printing Co. 1926

Michael D. Belis

Quartermaster Manual, QMC 16-3 August 1947 for Cemeteries Department of the Army Office of the Quartermaster General, Washington, D.C. 1947

Roster of Ohio Volunteers in the Service of the United States, War with Spain prepared under direction of Herbert B. Kingsley, Adjutant General of Ohio, Columbus, Ohio: J.L. Trauger, State Printer 1898

Report of the Adjutant General of the State of Maine for the Period of the World War 1917-1919 Volume II The Adjutant General State of Maine, Augusta, Maine 1929

Roster of the Men and Women who served in the Army or Naval Service (including the Marine Corps) of the United States or its Allies from the State of North Dakota in the World War 1917-1918 Volume 3 Adjutant General North Dakota 1931

The Santa Fe Magazine Volume XIII Number 5 April, 1919, Railway Exchange, Chicago 1919

SCHOOL of MINES and METALLURGY UNIVERSITY OF MISSOURI WAR RECORDS Missouri School of Mines, Rolla, Missouri 1920

Sergeant Wolinksi and the Great War Gary Mawyer 2016 ISBN: 153093169X, ISBN 13: 978-153093169

Service records: Connecticut Men and Women in the Armed Forces of the United States during World War, 1917-1920 State Armory Office of the Adjutant General, Hartford, Connecticut 1941

SOLDIERS ALL: PORTRAITS AND SKETCHES OF THE MEN OF

Soldiers Steadfast and Loyal

THE A. E. F. Joseph Cummings Chase, New York, George H. Doran Company 1920

SOLDIERS OF THE GREAT WAR Volume I W.M. Haulsee, F.G. Howe and A.C. Doyle, Soldiers Record Publishing Association, Washington, D.C. 1920

SOLDIERS OF THE GREAT WAR Volume II W.M. Haulsee, F.G. Howe and A.C. Doyle, Soldiers Record Publishing Association, Washington, D.C. 1920

SOLDIERS OF THE GREAT WAR Volume III W.M. Haulsee, F.G. Howe and A.C. Doyle, Soldiers Record Publishing Association, Washington, D.C. 1920

Sunset The Pacific Monthly Volume Forty-One July-December 1918

The Infantry Squad Part 1: How Did We get Here? Chris Raynor, NCO Journal, Army University Press March 19, 2018

THE TECHNE Vol. 1 No. 1 November, 1917 Published by the State Manual Training Normal College, Pittsburg, Kansas

THE TECHNE Vol. 1 No. 7 October, 1918 Published by the State Manual Training Normal College, Pittsburg, Kansas

THE UNITED STATES SHIP GREAT NORTHERN D. K. Romig, Eagle Press, Brooklyn, New York 1919

U.S. Army Uniforms of World War II Shelby Stanton, Stackpole Books, Mechanicsburg, Pennsylvania 1991

Michael D. Belis

With the Colors, from Whatcom, Skagit and San Juan Counties. An honor roll containing a pictorial record of the gallant and courageous men from northwestern Washington, U.S.A., who served in the World War, 1917-1918-1919 Louis Jacobin, Seattle, Washington: Press of Peters Pub. Co, 1921

UNITED STATES MILITARY ACADEMY PUBLICATIONS:

BIOGRAPHICAL REGISTER of the Officers and Graduates of the U.S. Military Academy George W. Cullum Ed. Edward S. Holden, Supplement, Volume IV 1890-1900 The Riverside Press, Cambridge, Massachusetts 1901

BIOGRAPHICAL REGISTER of the Officers and Graduates of the U.S. Military Academy George W. Cullum Ed. Charles Braden, Supplement, Volume V 1900-1910 Seeman & Peters Printers, Saginaw, Michigan 1910

BIOGRAPHICAL REGISTER of the Officers and Graduates of the U.S. Military Academy George W. Cullum Ed. Wirt Robinson, Supplement, Volume VI-A 1910-1920 Seeman & Peters Printers, Saginaw, Michigan 1920

BIOGRAPHICAL REGISTER of the Officers and Graduates of the U.S. Military Academy George W. Cullum Ed. Wirt Robinson, Supplement, Volume VI-B 1910-1920 Seeman & Peters Printers, Saginaw, Michigan 1920

BIOGRAPHICAL REGISTER of the Officers and Graduates of the U.S. Military Academy George W. Cullum Ed. WM. H. Donaldson, Supplement, Volume VII 1920-1930 R.R. Donnelley & Sons Company, The Lakeside Press Chicago, Illinois 1930

BIOGRAPHICAL REGISTER of the Officers and Graduates of the U.S. Military Academy George W. Cullum Ed. E. E. Farman, Supplement, Volume VIII 1930-1940 R.R. Donnelley & Sons Company, The Lakeside Press Chicago, Illinois 1940

BIOGRAPHICAL REGISTER of the Officers and Graduates of the U.S. Military Academy George W. Cullum Ed. Charles N. Branham, Supplement, Volume IX 1940-1950

Fiftieth Annual Report of the Association of Graduates of the United States Military Academy Seemann & Peters, Inc. Saignaw, Michigan 1919

The Howitzer 1911 The Hoskins Press, Philadelphia, Pennsylvania 1911

The Howitzer 1913 Chas. L. Willard Co., New York, N.Y. 1913

Official Register of the Officers and Cadets United States Military Academy United States Military Academy Printing Office, West Point, New York (1886, 1890, 1891, 1905, 1906, 1907, 1908, 1909, 1910, 1911, 1912, 1913, 1914, 1915, 1916, 1917)

NEWSPAPERS

Augusta Chronicle Augusta, Georgia Tuesday, July 5, 1921

Beckley Post-Herald Beckley, West Virginia Thursday, December 22, 1949

Beckley Post-Herald Beckley, West Virginia Monday, August 20, 1951

Beckley Post-Herald Beckley, West Virginia Tuesday, October 17, 1972

Michael D. Belis

Blackstone Valley Tribune Friday, September 28, 2018

Boston Globe Boston, Massachusetts Monday August 12, 1918

Boston Globe Boston, Massachusetts Tuesday, June 14, 1921

Boston Post Boston, Massachusetts Tuesday, November 19, 1918

Bridgeport Sunday Post Sunday, May 28, 1961

The Bridgeport Post Wednesday, July 12, 1961

The Brooklyn Daily Eagle Tuesday, June 4, 1940

Brooklyn Eagle Wednesday, February 20, 1946

California Examiner Los Angeles, California Tuesday, March 30, 1920

California Times Los Angeles, California Tuesday, March 30, 1920

Catholic Telegraph Cincinnati, Ohio Thursday, May 14, 1931

The Citizen-Advertiser Auburn, New York Friday, May 24, 1963

Dinuba California Sentinel Thursday, September 18, 1919

East Oregonian Friday, October 11, 1918 / Wednesday, October 10, 2018

El Paso Herald-Post Friday, August 1, 1952

Evening Public Ledger-Philadelphia Monday, August 5, 1918

Soldiers Steadfast and Loyal

Evening Public Ledger-Philadelphia Saturday, September 27, 1919

Evening Public Ledger-Philadelphia Friday, November 7, 1919

Evening Public Ledger-Philadelphia Tuesday, August 30, 1921

Fitchburg Daily Sentinel Monday, February 24, 1919

The Independent Long Beach, California Wednesday, December 16, 1953

The Independent Long Beach, California Thursday, June 30, 1960

Los Angeles California Examiner Tuesday, March 30, 1920

Los Angeles California Herald Tuesday, March 30, 1920

Los Angeles California Times Tuesday, March 30, 1920

Modesto California News Monday, August 4, 1919

New Orleans Item Tuesday, April 15, 1919

New Orleans Item Tuesday, September 30, 1919

The New York Post Friday, March 1, 1946

The New York Sun Monday, December 4, 1933

The New York Sun Wednesday, November 29, 1939

Norwich Bulletin Norwich, Connecticut Friday, November 29, 1918

Michael D. Belis

The Pantagraph Bloomington, Illinois Saturday March 14, 1987

Pittsburgh Post Gazette Wednesday, June 25, 1969

Reno Evening Gazette Wednesday, May 8, 1968

Rochester Democrat and Chronicle Thursday, August 28, 1919

San Francisco Bulletin Friday, October 17, 1919

The Saratogian Saratoga Springs, N.Y. Saturday, April 10, 1932

The Sea Coast Echo Tuesday, November 13, 2018

The Stars And Stripes Wednesday, September 20, 1950

The Stars And Stripes Sunday, December 31, 1950

The Stars And Stripes Friday, April 3, 1953

The Stars And Stripes Wednesday, January 20, 1954

The Stars And Stripes Friday, March 12, 1954

The Sun Wednesday, November 20, 1918

The Sun/Daily Herald Gulfport, Mississippi Sunday, November 29, 1981

The Troy Times Troy, N.Y. Saturday, April 30, 1932

Soldiers Steadfast and Loyal

OTHER SOURCES

American Battle Monuments Commission website: https://www.abmc.gov/

Ancestry.com https://www.ancestry.com/

Applications for Headstones for U.S. Military Veterans, 1925-1941. Microfilm publication M1916, 134 rolls. ARC ID: 596118. Records of the Office of the Quartermaster General, Record Group 92. National Archives at Washington, D.C.

Applications for Headstones, compiled 01/01/1925 - 06/30/1970, documenting the period ca. 1776 - 1970 ARC: 596118. Records of the Office of the Quartermaster General, 1774–1985, Record Group 92. National Archives and Records Administration, Washington, D.C.

Arlington National Cemetery Website https://www.arlingtoncemetery.mil/#/ Official U.S. Army website for Arlington National Cemetery

Arlington National Cemetery Website http://www.arlingtoncemetery.net/ A privately operated website by Michael Robert Patterson

THE ATTACK OF THE 39TH INFANTRY (U.S.) AS PART OF THE 33RD FRENCH DIVISION SOUTH OF FAVEROLLES, JULY 18-19, 1918 Major Henry Terrell Jr., Maneuver Center of Excellence Libraries, MCoE HQ Donovan Research Library, Fort Benning, Georgia website: https://www.benning.army.mil/Library/Donovanpapers/wwi/index.html

"The Battle of the Ourcq River" By Earl Starbuck A Thesis Submitted To The Faculty of the History Department and Graduate School At Lib-

Michael D. Belis

erty University In Partial Fulfillment of the Requirements for a Master of Arts in History May 2019 Liberty University website: https://digitalcommons.liberty.edu/cgi/viewcontent.cgi?article=1579&context=-masters

California Military History, California Militia and National Guard Unit Histories Volume 3. 1917-1941 California State Military History and Museums Program website: http://www.militarymuseum.org/historyCal-NG3.html

California State LibraryAncestry.com. California, WWI Soldier Photographs, 1917-1918 [database on-line]. Provo, UT, USA: Ancestry.com Operations, Inc. Original data: California, World War I Soldier Photographs. Sacramento, California: California State Library, California History Section.

California, Voter Registrations, 1900-1968 [database on-line]. Provo, UT, USA: Ancestry.com.

California, World War I Soldier Service Records. Microfilm publication, 28 rolls. California State Library, California History Section. Sacramento, California.

Catholic Heroes of the World War Scrapbook and Related Material Collection, The Catholic University of America, University Libraries website: https://libraries.catholic.edu/special collections/archives/collections/finding-aids/finding-aids.html?file=cathhero

COMPANY D, 47TH INFANTRY, 4TH DIVISION, NEAR SERGY IN THE AISNE-MARNE OFFENSIVE, JULY 29-AUGUST 1, 1918 Captain John W. Bulger, Maneuver Center of Excellence Libraries, MCoE

Soldiers Steadfast and Loyal

HQ Donovan Research Library, Fort Benning, Georgia website: https://www.benning.army.mil/Library/Donovanpapers/wwi/index.html

Connecticut Department of Health. Connecticut Death Index, 1949-2012 [database on-line]. Provo, UT, USA: Ancestry.com Operations, Inc., 2003.Original data: Connecticut Department of Health. Connecticut Death Index, 1949-2001. Hartford, CT, USA: Connecticut Department of Health.

Connecticut, Federal Naturalization Records, 1790-1996 [database on-line]. Lehi, UT, USA: Ancestry.com Operations, Inc., 2016. Original data: Naturalization Records. National Archives at Boston, Waltham, Massachusetts.

Connecticut WWI Military Questionnaires, 1919–1920. Connecticut State Library, Hartford, Connecticut.

The Diary of Henry Jetton Tudury: Mississippi's Most Decorated Doughboy of World War I: http://www.oocities.org/henry_tudury/diary.htm

Division of Passport Control: Emergency Passport Applications, 1906–1925 (Argentina thru Venezuela). NAI 1244183 A1 544. General Records of the Department of State, Record Group 59. National Archives, Washington D.C.

Doughboy Center website: http://www.worldwar1.com/dbc/ghq1arm.htm

Faid and Kasserine by Kenneth C. Haydon; The Restored Burlington Northern Depot & WWII Memorial Museum website: http://depothill.net/depot06.html

Michael D. Belis

Find A Grave website: https://www.findagrave.com/

Flickr website: https://www.flickr.com/photos/philip_wgtn/5173650417

GENERAL ORDERS AND BULLETINS War Department 1918 Government Printing Office, Washington, D.C. 1919; Hathi Trust Digital Library website: https://catalog.hathitrust.org/Record/000065622

Georgia Archives University System of Georgia Virutal Vault website: https://vault.georgiaarchives.org/digital/collection/p17154coll7

Georgia, Deaths Index, 1914-1940 [database on-line]. Provo, UT, USA: Ancestry.com Operations, Inc., 2011. "Georgia Deaths, 1914–1927." Index. FamilySearch, Salt Lake City, Utah, 2007. "Georgia Deaths, 1914–1927" and "Georgia Deaths, 1930," images, FamilySearch. Georgia Department of Health and Vital Statistics, Atlanta, Georgia.

GlobalSecurity.org website: https://www.globalsecurity.org/military/agency/army/militia-oh.htm

The Hall of Valor Project website: https://valor.militarytimes.com/

Harvard Advocates for ROTC website: https://www.advocatesforrotc.org/harvard/valor.html

Historical Register of National Homes for Disabled Volunteer Soldiers, 1866-1938; (National Archives Microfilm Publication M1749, 282 rolls); Records of the Department of Veterans Affairs, Record Group 15; National Archives, Washington, D.C.

Home of Heroes website: https://homeofheroes.com/

Indiana World War I Service Record Cards https://fromthepage.com/indianaarchives/indiana-wwi-service-record-cards

Initial Burial Plats for World War I American Soldiers November 6, 2018 By Brandi Oswald, Posted In Cartographic Records, Military, Uncategorized, World War I from the National Archives website The Unwritten Record: https://unwritten-record.blogs.archives.gov/2018/11/06/newly-digitized-series-initial-burial-plats-for-world-war-i-american-soldiers/

Interment Control Forms, 1928–1962. Interment Control Forms, A1 2110-B. Records of the Office of the Quartermaster General, 1774–1985, Record Group 92. The National Archives at College Park, College Park, Maryland.

Iowa, World War II Bonus Case Files, 1947-1954 [database on-line]. Provo, UT, USA: Ancestry.com Operations, Inc., 2014. Original data: WWII Bonus Case Files. State Historical Society of Iowa, Des Moines, Iowa.

Kentucky, Death Records, 1852-1965 [database on-line]. Lehi, UT, USA: Ancestry.com Operations Inc. 2007. Original data: Kentucky. Kentucky Birth, Marriage and Death Records – Microfilm (1852-1910). Microfilm rolls #994027-994058. Kentucky Department for Libraries and Archives, Frankfort, Kentucky.

Lineage and Honors Information Divisions and Brigades U.S. Army Center of Military History website: https://history.army.mil/html/forcestruc/lineages/branches/div/defaultDIV.htm

Lists of Incoming Passengers, 1917-1938. Textual records. 360 Boxes. NAI: 6234465. Records of the Office of the Quartermaster General, 1774-

Michael D. Belis

1985, Record Group 92. The National Archives at College Park, Maryland.

List of Mothers and Widows of American Soldiers, Sailors, and Marines Entitled to Make a Pilgrimage to War Cemeteries in Europe. Washington, D.C.: Government Printing Office, 1930. *Records Relating to Pilgrimages of Gold Star Mothers and Widows, 1930–1933.* NAI 6161915. Records of the Office of the Quartermaster General, 1774–1985, Record Group 92. The National Archives at Washington, D.C.

Lists of Outgoing Passengers, 1917-1938. Textual records. 255 Boxes. NAI: 6234477. Records of the Office of the Quartermaster General, 1774-1985, Record Group 92. The National Archives at College Park, Maryland.

"Louisiana World War I Service Records, 1917-1920." Database with images. FamilySearch. http://FamilySearch.org. Citing The Louisiana State Archives, Baton Rouge.

MACHINE GUN COMPANY, 39TH INFANTRY (4TH DIVISION) IN THE AISNE-MARNE OFFENSIVE (July 18--August 5, 1918) Major M.S. Eddy, Maneuver Center of Excellence Libraries, MCoE HQ Donovan Research Library, Fort Benning, Georgia website: https://www.benning.army.mil/Library/Donovanpapers/wwi/index.html

Mémoires des Hommes website: https://www.memoiredeshommes.sga.defense.gouv.fr/en/article.php?larub=80&titre=those-who-died-for-france-in-the-first-world-war

"Michigan, Census of World War I Veterans with Card Index, 1917-1919."

Soldiers Steadfast and Loyal

Database. FamilySearch. https://FamilySearch.org. Michigan Department of State, Lansing.

Military Enlistments (Montana), World War I Montana Adjutant General's Office Records 1889-1959 (RS 223) MONTANA MEMORY PROJECT: http://mtmemory.org/digital/custom/home/#/

Military Hall of Honor website http://www.militaryhallofhonor.com/

Minnesota Historical Society website: http://www.mnhs.org/search/people#/veteranrecords

Mississippi Department of Archives and History Series 1731: Mississippi World War I Statement of Service Cards, 1917-1919

Mississippi Encyclopedia website: https://mississippiencyclopedia.org/entries/world-war-i/

Missouri Digital Heritage website: Soldiers' Records: War of 1812 - World War I https://s1.sos.mo.gov/records/archives/archivesdb/soldiers/Default.aspx#soldierSearch

"Montana, Military Records, 1904-1918." Database. FamilySearch. https://FamilySearch.org. Montana State Historical Society, Helena.

Muster Rolls of the U.S. Marine Corps, 1798-1892. Microfilm Publication T1118, 123 rolls. ARC ID: 922159. Records of the U.S. Marine Corps, Record Group 127; National Archives in Washington, D.C.

National Archives Catalog: Card Register of Burials of Deceased American Soldiers, 1917 – 1922. Record Group 92: Records of the Office of the

Michael D. Belis

Quartermaster General, 1774 – 1985, National Archives Catalog website: https://catalog.archives.gov/id/6943087

National Cemetery Administration. U.S. Veterans' Gravesites, ca.1775-2006 [database on-line]. Provo, UT, USA: Ancestry.com Operations Inc. 2006. Original data: National Cemetery Administration. Nationwide Gravesite Locator.

NYNG Officer Service Cards prior to 1/1/1930. Saratoga Springs, New York: New York State Military Museum

New York State Abstracts of World War I Military Service, 1917–1919. Adjutant General's Office. Series B0808. New York State Archives, Albany, New York

New York State Military Museum Ancestry.com. New York, Record of Award Medal, 1920-1991 [database on-line]. Provo, UT, USA: Ancestry.com Operations, Inc., 2013 Original data: Record of Award of Medal, 1920–1991. New York State Military Museum, Saratoga Springs, New York.

Oflag 64 Remembered website: http://oflag64.us/index.html

"Ohio, World War I Statement of Service Cards, 1914-1919." Database. FamilySearch. https://FamilySearch.org. The Ohio Adjutant General's Office, Columbus.

OPERATIONS OF COMPANY E 58TH INFANTRY (4TH DIVISION), IN THE MEUSE-ARGONNE OFFENSIVE, SEPTEMBER 26 - OCTOBER 9, 1918 (Personal experience of a Company Commander) Captain John E. Hull, Maneuver Center of Excellence Libraries, MCoE HQ

Soldiers Steadfast and Loyal

Donovan Research Library, Fort Benning, Georgia website: https://www.benning.army.mil/Library/Donovanpapers/wwi/index.html

Passenger Lists of Vessels Arriving at New York, New York, 1820-1897. Microfilm Publication M237, 675 rolls. NAI: 6256867. Records of the U.S. Customs Service, Record Group 36. National Archives at Washington, D.C.

Passport Applications, January 2, 1906–March 31, 1925. NARA Microfilm Publication M1490, 2740 rolls. General Records of the Department of State, Record Group 59. National Archives, Washington, D.C.

Pennsylvania, Death Certificates, 1906-1966 [database on-line]. Provo, UT, USA: Ancestry.com Operations, Inc., 2014. Original data: Pennsylvania (State). Death certificates, 1906–1963. Series 11.90 (1,905 cartons). Records of the Pennsylvania Department of Health, Record Group 11. Pennsylvania Historical and Museum Commission, Harrisburg, Pennsylvania.

Pennsylvania Historical & Museum Commission website: https://www.phmc.pa.gov/Archives/Research-Online/Pages/World-Wars.aspx

Pennsylvania War History Commission

Pennsylvania, WWI Veterans Service and Compensation Files, 1917-1919, 1934-1948 [database on-line]. Provo, UT, USA: Ancestry.com Operations, Inc., 2015. Original data: World War I Veterans Service and Compensation File, 1934–1948. RG 19, Series 19.91. Pennsylvania Historical and Museum Commission, Harrisburg Pennsylvania.

Records of the American Jewish Committee—Office of Jewish War Records, un-

Michael D. Belis

dated, 1918–1921; I-9; boxes 1-22. Newton Centre, MA and New York, NY: American Jewish Historical Society.

Register of Enlistments in the U.S. Army, 1798-1914; (National Archives Microfilm Publication M233, 81 rolls); Records of the Adjutant General's Office, 1780's-1917, Record Group 94; National Archives, Washington, D.C.

Returns from Regular Army Infantry Regiments, June 1821–December 1916. NARA microfilm publication M665, rolls 1–244, 262-292, 297–300 of 300. Records of the Adjutant General's Office, 1780's–1917, Record Group 94, and Records of United States Regular Army Mobile Units, 1821–1942, Record Group 391. National Archives and Records Administration, Washington, D.C.

77th Artillery Regiment Association website: http://www.77fa.org/index.html

Social Security Death Index, Master File. Social Security Administration.

Society of the Fifth Division United States Army website: http://www.societyofthefifthdivision.com/

Sons of the American Revolution Membership Applications, 1889-1970. Louisville, Kentucky: National Society of the Sons of the American Revolution. Microfilm, 508 rolls.

State of New Jersey Department of State, New Jersey State Archives website: https://www.nj.gov/state/archives/index.html

Soldiers Steadfast and Loyal

Stewart Bell Jr. Archives Handley Regional Library 100 West Piccadilly Street, Winchester, VA 22601

Texas, Death Certificates, 1903-1982 [database on-line]. Provo, UT, USA: Ancestry.com Operations, Inc., 2013. Original data: Texas Department of State Health Services. Texas Death Certificates, 1903–1982. Austin, Texas, USA.

"Texas, World War I Records, 1917-1920." Database with images. FamilySearch. http://Familysearch.org. Texas Military Forces Museum, Austin.

The Institute of Heraldry website: https://tioh.army.mil/

United States of America, Bureau of the Census. Twelfth Census of the United States, 1900. Washington, D.C.: National Archives and Records Administration, 1900. T623, 1854 rolls.

United States of America, Bureau of the Census. Thirteenth Census of the United States, 1910 (NARA microfilm publication T624, 1,178 rolls). Records of the Bureau of the Census, Record Group 29. National Archives, Washington, D.C.

United States of America, Bureau of the Census. Fourteenth Census of the United States, 1920. (NARA microfilm publication T625, 2076 rolls). Records of the Bureau of the Census, Record Group 29. National Archives, Washington, D.C.

United States of America, Bureau of the Census. Fifteenth Census of the United States, 1930. Washington, D.C.: National Archives and Records Administration, 1930. T626, 2,667 rolls.

Michael D. Belis

United States of America, Bureau of the Census. Sixteenth Census of the United States, 1940. Washington, D.C.: National Archives and Records Administration, 1940. T627, 4,643 rolls.

"United States, Veterans Administration Master Index, 1917-1940." Database. FamilySearch. https://FamilySearch.org. Citing NARA microfilm publication 76193916. St. Louis: National Archives and Records Administration, 1985.

The United States World War One Centennial Commission website: https://www.worldwar1centennial.org/index.php/commemorate/family-ties/documenting-doughboys/2223-graves-registration-card-registers-1917-22.html

U.S., Adjutant General Military Records, 1631-1976 [database on-line]. Provo, UT, USA: Ancestry.com Operations, Inc., 2011

U.S. Army Medical Department Office of Medical History website: https://history.amedd.army.mil/booksdocs/wwi/adminamerexp/chapter24.html

U.S. City Directories, 1822-1995 [database on-line]. Provo, UT, USA: Ancestry.com Operations, Inc., 2011.

U.S. National Cemetery Interment Control Forms, 1928-1962 [database on-line]. Provo, UT, USA: Ancestry.com Operations, Inc. Original Source: Interment Control Forms, 1928–1962. Interment Control Forms, A1 2110-B. Records of the Office of the Quartermaster General, 1774–1985, Record Group 92. The National Archives at College Park, College Park, Maryland.

Soldiers Steadfast and Loyal

U.S. Naturalization Records Indexes, 1794-1995 [database on-line]. Provo, UT, USA: Ancestry.com Operations, Inc., 2007. Original data: National Archives and Records Administration (NARA) Microfilm Publications; Records of District Courts of the United States, Record Group 21; National Archives, Washington, D.C.

U.S., World War I Draft Registration Cards, 1917-1918 [database on-line]. Provo, UT, USA: Ancestry.com Operations Inc. 2005. Original Source: *World War I Selective Service System Draft Registration Cards, 1917-1918.* Washington, D.C.: National Archives and Records Administration. M1509, 4,582 rolls. Imaged from Family History Library microfilm.

U.S., World War II Draft Registration Cards, 1942 [database on-line]. Lehi, UT, USA: Ancestry.com Operations, Inc., 2010. Original Source: United States, Selective Service System. Selective Service Registration Cards, World War II: Fourth Registration. Records of the Selective Service System, Record Group Number 147. National Archives and Records Administration.

Veterans Affairs, Department of, World War I Service Statement Cards, 1917-1919, Washington State Archives, Digital Archives, http://digitalarchives.wa.gov

War Department General Orders No. 66 May 21, 1919 University of Massachusetts Amherst website: https://credo.library.umass.edu/view/full/mums312-b219-i239

War Relics Forum website: http://www.warrelics.eu/forum/orders-medals-decorations/distinguished-service-cross-pair-679822/

World War I Veterans Service and Compensation File, 1934–1948. RG 19,

Michael D. Belis

Series 19.91. Pennsylvania Historical and Museum Commission, Harrisburg Pennsylvania.

PHOTO CREDITS:

Samuel P. Adkisson - *The Santa Fe Magazine Volume XIII Number 5* April, 1919 Railway Exchange, Chicago 1919 pp35

Tom F. Barto - *Sunset The Pacific Monthly* December 1918 Vol. 41 No. 6 pp14

Elwyn L. Berwick - *SOLDIERS OF THE GREAT WAR Volume I* W.M. Haulsee, F.G. Howe and A.C. Doyle, Soldiers Record Publishing Association Washington, D.C. 1920 pp126

Guy W. Boardman - *SOLDIERS OF THE GREAT WAR Volume I* W.M. Haulsee, F.G. Howe and A.C. Doyle, Soldiers Record Publishing Association Washington, D.C. 1920 pp123

Frank C. Bolles - *The History of the 39th U.S. Infantry During the World War* by Richard Barrett Cole and Barnard Eberlin, New York, Press of J.D. McGuire 1919 pp9

Gustav J. Braun - *The Doughboy* The Infantry School, Camp Benning Georgia 1922

Charles F. Carbaugh - Nettie Carbaugh Collection, Stewart Bell, Jr. Archives Room, Handley Regional Library, Winchester, VA

Marion H. Cardwell - *The Fifty-Eighth Infantry in THE WORLD WAR* George L. Morrow, 58th Infantry History Association 1919 pp270

Soldiers Steadfast and Loyal

Arthur I. Clark - *The History of the 39th U.S. Infantry During the World War* by Richard Barrett Cole and Barnard Eberlin, New York, Press of J.D. McGuire 1919 pp131

Joseph T. Clement - *Passport Applications, January 2, 1906–March 31, 1925*. NARA Microfilm Publication M1490, 2740 rolls. General Records of the Department of State, Record Group 59. National Archives, Washington, D.C. 1922

Clinton Day - *The Fifty-Eighth Infantry in THE WORLD WAR* George L. Morrow, 58th Infantry History Association 1919 pp272

Walter H. Detrow - Courtesy of Gary A. Mitchell

Albert Dietz - *INDIANA HISTORICAL COLLECTIONS VOLUME XVIII Indiana Book of Merit: Official Individual Decorations and Commendations Awarded to Indiana Men and Women for Services in the World War* Harry A. Rider, Indianapolis: The Historical Bureau, Indiana Library and Historical Dept., 1932

Frank J. Downs - *The Fifty-Eighth Infantry in THE WORLD WAR* George L. Morrow, 58th Infantry History Association 1919 pp271

Charles B. Duncan - *SOLDIERS OF THE GREAT WAR Volume III* W.M. Haulsee, F.G. Howe and A.C. Doyle, Soldiers Record Publishing Association Washington, D.C. 1920 pp213

Peter W. Ebbert - *The Fifty-Eighth Infantry in THE WORLD WAR* George L. Morrow, 58th Infantry History Association 1919 pp167

Michael D. Belis

Andrew Edmiston - *Charleston Daily Mail* Wednesday November 4, 1936 Charleston, West Virginia

Emil J. Eklund - *The Fifty-Eighth Infantry in THE WORLD WAR* George L. Morrow, 58th Infantry History Association 1919 pp270

Edwin A. Elliott –Amon Carter Papers, MS 014, Box 1, Special Collections, Mary Couts Burnett Library, Texas Christian University

Earl R. Fretz - *Memoirs of the Harvard Dead in the War Against Germany Volume V* M.A. DeWolfe, Harvard University Press, Cambridge, Mass., 1924 pp277

David S. Grant - *Passport Applications, January 2, 1906–March 31, 1925.* NARA Microfilm Publication M1490, 2740 rolls. General Records of the Department of State, Record Group 59. National Archives, Washington, D.C. 1915

Glenn Grove - *HEADQUARTERS Army and Navy LEGION OF VALOR OF THE U.S. OF AMERICA General Orders (1942-1943) Volume 53, No. 4* Pittsburg, Pa., January 26, 1943

Harold Lamonte Hall - *SOLDIERS OF THE GREAT WAR Volume III* W.M. Haulsee, F.G. Howe and A.C. Doyle, Soldiers Record Publishing Association Washington, D.C. 1920 pp390

Arthur M. Hamilton - *The Fifty-Eighth Infantry in THE WORLD WAR* George L. Morrow, 58th Infantry History Association 1919 pp272

William H. Hammond - *The Doughboy* The Infantry School, Fort Benning Georgia 1924

James W. Hanbery - State Manual Training Normal College yearbook, Pittsburg, Kansas 1915

William Herren - *The Fifty-Eighth Infantry in THE WORLD WAR* George L. Morrow, 58th Infantry History Association 1919 pp271

Samuel H. Houston - *The Fifty-Eighth Infantry in THE WORLD WAR* George L. Morrow, 58th Infantry History Association 1919 pp40

William L. Hunter - *The Fifty-Eighth Infantry in THE WORLD WAR* George L. Morrow, 58th Infantry History Association 1919 pp272

Thomas W. Kearns - *Boston Globe* Boston, Massachusetts Monday August 12, 1918

Max Kos - *INDIANA HISTORICAL COLLECTIONS VOLUME XVIII Indiana Book of Merit: Official Individual Decorations and Commendations Awarded to Indiana Men and Women for Services in the World War* Harry A. Rider, Indianapolis: The Historical Bureau, Indiana Library and Historical Dept., 1932

Ralph E. Ladue - *The Brooklyn Daily Eagle* Tuesday June 4, 1940 Brooklyn, New York pp18

John Legnosky - *The Fifty-Eighth Infantry in THE WORLD WAR* George L. Morrow, 58th Infantry History Association 1919 pp270

Robert A. Madden - *Indiana World War Records GOLD STAR HONOR ROLL A Record of Indiana Men and Women who died in the service of the United States and the Allied Nations in the World War 1914-1918* Indiana Historical Commission Indianapolis 1921 pp405

Michael D. Belis

Robert G. Marshall - *The Fifty-Eighth Infantry in THE WORLD WAR* George L. Morrow, 58th Infantry History Association 1919 pp270

Henry F. Martin - *Passport Applications, January 2, 1906–March 31, 1925.* NARA Microfilm Publication M1490, 2740 rolls. General Records of the Department of State, Record Group 59. National Archives, Washington, D.C. 1920

Edward McAndrew - *Indiana World War Records GOLD STAR HONOR ROLL A Record of Indiana Men and Women who died in the service of the United States and the Allied Nations in the World War 1914-1918* Indiana Historical Commission Indianapolis 1921 pp300

Howard C. McCall - *Passport Applications, January 2, 1906–March 31, 1925.* NARA Microfilm Publication M1490, 2740 rolls. General Records of the Department of State, Record Group 59. National Archives, Washington, D.C. 1916

George C. McCelvey - *The Sphinx*, The Citadel, Charleston, S.C. 1911

Manton C. Mitchell - *The History of the 39th U.S. Infantry During The World War* by Richard Barrett Cole and Barnard Eberlin, New York, Press of J.D. McGuire 1919 pp14

Martin Nelson - *The Fifty-Eighth Infantry in THE WORLD WAR* George L. Morrow, 58th Infantry History Association 1919 pp271

Francis K. Newcomer - *The Howitzer* U.S. Military Academy 1913 pp107

John H. Norton - *Fiftieth Annual Report of the Association of Graduates of the*

Soldiers Steadfast and Loyal

United States Military Academy Saignaw, Michigan Seemann & Peters, Inc. 1919 pp150

Robert W. Norton - *The History of the 39th U.S. Infantry During The World War* by Richard Barrett Cole and Barnard Eberlin, New York, Press of J.D. McGuire 1919 pp52

Paul J. Pappas - *The History of the 39th U.S. Infantry During The World War* by Richard Barrett Cole and Barnard Eberlin, New York, Press of J.D. McGuire 1919 pp135

James K. Parsons - Naval Historical Collection, U.S. Naval War College - "Naval War College Staff and Classes of 1925" 1925

Donald A. Pegg - *INDIANA HISTORICAL COLLECTIONS VOLUME XVIII Indiana Book of Merit: Official Individual Decorations and Commendations Awarded to Indiana Men and Women for Services in the World War* Harry A. Rider, Indianapolis: The Historical Bureau, Indiana Library and Historical Dept., 1932

James J. Pirtle – *The 1930 Arbutus Volume Thirty Seven* Indiana University 1930

Richard G. Plumley - *The History of the 39th U.S. Infantry During the World War* by Richard Barrett Cole and Barnard Eberlin, New York, Press of J.D. McGuire 1919 pp52

Benjamin A. Poore - *U.S. Army Signal Corps Collection, National Archives, 19023 [1087G8]*

Ernest R. Potter - *The History of the 39th U.S. Infantry During the World War* by Richard Barrett Cole and Barnard Eberlin, New York, Press of J.D. McGuire 1919 pp131

Fred N. Rapp - *SOLDIERS OF THE GREAT WAR Volume II* W.M. Haulsee, F.G. Howe and A.C. Doyle, Soldiers Record Publishing Association Washington, D.C. 1920 pp182

Carl Rasmussen - *Passport Applications, January 2, 1906–March 31, 1925.* NARA Microfilm Publication M1490, 2740 rolls. General Records of the Department of State, Record Group 59. National Archives, Washington, D.C 1920

Lowell H. Riley - *The Fifty-Eighth Infantry in THE WORLD WAR* George L. Morrow, 58th Infantry History Association 1919 pp167

James H. Roberts - *The History of the 39th U.S. Infantry During The World War* by Richard Barrett Cole and Barnard Eberlin, New York, Press of J.D. McGuire 1919 pp135

Alvan C. Sandeford - *The Howitzer* U.S. Military Academy 1911

Anthony F. Shedlock - *The Fifty-Eighth Infantry in THE WORLD WAR* George L. Morrow, 58th Infantry History Association 1919 pp271

William Shemin - U.S. Army photo

Albert B. Simpson - *The Georgia State Memorial Book* Bert E. Boss, American Memorial publishing Company, 1921

Ralph Slate - *The History of the 39th U.S. Infantry During the World War* by Richard Barrett Cole and Barnard Eberlin, New York, Press of J.D. McGuire 1919 pp40

Rutherford H. Spessard - *The Fifty-Eighth Infantry in THE WORLD WAR* George L. Morrow, 58th Infantry History Association 1919 pp134

Thomas Vander Veen - *Division of Passport Control: Emergency Passport Applications, 1906–1925 (Argentina thru Venezuela)*. NAI 1244183 A1 544. General Records of the Department of State, Record Group 59. National Archives, Washington D.C.

Harrison B. Webster - *HARVARD CLASS OF 1905 FOURTH REPORT June 1920* Plimpton Press, Norwood, Massachusetts 1920

ABOUT THE AUTHOR

Michael D. Belis served as a rifleman in the United States Army's 1st Battalion 22nd Infantry Regiment of the 4th Infantry Division in the Republic of Vietnam in 1970 and is a proud holder of the Combat Infantryman Badge. He is a member of the 4th Infantry Division Association and the 22nd Infantry Regiment Society.

From 1999-2014, he was a historical advisor to 1st Battalion 22nd Infantry while the Battalion was part of the 4th Infantry Division (Mechanized) based out of Fort Hood, Texas and then Fort Carson, Colorado and contributed an extensive number of historical artifacts of the Regiment to the Battalion's Regimental Room during those years. Since 2014, he has been working in a historical advisory capacity to 2nd Battalion 22nd Infantry, 10th Mountain Division stationed at Fort Drum, New York and has also contributed numerous artifacts of the Regiment to that Battalion for their displays at their home base.

He is the Historian for the 22nd Infantry Regiment Society and for over twenty years has written articles for the 22nd Infantry Regiment Society newsletter and for other projects by the Society. He has written articles for the 4th Infantry Division Association publication "The Ivy Leaves" and for the Division Association's website and Facebook pages. He is the author of the book *Earl William "Lum" Edwards "Mr. 22nd Infantry"* a biography of a 4th Infantry Division officer in World War Two. He is also the webmaster and one of the four founding members of the 1st Battalion 22nd Infantry website at www.1-22infantry.org.

In 2012, the title of Distinguished Member of the Regiment (DMOR) was conferred upon him by the Secretary of the Army and his name was added to the roll of the Distinguished Members of the Regiment of the United States 22nd Infantry.

He lives with his wife, Margaret, and their cats Bijou, Scottie, and Elliot in the heart of Cajun Country near Carencro, Louisiana.

www.ingramcontent.com/pod-product-compliance
Lightning Source LLC
Chambersburg PA
CBHW021147230426
43667CB00006B/281